DYES AND DRUGS

New Uses and Implications

DYES AND DRUGS

New Uses and Implications

Harold H. Trimm, PhD, RSO

Chairman, Chemistry Department, Broome Community College;
Adjunct Analytical Professor, Binghamton University,
Binghamton, New York, U.S.A.

William Hunter Jr.

Olean General Hospital, Olean, New York, U.S.A.

Apple Academic Press

TORONTO NEW JERSEY

Dyes and Drugs: New Uses and Implications

First Published in the Canada, 2011
Apple Academic Press Inc.
3333 Mistwell Crescent
Oakville, ON L6L 0A2
Tel. : (888) 241-2035
Fax: (866) 222-9549
E-mail: info@appleacademicpress.com
www.appleacademicpress.com

First issued in paperback 2021

The full-color tables, figures, diagrams, and images in this book may be viewed at www.appleacademicpress.com

ISBN 13: 978-1-77463-256-7 (pbk)
ISBN 13: 978-1-926692-62-3 (hbk)

Harold H. Trimm, PhD, RSO
William Hunter Jr.

Cover Design: Psqua

Library and Archives Canada Cataloguing in Publication Data
CIP Data on file with the Library and Archives Canada

CONTENTS

ACKNOWLEDGMENTS AND HOW TO CITE

The chapters in this book were previously published in various places and in various formats. By bringing these chapters together in one place, we offer the reader a comprehensive perspective on recent investigations into this important field.

We wish to thank the authors who made their research available for this book, whether by granting permission individually or by releasing their research as open source articles or under a license that permits free use provided that attribution is made. When citing information contained within this book, please do the authors the courtesy of attributing them by name, referring back to theiroriginal articles, using the citations provided at the end of each chapter.

INTRODUCTION

The science of chemistry is so broad that it is normally broken into fields or branches of specialization. The manufacture of drugs and dyes is one of the most practical industrial applications of chemistry.

The development of dyes and medicines has always been connected. Scientists working to discover and synthesize both dyes and drugs developed the organic, synthetic chemistry that gave birth to the pharmaceutical industry at the end of the nineteenth century and the beginning of the twentieth. In fact, it was the accidental discovery of the first synthetic dye by Henry Perkin in England in 1856 that ushered in the age of synthetics. Perkin had been trying to synthesize the antimalarial drug quinine when he discovered a mauve dye later popularized by Queen Victoria. On the other side of this intimate connection between drugs and dyes, nineteenth-century chemists noticed the antibacterial properties some dyes exhibited in test tubes and then investigated their uses as therapeutic agents. Many proved to be quite useful. Paul Ehrlich, the German drug developer, systematically studied aniline dyes and modified them for use as drugs

Research during the early years of dye synthesis produced compounds such as methylene blue and acriflavine, which were used as biological stains. The selectivity of such compounds for "non-economic" cells such as pathogenic bacteria or tumor cells gave rise both to the principle of selective toxicity and to the development of many modern drugs. The use of dyes in therapy is again gaining credence today, given the efficacy of light-activated drugs based on dye molecules against drug-resistant organisms such as MRSA. In addition, because of the ongoing prevalence of "superbugs", older drugs developed from dye chromophores may again be of use in medical settings.

It is the intent of this collection to present the reader with a broad spectrum of articles in the particular area of drugs and dyes, thereby demonstrating key developments in this rapidly changing field.

— Harold H. Trimm, PhD, RSO

Solvent Effect on the Spectral Properties of Neutral Red

Muhammad A. Rauf, Ahmed A. Soliman and Muhammad Khattab

ABSTRACT

Background

The study was aimed at investigating the effect of various solvents on the absorption spectra of Neutral Red, a dye belonging to the quinone-imine class of dyes. The solvents chosen for the study were water, ethanol, acetonitrile, acetone, propan-1-ol, chloroform, nitrobenzene, ethyleneglycol, acetic acid, DMSO and DMF.

Results

The results have shown that the absorption maxima of dyes are dependent on solvent polarity. In non-hydrogen-bond donating solvents, solvation of dye molecules probably occurs via dipole-dipole interactions, whereas in hydrogen-bond donating solvents the phenomenon is more hydrogen bonding in nature. To estimate the contribution of the different variables on the wave number of the Neutral Red dye, regression analyses using the ECW model were compared with the π scale model. This showed that the unified scale for estimating the solvent effect on the absorption of the Neutral Red dye is more adopted and more applicable than the π* scale model.*

Conclusion

Absorption maxima of dyes are dependent on solvent polarity. Solvation of dye molecules probably occurs via dipole-dipole interactions in non-hydrogen-bond donating solvents, whereas in hydrogen-bond donating solvents the phenomenon is more hydrogen bonding in nature. The unified scale for estimating the solvent effect on the absorption of Neutral Red dye is more adopted and more applicable than the π scale model. This may be due to complications from both π-π* charge transfer interactions and incomplete complexation of the solute; these effects are averaged out in the derived β and π parameters and thus limit their applicability.*

Background

It is very well known that chemical processes are influenced by the properties of solvents in which they are carried out. These include the dipole moment, dielectric constant, and refractive index values. The most important property in this regard is the solvent polarity which can change the position of the absorption or emission band of molecules by solvating a solute molecule or any other molecular species introduced into the solvent matrix. A number of literature citations are available for the studies of simple organic molecules with regard to their interactions in different solvents [1-3]. Dye molecules on the other hand are complex organic molecules which might carry charge centers (as an integral part of their structure or because they are derived salts) and are thus prone to absorption changes in various media [4,5]. The structural complexity of dye molecules has drawn attention of many workers to understand their behavior in various media [6-8]. These changes are important to understand various physical-organic reactions of these macromolecules which have become important in different fields of pure and applied chemistry such as synthetic chemistry, extraction of dyes from solution, photodynamic therapy and chelation processes [9-11].

Experimental

The dye used in this work was purchased from Sigma Chemicals and used as such. The solvents used in this work namely propan-1-ol, chloroform, acetonitrile, DMSO, ethanol, acetone, and DMF were purchased from Merck or Fluka and had a purity of > 99%. They were kept over molecular sieves 5Å prior to their use in this work. Stock solution of the dye was made in all these solvents and then diluting them appropriately with a given solvent. Absorption spectrum of each dye solution (1×10^{-5} M) was recorded on a CARY 50 UV/VIS spectrophotometer, using a 1 cm quartz cell.

Results and Discussion

The present study deals with the solvent effect on the absorption spectra of Neutral Red (NR), which belongs to Quinone-Imine class of dyes. The molecular structure of this dye is shown in figure 1. Absorption spectrum of the dye solution was recorded in different solvents with the aim to probe the effects of various solvents and correlate various absorption parameters to dye spectra in various solvents. For this purpose solvents of different types were selected, firstly the non hydrogen-bond donating solvents (also called as non-HBD type of solvents) such as acetone, acetonitrile, nitrobenzene, DMF and DMSO; and secondly the hydrogen-bond donating solvents (also called as HBD type solvents) such as water, ethanol, acetic acid, ethyleneglycol, chloroform and propan-1-ol. The values of λ_{max} of Neutral Red in these solvents are given in Table 1. One can see from this table that the absorption maximum of the dye is affected by solvent type and has a maximum shift of $\Delta\lambda$ = 22 nm for the solvents used in this work. Thus this change in spectral position can be used as a probe for various types of interactions between the solute and the solvent.

Table 1. Absorption maxima of Neutral Red (NR) in various solvents and selective Kamlet-Taft solvent properties.

Solvent	λ(nm)	π*	α	β	ε	n
Water	527	1.09	1.17	0.18	80.1	1.3330
Propan-1-ol	538	0.52	0.84	0.90	20.33	1.3856
Chloroform	522	0.58	0.44	0.0	4.81	1.4429
Acetonitrile	531	0.75	0.19	0.31	38.80	1.3440
DMSO	546	1.0	0.0	0.76	47.2	1.4790
Ethanol	537	0.54	0.83	0.77	24.55	1.3614
Acetone	525	0.71	0.08	0.48	20.70	1.3590
DMF	540	0.88	0.0	0.69	38.3	1.4305
Acetic acid	534	0.64	1.12	-	6.2	1.3710
Ethyleneglycol	539	0.92	0.9	0.52	38.7	1.4310
Nitrobenzene	544	1.01	0.0	0.39	32.0	1.5510

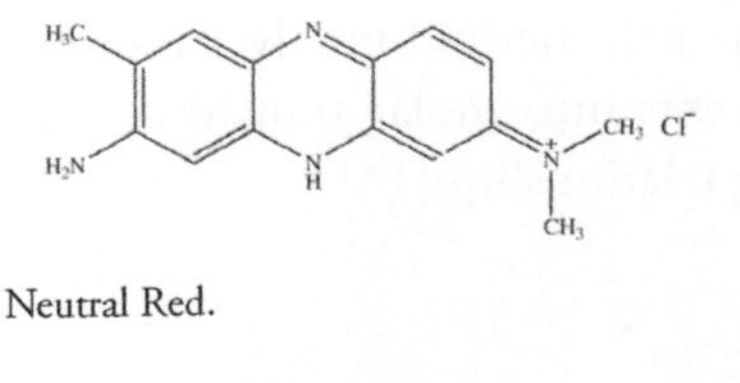

Figure 1. Chemical Structure of Neutral Red.

The analysis of solvent effect on spectral properties of dye solutions were carried out by using the spectral position in above mentioned solvents and correlating these with the Kamlet-Taft solvent properties namely, π^*, α, β, n and ε, obtained from the literature [12,13]. Since the shift in λmax values with solvent type reflects dye-molecule interactions, an attempt was made to study this phenomenon in detail. Table 1 shows the essential solvent parameters required in this study along with the absorption maxima for each dye in these solvents. The spectral position of dye in various solvents has revealed interesting results. Since all the solvents used in this work were polar in nature, one would expect that the dye would bind more strongly to a more polar solvent and thus cause the spectra to shift to lower wavelengths. However, this is not seen from our results as λmax is lowest in the case of chloroform. This is due to the reason that all other solvents used in this work are more polar than chloroform and can engage more strongly in a solvent-solvent type of interaction. Thus their ability to interact with the dye molecules becomes less. On the other hand, chloroform which is less polar, can interact with the dye molecule in terms of dipole-dipole interactions, thereby resulting in a net stabilization of the ground state of the dye molecule, and hence one sees a hypsochromic shift in the spectrum in this solvent. On the other hand, the λmax value is shifted to lower energies in highly polar solvents such as DMSO because of strong solvent-solvent interaction or the specific interaction between the solvent and hydrogen from NH2 group in the dye molecule.

A plot of λmax versus the dielectric constant values in various non-HBD and HBD solvents is shown in figure 2A and 2B. It can be seen from this figure that with increasing dielectric constant values, the spectrum is shifted to higher wavelength. The spectral changes observed in water were quite distinct from those in other solvents. The λmax of dye solution in water was found at lower wavelengths as compared to in other solvents although its dielectric constant is the highest among these solvents. This might be due to the formation of strong hydrogen bond between dye and water molecule. Thus different phenomena are present in various media. An increase in λmax values with π^* (dipolarity/polarizability) as shown in figure 3 also indicates that dye interaction becomes different with increasing capability of a given solvent to form H bonds in solution.

The absorption data of dyes in various solvent was also analyzed in terms of various polarity scales. The first method involves the transformation of λmax (nm) of dyes in various solvents into molar transition energies {ET(dye), kcal/mole} by using the following relationship [14]

A:

B:

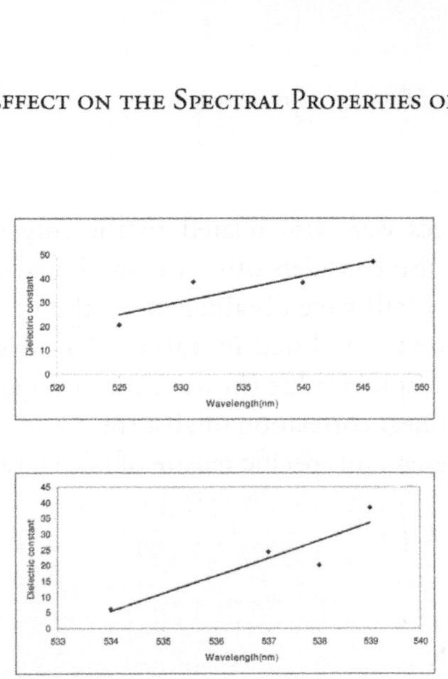

Figure 2. (A) Absorption shift of dye solution as a function of dielectric constant in non-hydrogen-bond donating solvents, (B) Absorption shift of dye solution as a function of dielectric constant in hydrogen bonding solvents.

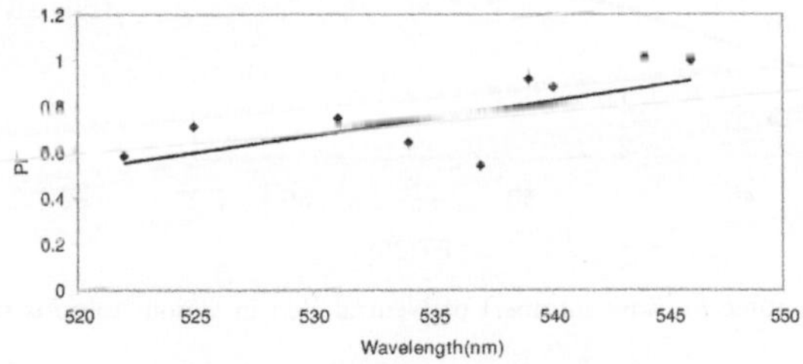

Figure 3. Absorption shift of dye solution as a function of solvent polarizability (π^*).

$$ET(dye) = 28{,}591/\lambda max. \quad (1)$$

The ET(dye) values signify transition energy which also reflects the stabilization of the dye in its ground state in a given solvent. This may be due to either hydrogen bond formation or dye-solvent interaction. Therefore, ET(dye) provides a direct empirical measure of dye solvation behavior. Again, from Table 2 one can see that ET(dye) is maximum in the case of chloroform as compared to other solvents. The rationale behind this is the same as described previously.

Table 2. Empirical parameters of solvent polarity

Solvent	E_T (30) kcal/mol	E_T (NR) kcal/mol	$f(\varepsilon, n)$	$g(n)$	$\phi(\varepsilon, n)$
Water	62.8	54.25	0.91363	0.226851	1.367334
Propan-1-ol	50.7	53.14	0.77905	0.262207	1.303468
Chloroform	39.1	54.77	0.37245	0.300187	0.972823
Acetonitrile	55.4	53.84	0.86602	0.234279	1.334575
DMSO	45.1	52.36	0.84132	0.323779	1.488877
Ethanol	51.9	53.24	0.81293	0.245993	1.304918
Acetone	39.1	54.45	0.79028	0.244380	1.279042
DMF	43.2	52.95	0.83944	0.292021	1.423484
Acetic acid	51.7	53.54	0.49972	0.254245	1.004628
Ethyleneglycol	56.3	54.77	1.11158	0.292350	1.696281
Nitrobenzene	41.2	52.55	0.78171	0.369957	1.521691

The absorption values were also related to the solvent polarity parameter, namely ET(30), which also considers other interactions besides those of specific nature. The values of ET(30) were obtained from the literature for various solvents used in this work and are listed in Table 2[14]. Figure 4 shows the correlation between the absorption value (in wave number) and ET(30) for the dye studied in this work. A linear correlation of absorption energy covering a range of ET(30) indicates the presence of specific nature of interactions between the solute and solvents.

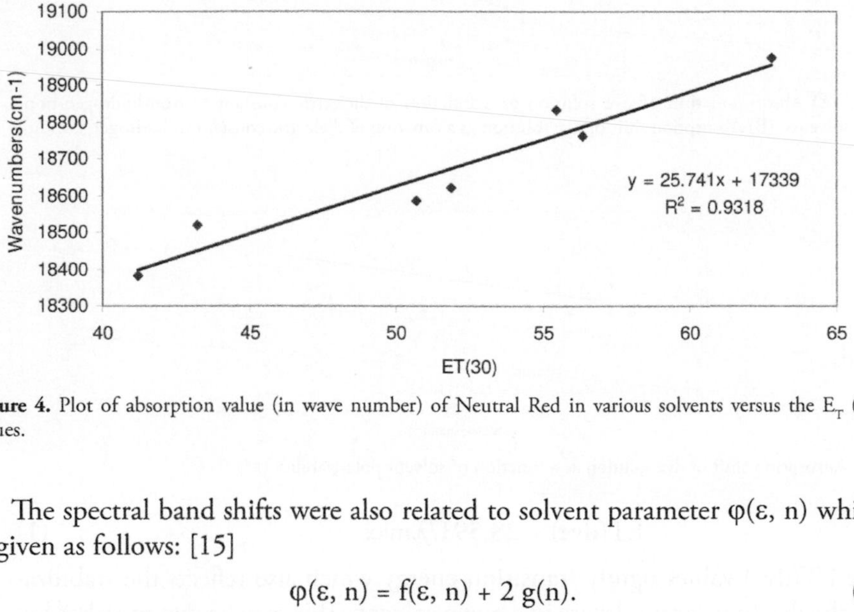

Figure 4. Plot of absorption value (in wave number) of Neutral Red in various solvents versus the E_T (30) values.

The spectral band shifts were also related to solvent parameter φ(ε, n) which is given as follows: [15]

$$\varphi(\varepsilon, n) = f(\varepsilon, n) + 2\,g(n). \tag{2}$$

The function takes into account two important properties of the solvents namely the dielectric constant and the refractive index and is a sum of two independent terms namely f(ε, n) and g(n) which are given as follows

$$f(\varepsilon, n) = [(2n2+1)/(n2+2)]\ [\{(\varepsilon - 1/\varepsilon + 2)\} - \{(n2-1)/(n2+2)\}] \tag{3}$$

$$g(n) = 3/2\ [(n4-1)/(n2+2)2] \tag{4}$$

where, ε is the dielectric constant and n is the refractive index and both these quantities reflect the freedom of motion of electrons in the solvent and the dipole moment of the molecules. Specific solvent effects occur by interactions of the solvent and the chromophores. Figure 5A and 5B shows the trend when the spectral position (λ_{max}) of the dyes in various solvents (non-protic and protic) were plotted against the solvent polarity parameter φ(ε, *n*).

A:

B:

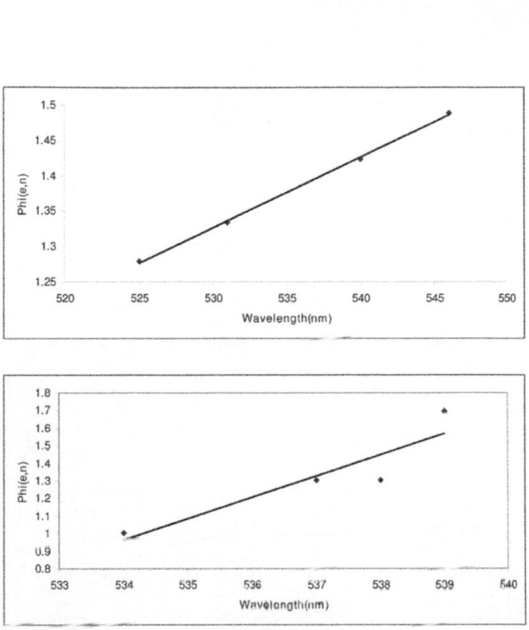

Figure 5. (A) Absorption shift of dye solution in non-hydrogen-bond donating solvents as a function of solvent polarity parameter $\varphi(\varepsilon, n)$, (B) Absorption shift of dye solution in hydrogen bonding solvents as a function of solvent polarity parameter $\varphi(\varepsilon, n)$.

Since dye spectra are also influenced by the presence of a co-solvent, some studies on selective mixtures of solvents were also carried out and their results are hereby discussed. For these studies, two sets of solvents were selected. The first solvent mixture selected was ethanol and water mixture, which belongs to the HBD type of solvents, whereas, the second solvent mixture consists of acetone and ethanol, which belong to the HBD-non HBD type of solvents. The mixtures were prepared in various mole fractions containing a fixed amount of dye.

In ethanol-water mixtures, the spectra were found to shift towards red with increasing mole fraction of ethanol. These results show that the dye cation is preferentially solvated by the alcoholic component in all mole fractions in aqueous mixtures with ethanol. It is well known that water makes strong hydrogen-bonded nets in the water-rich region, which are not easily disrupted by the cosolvent [13]. This can explain the strong preferential salvation by the alcoholic component in this region since water preferentially interacts with itself rather than with the dye. In the alcohol-rich region, the alcohol molecules are freer to interact with the water and with the dye, since their nets formed by hydrogen bonds are weaker than in water. In this situation, the alcohol molecules can, to a greater or lesser extent, interact with water through hydrogen bonding. A change in the ET(NR) as a function of mole fraction of water in water-ethanol mixture is shown in figure 6A. The ET(NR) in the mixture was calculated by the method given in the literature [5].

A:

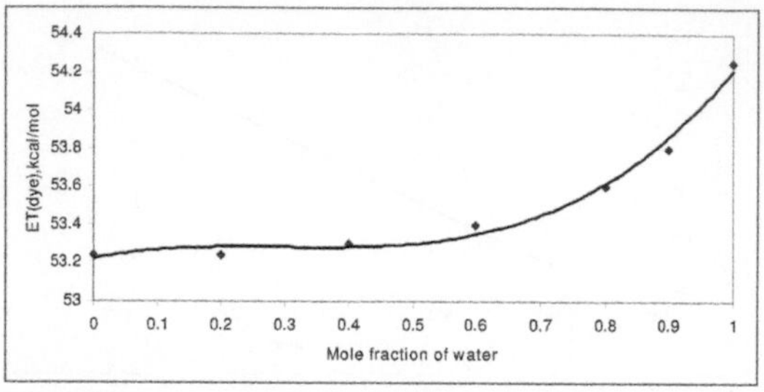

B:

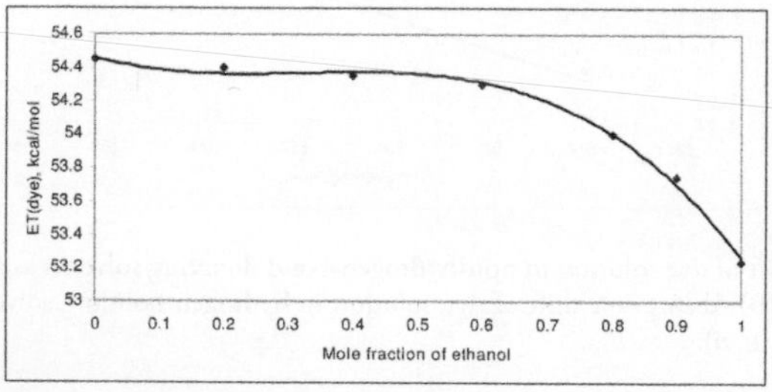

Figure 6. (A) E_T(NR) as a function of mole fraction of water in water-ethanol mixture, (B) E_T(NR) as a function of mole fraction of ethanol in acetone- ethanol mixture.

On the other hand, in the case of acetone and ethanol mixtures, the spectra also shifted towards red with increasing amount of ethanol. The solvation of dye in non-HBD type of solvent mainly occurs through charge-dipole type of interaction, whereas in HBD type of solvent, the interaction also occurs by hydrogen bonding besides the usual ion-dipole interaction. In this situation, the methyl groups of acetone are responsible for the solvation of the dye. Thus in these solvent mixtures, increasing the amount of HBD solvent (ethanol in this case) shall break these interactions with the dye molecule, thereby shifting the spectra towards red. A similar behavior is reported in the literature for some other probe molecule [5]. Figure 6B shows the change in ET(NR) as a function of mole fraction of ethanol in acetone-ethanol mixture.

Great attention has been paid to the problem of solvent effect on spectral, chemical and reactivity data [16]. Kamlet considered the total solvent effect to be composed of three independent contributions; solvent polarity (π^*), acidity (α) and basicity (β) for hydrogen bond acceptor (HBA) solvents. These contributions are gathered in one equation as follows:

$$\nu = \nu o + s\pi^* + a\alpha + b\beta. \quad (5)$$

where ν is the wave number at maximum absorption, *s*, *a* and *b* are regression factors, whose values depend on the extent of contribution of each solvent parameter (π^*, α, β) to the predicted values ν'.

A unified scale of solvent polarities, taking into account both the non-specific and specific donor acceptor interactions of these solvents with solute probes, was introduced by Drago and co-workers and shown in its mathematical form below [17]:

$$\Delta\chi = E^*A\,EB + C^*ACB + S'P + W \tag{6}$$

where Δχ or χ is used as a value of physicochemical property measured in the specific solvent polarity; P is a measure of the susceptibility of the solute probe to solvation and W is the value of Δχ at S' = 0 value which provides a scale of non-specific solvating ability. Similarly, E_B and C_B are solvent parameters which are reported for the donors reacting with a wide range of acceptors in solvents of poor solvation property [18].

The unified scale of solvent polarities was expanded to include the very important class of polar hydrogen bonding solvents. Since these solvents are capable of undergoing both non-specific and specific donor-acceptor interactions with donor solute probes, the above equation modifies to:

$$\Delta\chi = E'A\,E^*B + C'AC^*B + S'P + W. \tag{7}$$

The prime values represent parameters that are consistent with enthalpy based parameters of the ECW model [18-21].

To estimate the contribution of the different variables on the wave number of the Neutral Red dye; the coefficients in equations (5) and (7) were estimated using a multiple linear regression analyses (where Δχ in equation 7 is used as the wave number (ν) of the dye in the different solvents), and the results are shown in Tables 3, 4 and 5. The high R2 value for the regression using ECW model compared with that of the π* scale model reflect the fact that the unified scale for estimating the solvent effect on the absorption of Neutral Red dye is more adopted and more applicable than the π* scale model. This may be due to complications from both π-π* charge transfer interactions and incomplete complexation of the solute; these effects are averaged out in the derived β and π parameters and thus limit their applicability [22]. DMF and DMSO were incompatible and this could be attributed to complications in the specific interactions which may arise due to a variety of bonding sites of different extent in these solvents.

Table 3. Neutral Red absorption (cm-1) in various solvents and some selective solvent properties

Solvent	$\nu(cm^{-1})$	E'_A	C'_A	S'
Water	18975.33	1.91	1.78	3.53
Propan-1-ol	18587.36	1.28	0.83	2.66
Chloroform	19157.09	1.56	0.44	1.74
Acetonitrile	18832.39			
DMSO	18315.02			
Ethanol	18621.97	1.33	1.23	2.8
Acetone	19047.62			
DMF	18518.52			

Table 4. π*-scale parameters of Neutral Red

Effect	Coefficient	Std Error	Std Coef	Tolerance	t	P(2 Tail)	R^2	$R^2_{adjusted}$
	19473.838 (ν_o)	278.232	0	.	69.991	0	0.830	0.702
π*	-438.856 (s)	288.061	-0.322	0.952	-1.523	0.202		
α	63.216 (a)	135.746	0.098	0.969	0.466	0.666		
β	-805.858 (b)	191.874	-0.888	0.953	-4.2	0.014		

Table 5. ECW parameters of the Neutral Red as a new probe

Effect	Coefficient	Std Error	Std Coef	Tolerance	t	P(2 Tail)	R^2
	18220.243 (W)	0	0	.	.	.	1.00
E'_A	976.912 (E^*_B)	0	1.013	0.501	.	.	
C'_A	90.525 (C^*_B)	0	0.187	0.042	.	.	
S'	-360.325 (P)	0	-0.956	0.05	.	.	

Conclusion

Absorption maxima of dyes are dependent on solvent polarity. Solvation of dye molecules probably occurs via dipole-dipole interactions in non-hydrogen-bond donating solvents, whereas in hydrogen-bond donating solvents the phenomenon is more hydrogen bonding in nature. The unified scale for estimating the solvent effect on the absorption of Neutral Red dye is more adopted and more applicable than the ð π* scale model. This may be due to complications from both πð- π* charge transfer interactions and incomplete complexation of the solute; these effects are averaged out in the derived β and π parameters and thus limit their applicability.

Competing Interests

The authors declare that they have no competing interests.

References

1. Reichardt C: Solvents and solvent effects in organic chemistry. 2nd edition. VCH, New York, USA; 1991.
2. Ikram M, Rauf MA, Jabeen Z: Solvent effect on Si-H stretching bands of substituted silanes. Spectrochim Acta, Part A 1994, 50:337–342.
3. Mishra A, Behera RK, Behera PK, Mishra BK, Behera GB: Cyanines during the 1990s: A Review. Chem Rev 2000, 100:1973–2011.
4. Oliveira CS, Bronco KP, Baptista MS, Indig GL: Solvent and concentration effects on the visible spectra of tri-para-dialkylamino-substituted triarylmethane dyes in liquid solutions. Spectochim Acta, Part A 2002, 58:2971–2982.
5. Bevilaqua T, Goncalves TF, Venturini CG, Machado VG: Solute-solvent and solvent-solvent interactions in the preferential solvation of 4-[4-(dimethylamino) styryl]-1-methylpyridinium iodide in 24 binary solvent mixtures. Spectochim Acta, Part A 2006, 65:535–542.
6. El-Kemary MA, Khedr RA, Etaiw SH: Fluorescence decay of singlet excited-state of safranine T and its interaction with ground-state of pyridinthiones in micelles and homogeneous media. Spectochim Acta, Part A 2002, 58:3011–3014.
7. Ishikawa M, Ye JY, Maruyama Y, Nakatsuka H: Triphenylmethane Dyes Revealing Heterogeneity of Their Nanoenvironment: Femtosecond, Picosecond, and Single-Molecule Studies. J Phys Chem A 1999, 103:4319–4331.
8. Jedrzejewska B, Kabatc J, Paczkowski J: Dichromophoric hemicyanine dyes. Synthesis and spectroscopic investigation. Dyes Pigm 2007, 74:262–268.
9. Rauf MA, Akhter Z, Kanwal S: Photometric studies of the complexation of Sudan Red B with Mn+2 and Fe+3 ions. Dyes Pigm 2004, 63:213–215.
10. Gomez ML, Previtali CM, Montejano HA: Photophysical properties of safranine O in protic solvents. Spectrochim Acta, Part A 2004, 60:2433–2439.
11. Souza HM, Bordin JO: Strategies for prevention of transfusion-associated Chagas' disease. Trans Med Rev 1996, 10:161–170.
12. Kamlet JM, Abboud JLM, Abraham MH, Taft RW: Linear solvation energy relationships. 23. A comprehensive collection of the solvatochromic parameters, .pi.*, .alpha., and .beta., and some methods for simplifying the generalized solvatochromic equation. J Org Chem 1983, 48:2877–2887.
13. Marcus Y: The properties of organic liquids that are relevant to their use as solvating solvents. Chem Soc Rev 1993, 22:409–416.

14. Reichardt C: Solvatochromic Dyes as Solvent Polarity Indicators. Chem Rev 1994, 94:2319–2358.
15. Kawski A: On the Estimation of Excited-State Dipole Moments from Solvatochromic Shifts of Absorption and Fluorescence Spectra. Z Naturforsch, A: Phys Sci 2002, 57A:255–262.
16. Kamlet JM, Abboud JL, Taft RW: The solvatochromic comparison method. 6. The .pi.* scale of solvent polarities. J Am Chem Soc 1977, 99:6027–6038.
17. Drago RS, Hirsch MS, Ferris DC, Chronister CW: A unified scale of solvent polarities for specific and non-specific interactions. J Chem Soc Perkin Trans 2 1994, 2:219–230.
18. Drago RS: The interpretation of reactivity in chemical and biological systems with the E and C model. Coord Chem Rev 1980, 33:251–277.
19. Drago RS, Ferris DC, Wong N: A method for the analysis and prediction of gas-phase ion-molecule enthalpies. J Am Chem Soc 1990, 112:8953–8961.
20. Drago RS, Vogel GC: Interpretation of spectroscopic changes upon adduct formation and their use to determine electrostatic and covalent (E and C) parameters. J Am Chem Soc 1992, 114:9527–9532.
21. Drago RS, Vogel GC, Dadmum A: Addition of new donors to the E and C model. Inorg Chem 1993, 32:2473–2479.
22. Soliman AA: Effect of solvents on the electronic absorption spectra of some salicylidene thioschiff bases. Spectrochim Acta, Part A 1997, 53:509.

CITATION

Rauff MA, Soliman AA, and Khattab M. Solvent Effect on the Spectral Properties of Neutral Red. Chemistry Central Journal 2008, 2:19. doi:10.1186/1752-153X-2-19.

The Degradation of Organic Dyes by Corona Discharge

S. C. Goheen, D. E. Durham, M. McCulloch and W. O. Heath

ABSTRACT

Several dyes in water were individually exposed to corona discharge. Light absorbance decreased for all organic dyes with time. Absorbance losses with methylene blue, malachite green, and new coccine were studied. The loss of color was followed using an in situ colorimeter and the effects of varying the current, voltage, gas phase, stirring rates, salinity, and electrode spacing were investigated. The highest reaction rates were observed using the highest current, highest voltage (up to 10kV), highest stirring rate, lowest salinity, smallest electrode spacing, and an environment containing enhanced levels of oxygen. Current was higher in the presence of nitrogen than in the presence of oxygen (for the same voltage), but the reaction of methylene blue did not proceed unless oxygen was present. These results help identify conditions using corona discharge in which dyes, and potentially other organics, can be destroyed.

Background

Relatively little research has focused on the destructive capabilities of corona discharge for organic or biochemicals. Polychlorinated biphenyls (PCBs) and other halogenated organics may be easily broken down into a mixture of harmless chemicals using superoxide (O_2^-) generated in solution [I].

The oxidizing environment caused by corona discharge is believed to be potent, generating electrons, negative molecular ions (including ions of nitrogen and oxygen), ozone and ultraviolet light [2]. Corona discharge has been reported to destroy aerosolized bacterial spores of Bacillus thermophillius, reducing the spore concentration >99% [2]. These authors further proposed that corona discharge might also fragment both harmful bacterial spores and toxic organic materials. Two articles have reported the oxidation of biomolecules in the laboratory using corona discharge. One of these reported the oxidation of NADH and NADPH [3]; the other reported the rapid destruction of hemoglobin [4].

Corona discharge has also been shown to oxidize a polyethylene surface and induce the graft polymerization of acrylamide [5], and to oxidize inorganic molecules. One report claimed that mixtures of SF6 and water subjected to point-to-plane 50 Hz ac corona discharges produced only stable gaseous by-products of SOF2 and SO2F2 as detected by GC or GC-MS [6]. Another study reported the oxidation of copper, silver, and gold foils [7].

Oxygen free radicals and related species have been shown to effectively destroy several organic hazardous species. A reaction process that breaks down industrial solvents, pesticides, dioxins, PCB's, and munitions chemicals into smaller, safer products has been demonstrated using titanium dioxide as a catalyst and the sun's ultraviolet light. In this process, hydroxyl radicals and peroxide ions are believed to break down the hazardous organics into water, carbon dioxide, and some very dilute acids [8]. Others have shown that a reaction involving the superoxide ion (O_2^-) will convert PCB's into bicarbonate of soda and halide ions. This is accomplished via the in situ electrolytic reduction of dissolved oxygen [I]. The same authors suggested that O_2^- generated in solution can react with PCB's to form HCl and carbonic acid.

Reaction rates, kinetics, and mechanisms of the effects from corona discharge have also been studied. SOx and NOx have been successfully removed from iron-ore sintering machine flue gas using high energy electrons in the presence of NH3. Oxidation by corona discharge creates ammonium sulfate and ammonium nitrate precipitates [9,10]. Similar tests have been run on synthetic coal-fired boiler flue gases [10]. Sulfur dioxide has been converted to sulfuric acid mist by corona discharge. The rate-determining step was the formation of atomic oxygen by electrical discharge. The concentration of SO2 was 500-3000 ppm in a flowing humid air mixture. Optimum reaction rates occurred at 70% relative humidity and above a 15% oxygen concentration [11]. SO2 removal was enhanced when the liquid surface was subjected to corona discharge, with the bulk of the acid deposition within the chamber. Gas-phase oxidation of SO2 in the absence of water was minimal [12].

A negatively charged electrode placed in the gas phase above grounded water will generate superoxide (O_2^-) [13] and ozone (O_3) [8]. In the presence of iron salts, superoxide can also be converted to hydrogen peroxide and the highly reactive hydroxyl radical [14]. Reports indicate that ion species created by corona discharge in air at atmospheric pressure also include CO_3^-, $CO_3^-(H_2O)x$, (x = 1, 2), O^-, NO_2^-, O_3, O_3^- NO+, NO_2+, N_2 and H+(H_2O)n (n≤9) [15–17] . These ions and radicals are highly unstable and may react with a broad spectrum of organics. The present report demonstrates the ability of corona discharge to bleach various organic dyes. These reactions may become useful as indicators of corona-stimulated reactions.

Materials

A high voltage power supply (Kilovolt Corporation, Hackensack, NJ) was used with an output of up to ca. 15kV. A dual-pen strip chart recorder (Soltec, San Fernando, CA), a PC600 colorimeter (Brinkman Instruments, Cleveland, OH) for constant on-line monitoring, and a picoammeter (Keithley Instruments, Cleveland, OH) were also used to record the current between the two electrodes. Initial spectrophotometric results were obtained using a Varian DMS200 UV-Vis spectrophotometer (Varian Instruments, Sunnyvale, California). The reaction chamber consisted of a one liter wide mouth flask with 6 access ports (Figure 1). The ports were fitted with ground glass stoppers. Each port had access-tube options such that electrodes or gasses could be introduced near the bottom, middle, or top of the reaction chamber (Figure 1). A stainless steel anode with a sharpened point was placed above the water level, inserted through one of the six ports of the reaction vessel. The platinum cathode (ground) was insulated through the gas phase of the chamber by a glass tube, but exposed at the bottom of the vessel. It was inserted into the vessel through another port at the top. The electrodes were insulated using glass tubes which were connected by ground glass joints to help seal the vessel from external air. The cathode was inserted through a glass tube which protruded into the liquid phase all the way to within about 1/4" of the bottom of the flask. The anode was connected to the power supply and served as the ion source. Efforts have been made to minimize the number of construction materials present in the reaction chamber. The main substance with which material came into contact was glass, which appeared to have been resistant to degradation.

Teflon stirrer bars were used. The stainless steel electrodes appeared to corrode in the reaction vessel, while the platinum (grounded) electrode did not.

Reagents used were methylene blue (Kodak, Rochester, NY), Malachite Green (Sigma Chemical, St. Louis, MO), New Coccine (Aldrich Chemical, Milwaukee, WI) , and silicic acid (Mallinckrodt, Paris, KY) . Water was purified using a Milli Q water purification system (Millipore Corp., Bedford, MA).

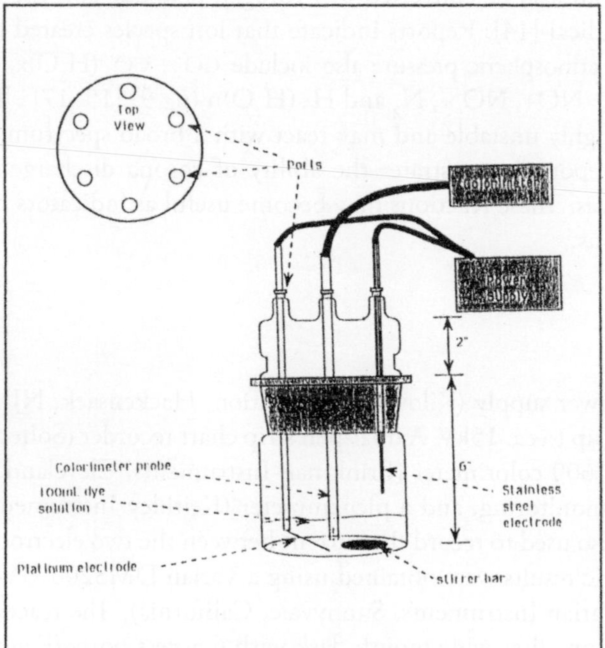

Figure 1. Reaction Vessel. A glass two liter reaction vessel was used for all experiments. 100ml of dye in Milli-Q deionized water was placed in each vessel. A dc power supply provided up to 15kV between the electrodes, with the negatively charged corona above the water surface. A teflon magnetic stirrer bar was used with continuous, vigorous stirring in all cases. The vessel was kept sealed during most experiments.

Methods

A concentration of 6 μg/mL of dye in water was used in all cases. This concentration was chosen because it was found to show best results without overloading the sensitivity of the detector. Absorbance was measured continuously using a submersible fiber optic cell, supplied with the PC600 colorimeter. Current between the two electrodes (stainless steel and platinum) was measured continuously, using a picoammeter. Typically, 5 – 15,000 Volts were used which resulted in a current of Ca. 10 to 50 μA. Current fluctuations of Ca. 10% were often present, largely due to rigorous stirring. When the magnetic stirrer bar was removed, or the stirrer was turned off, current fluctuated only slightly. All experiments were carried out with one control. Controls consisted of aliquots of mixed solutions of dye which were stirred adjacent to the exposed fractions, but without current between the electrodes.

Results

Several experiments were carried out exposing methylene blue, malachite green, and other dyes to corona discharge in the presence of air, using the reaction vessel shown in Figure 1. All dyes were dissolved in water as described in Methods. In the presence of air and at ambient temperature, methylene blue lost most of its color (absorbance) within four hours of exposure to 10 kV and 20 μA or corona discharge. Figure 2 shows the change in absorbance of a 6.0 μg/mL solution of methylene blue at 664 nm both with and without (control) exposure to corona discharge. Approximately half of that absorbance was lost within one hour under these conditions. There was little or no change in absorbance after 4.5 hours of exposure to corona discharge. The experiment was ended after 24 hours.

Malachite green (Figure 3) and new coccine (Figure 4) behaved in a similar manner in the presence of corona discharge and under the same conditions. Experiments were also performed under different atmospheric conditions (in the presence of N2, O2, He, individually) to determine their effect on reaction rates. Results from these experiments indicated dyes reacted rapidly only in the presence of oxygen and corona discharge to form colorless solutions (Figures 2-4). For these and in subsequent experiments, methylene blue was chosen rather than all three dyes.

Experiments were performed to measure the effect of current on bleaching rate of methylene blue (Figure 5). Current was adjusted by changing the voltage. Reaction rates were measured as a function of current by following the loss of color. The reaction rate increased with increasing current. The rate of this increase was proportional to the current.

There was little or no reaction in the presence of helium, a substantial reaction (faster than air) in the presence of oxygen, and no reaction in the presence of nitrogen (Figure 5). The current between the electrodes (at 10 kV dc power supplied) fluctuated wildly in the presence of helium, was constant in the presence of oxygen, as well as in the presence of nitrogen, the current was ca. 150 pA, significantly higher than in air, at a constant voltage of 5 kV. Fluctuations in current coincided with the motion of the aqueous solution during stirring. An experiment with 0.15 M NaC1 in the methylene blue solution indicated that NaC1 substantially slowed the bleaching reaction (Figure 5).

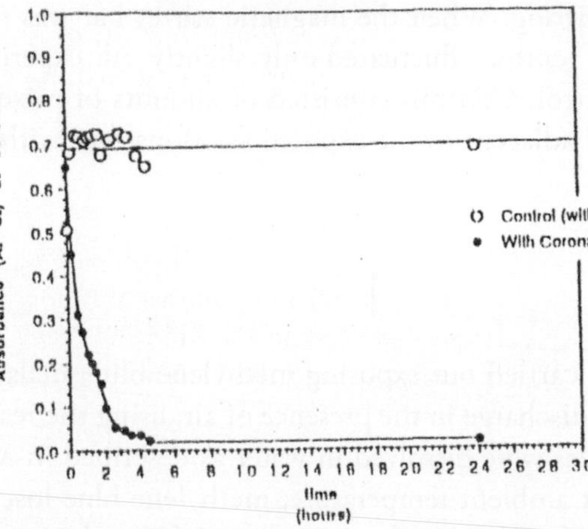

Figure 2. Methylene Blue at 664.4 nm. The change in absorbance at 664 nm of a 100 mg/mL solution of methylene blue with and without exposure to corona discharge is shown. The exposure chamber used for these experiments is shown in Figure 1. A 10 kV potential and 20 μA current were used for the corona exposed solution. More than half the absorbance was lost in the first hour of exposure. Little additional change in absorbance occurred between 4.5 and 24 hours of exposure to corona discharge. No significant effects were observed in the absence of corona discharge.

In the flasks in which solutions were exposed to corona, water directly under the electrode was disturbed, even in the absence of mechanical stirring. The disturbance appeared like a dimple, similar to the effect of blowing a small stream of air directly down on the water surface. We referred to this effect as the "corona wind."

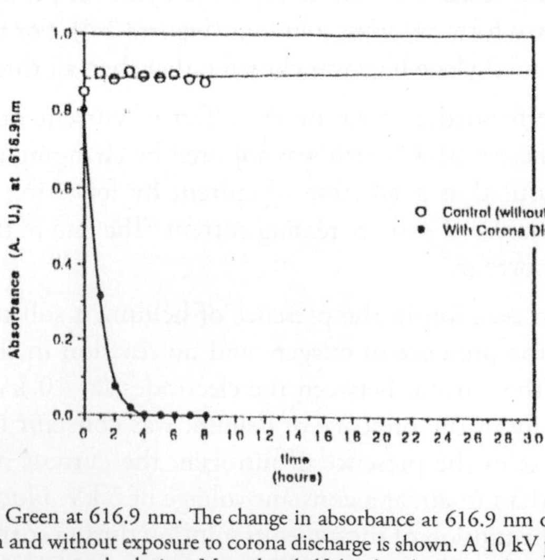

Figure 3. Malachite Green at 616.9 nm. The change in absorbance at 616.9 nm of a 100 mg/mL solution of malachite green with and without exposure to corona discharge is shown. A 10 kV potential and 20 μA current were used for the corona exposed solution. More than half the absorbance was lost in the first hour of exposure. Little additional change in absorbance occurred between 4.5 and 24 hours of exposure to corona discharge. No significant effects were observed in the absence of corona discharge.

Discussion and Conclusion

Many efforts have been made to explore the efficiency of corona emitters. For example, it has been shown that a triangle-shaped point-to-plane geometry electrostatic dc corona device is more efficient than a point-to-plane dc corona device with a rectangular cross section [18]. A semi-empirical mathematical model was used to describe the electrical characteristics and to refine estimates on the optimum shape of the device [18]. Experiments have also shown that a greater voltage was required to generate corona discharges when the anode radius was increased [19].

Furthermore, when cylindrical hollow electrodes generate corona discharge, the temperature of the electrodes has been shown to increase [20]. In the present study, the point-to-plane geometry was chosen with the aqueous surface acting as the (grounded) plane. The pointed electrode was solid, not hollow, and composed of stainless steel.

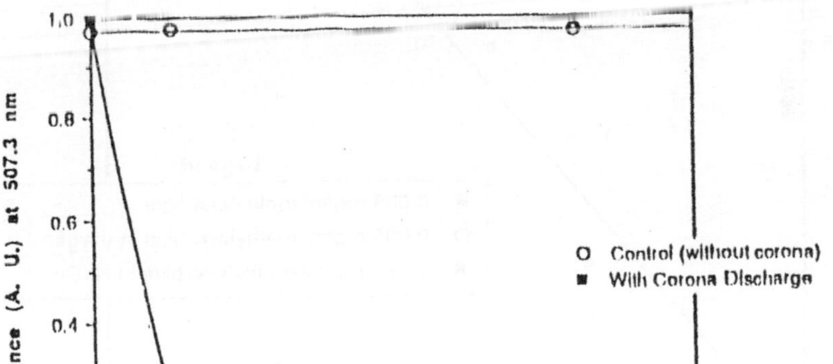

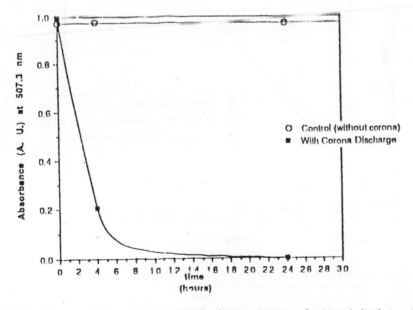

Figure 4. New Coccine at 503.3 nm. The change in absorbance at 503.3 nm of a 100 mg/mL solution of new coccine with and without exposure to corona discharge is shown. A 10 kV potential and 20 μA current were used for the corona exposed solution. More than half the absorbance was lost in the first hour of exposure. Little additional change in absorbance occurred between 4.5 and 24 hours of exposure to corona discharge. No significant effects were observed in the absence of corona discharge.

In the presence of gases other than air, the current, but not the reaction rate, was greater in the presence of nitrogen. Corona discharges have previously been shown to be enhanced in a nitrogen atmosphere [21]. However, the present study demonstrates that current and composition of gas phase both influence the rate of the reaction, and that oxygen is involved.

Previous attempts have been made to develop reactions using electrical oxidation of gas pollutants. One of these was an electrochemical flow reactor (essentially an electrostatic precipitator with a catalyst) developed to oxidize a large volume of gas with hydrocarbon pollutants. However, the reactor apparently precipitated the hydrocarbons to the inner wall (outer electrode) of the reactor in the presence of corona discharge [22]. In the present study, no precipitates or dye deposits were observed in the reaction vessel after bleaching had taken place.

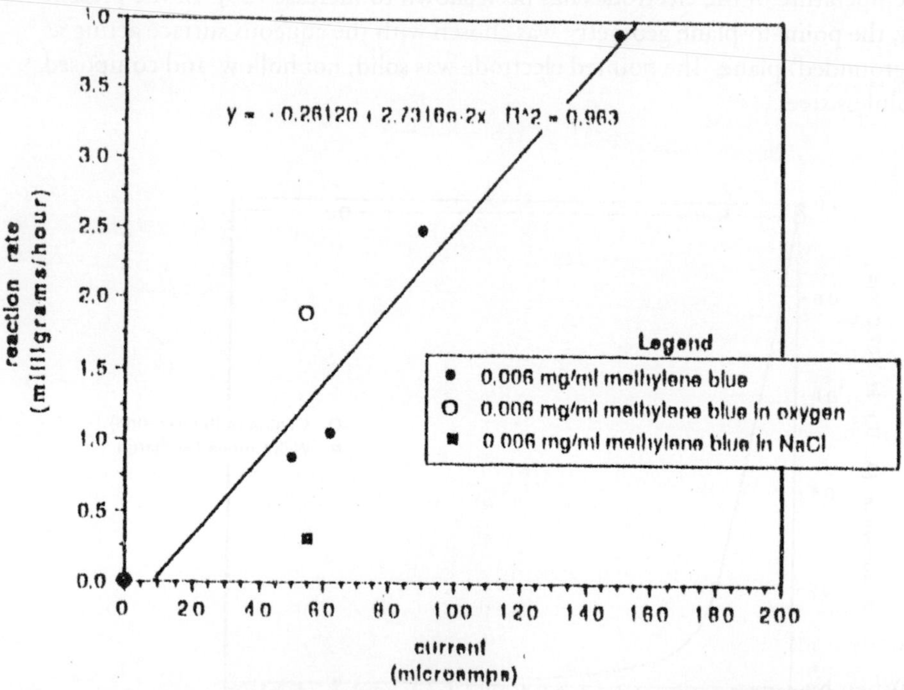

Figure 5. Effect of Current on Reaction Rate. Figure 5 shows the initial reaction rate of methylene blue when exposed to corona discharge versus current (between the two electrodes of Figure 1). Conditions were the same as for Figure 2. The dark circles were from experiments in ambient air, using only water as the solvent. The open circle represents the reaction rate in the presence of oxygen. The dark square represents the reaction rate when 0.15 NaC1 was used in the solvent.

Corona discharge is known to be responsible for the generation of superoxide [13], singlet oxygen [3], and ozone [8] in the presence of water. The observed loss of dye absorbance may have been due to the destruction of dyes by any of a number of mechanisms. However, when hemoglobin was exposed to corona discharge in a similar system, ozone was suspected as the prime oxidizing component [4]. In the present studies, ozone or other species such as free radicals of oxygen and hydroxyl radical (OH-) could have been responsible for the observed reaction.

The reactions observed in this study were remarkable in that after the reaction was complete, no absorbance remained within the absorbance range 200 to 900 nm. This is similar to the effect observed previously for hemoglobin [4]. In that case, reaction products appeared to have been much lower in molecular weight than hemoglobin. Reaction products were not identified in either this or the earlier study

These experiments were undertaken to explore methods for optimizing degradation reactions in the presence of corona discharge. The advantage of using dyes is that the reactions could be monitored continuously by colorimetry. Effects were observed when NaC1 was added to the solvent, as well as when the gas phase was displaced with oxygen, helium, and nitrogen. In addition, we observed the reactions to be highly dependent upon stirring rate. The influence of current on the reaction rate (Figure 5) was possibly influenced by the composition of the ambient air. While no reaction took place in a nitrogen atmosphere, the current was very high. Oxygen reduced the current, but increased the reaction rate. No reaction took place in the presence of helium, but in that case current was also low. Voltage alone was poorly correlated with the reaction rate although higher currents generally gave faster bleaching. Furthermore, the electrode spacing to the water surface greatly influenced current at the same voltage. This effect was most remarkable with rigorous stirring, causing current to fluctuate with the motion of the water surface. Attempts were made to control the spacing, but under vigorous stirring conditions, current still fluctuated. An average current was used for the data presented here. An explanation for poor correlations between current and voltage could have been poor control of the gas phase consumption. Small changes in humidity, temperatures, and particulates, as well as the geometry of the electrode tip may have influenced current at a fixed voltage. These variables are currently under investigation.

When oxygen was continuously blown into the reaction vessel, the bleaching of methylene blue proceeded more slowly than when there was no replenishment of oxygen. This suggests that species generated from oxygen, not oxygen alone, were responsible for the reaction. Lower flow rates of oxygen gave faster reactions. When we added oxygen near the end of a reaction, the rate was again diminished. This suggests that the reactions either required relatively small amounts of oxygen or adding fresh oxygen displaced the reaction species.

Reactions between species generated by corona discharge and the organic dyes used in this study were increased by increasing current, adding oxygen, and vigorous stirring. The reaction with methylene blue was also decreased by adding NaCl. These conditions may also enhance reactions with hazardous organic species in the presence of corona discharge. No reaction was observed when the vessel was filled with nitrogen, although current, corona glow, and the corona wind were enhanced. The diminished reaction in the presence of NaCl may have been due to the ability of Na+ to either scavenge superoxide anions, or NaCl to protect the dye by coating the dye molecular surface. Oxygen is clearly required for the dye to react with species generated by corona discharge. This supports the concept that either free radicals of oxygen or ozone are involved in the dye bleaching reaction.

Acknowledgements

The authors thank Richard Richardson for his assistance in designing the circuitry for this project, and the U. S. Department of Energy for funding this research.

References

1. Sugimoto, H., S. Matsumoto, D. T. Sawyer, 1988. "Degradation and Dehalogenation of Polychlorobiphenyls and Halogenated Aromatic Molecules by Superoxide Ion and by Electrolytic Reduction." Environmental Science and Technology, 22 (10) :1182–1186.
2. Hoenig, S. A., G. T. Sill, L. M. Kelley, and K. J. Garvey, 1980. "Destruction of Bacteria and Toxic Organic Chemicals by a Corona Discharge." Air Pollution Control Association Journal, 30 (3):277–278.
3. Bissell, M. G., S. C. Goheen, E. C. Larkin and G. A. Rao, 1983. "Oxidation of Reduced Nicotinamide Adenine Dinucleotide by Negative Air Ions." Biochem. Archives, 1:231–238.
4. Goheen, S. C., M. G. Bissell, G. A. Rao and E. C. Larkin, 1985. "Destruction of Human Hemoglobin in the Presence of Water and Negative Air ions Generated by Corona Discharge." Intl. J. Biometerol., 29: 353–359.
5. Iwata, H., A. Kishida, M. Suzuki, Y. Hata, and Y. Ikada, 1988. "Oxidation of Polyethylene Surface by Corona Discharge and the Subsequent Graft Polymerization." Journal of Polymer Science, Part A, 26(12):3309–3322.
6. Derdouri, A., J. Casanovas, R. Hergli, R. Grob, and J. Mathieu, 1989. "Study of the Decomposition of Wet SF6, Subjected to 50 Hz AC Corona Discharges." Journal of Applied Physics, 65 (5):1852–1857.

7. Bigelow, R. W., 1988. "An XPS Study of Air Corona Discharge-Induced Corrosion Products at Cu, Ag, and Au Ground Planes." Applied Surface Science, 32 (1-2) :122–140.
8. Goheen, S. C., E. C. Larkin and M. G. Bissell, 1984. "Ozone Produced by Corona Discharge in the Presence of Water." Intl. J. Biometerol, 28:157–162.
9. Kawamura K. and V. H. Shui, 1984 . "Pilot Plant Experience in Electron-Beam Treatment of Iron-Ore Sintering Flue Gas and Its Application to Coal Boiler Flue Gas Cleanup." Physical Chemistry, 24(1):117–127.
10. Clements, J. S., A. Mizuno, W. C. Finney, and R. H. Davis, 1989. "Combined Removal of SO 2, NO x, and Fly Ash from Simulated Flue Gas Using Pulsed Streamer Corona." IEEE Transactions on Industry Applications, 25(i) :62–69.
11. Matteson, M. J., H. L. Stringer, and W. L. Busbee. 1972. "Corona Discharge Oxidation of Sulfur Dioxide." Environmental Science & Technology 6(10) :895–901.
12. Vasishtha, N. and A. V. Someshwar, 1988. "Absorption Characteristics of Sulfur Dioxide in Water in the Presence of a Corona Discharge." Industrial and Engineering Chemical Research, 27(7):1235–1241.
13. Kellogg, E. W., M. G. Yost, N. Barthakur and A. P. Kreuger, 1979. "Superoxide Involvement in the Bactericidal Effects of Negative Air Ions on Staphylococcus albus." Nature, 281:400–401.
14. Halliwell, B., 1981. "The Biological Effects of the Superoxide Radical and its Products" in: Clinical Resperatory Physiology, J. Hakin and M. Torres, ed., Pergamon Press:Oxford, pp. 21–28.
15. Williams, E. M., 1984. The Physics and Technology of Xerographic Processes, New York:Wiley-Interscience.
16. Shahin, M. M., 1969. "Nature of Charge Carriers in Negative Coronas." Appl. Opt. Suppl., 3:106–110.
17. Shahin, M. M., 1971. "Characteristics of corona Discharge and Their Application to Electrophotography." Phot. Sci. Eng., 15:322–328.
18. Yamamoto, T., P. A. Lawless, L. E. Sparks, 1989. "Triangle-shaped DC Corona Discharge Device for Molecular Decomposition." IEEE Transactions on Industrial Applications, 25(4):743–749.
19. Nashimoto, K., 1988. "Silicon Oxide Projections Grown By Negative Corona Discharge." Japanese Journal of Applied Physics, 27(6):892–898.
20. Chang, J. S.and I. Maezono, 1988. "The Electrode Surface Temperature Profile in a Corona Discharge." Journal of Physics, D: Appl. Phys., 21(6):1023–1024.

21. Maezono, I. and J. S. Chang, 1988. "Flow-Enhanced Corona Discharge -The Corona Torch." Journal of Applied Physics, 64 (6):3322–3324.

22. Kipp, E., B. K. A. Shelstad, and G. S. P. Castle, 1973."An Electrochemical Flow Reactor for Oxidation of Hydrocarbon Pollutants." Canadian Journal of Chemical Engineering, 1(8):494–498.

CITATION

Goheen SC, Durham DE, McCulloch M, and Heath WO. The Degradation of Organic Dyes by Corona Discharge. Presented at the Chemical Oxidation: Technology for the 90's Conference, Feb. 19–21, 1992, Nashville, TN. http://www.osti.gov/scitech/servlets/purl/10135257. US. Government publication.

UV-Vis Spectrophotometrical and Analytical Methodology for the Determination of Singlet Oxygen in New Antibacterials Drugs

Tamara Zoltan, Franklin Vargas and Carla Izzo

ABSTRACT

We have determined and quantified spectrophotometrically the capacity of producing reactive oxygen species (ROS) as $^1O_2$ during the photolysis with UV-A light of 5 new synthesized naphthyl ester derivates of well-known quinolone antibacterials (nalidixic acid (1), cinoxacin (2), norfloxacin (3), ciprofloxacin (4) and enoxacin (5)). The ability of the naphthyl ester derivatives (6–10) to generate singlet oxygen were detecting and for the first time quantified by the histidine assay, a sensitive, fast and inexpensive method.

The following tendency of generation of singlet oxygen was observed: compounds 7>10>6>8>9>> parent drugs 1–5.

Keywords: histidine assay, singlet oxygen, anti-bacterial, photochemical activity, quinolone, reactive oxygen species

Introduction

Quinolones are antibacterial agents, whose pharmacological action involves the inhibition of an enzyme (bacterial topoisomerase DNA gyrase) that controls the shape of DNA (Sauvaigo et al. 2001). A major side effect of these drugs is skin photosensitization (Vargas et al. 2003; Vargas and Rivas, 1997; Tokura 1998; Naldi et al. 1999; Arata et al. 1998). This has stimulated the photophysical, photochemical and photobiological studies on a large number of quinolones, such as nalidixic acid (1), cinoxacin (2), norfloxacin (3), ciprofloxacin (4) and enoxacin (5) and their new naphthyl ester derivatives (6–10) respectively (Fig. 1).

Quinolones undergo a variety of photochemical processes such as decarboxylation, defluorination, oxidation of an amino substituent at C-7, generation of singlet oxygen and production of superoxide (Vargas et al. 2006 and 1991; Vargas 1997; Fasani et al. 1998; Martinez et al. 1997). These oxygenated species (1O2 and .-O2) can alter the oxidant - antioxidant balance of the biological system through different processes broadly studied as the lipid peroxidation, hemolysis of erythrocytes and damages to neutrophils and DNA (photo-genotoxicity). Singlet oxygen can be generated inside cells by photosensitization and can react efficiently with DNA. An understanding of the genotoxic potential of singlet oxygen requires insight into the following parameters: generation of singlet oxygen and bioavailability to DNA; the reactivity of singlet oxygen with DNA and the nature of the modifications induced; mutagenecity and repair of the DNA modifications induced; and secondary products generated from singlet oxygen that could account for any indirect mutagenecity.

On the other hand, recent joint efforts of physicists, chemists, and physicians resulted in a significant progress in photodynamic therapy (PDT) of malignant tumors and some non-oncological diseases (Privalov et al. 2002; Dougherty 2002). This therapy involves the administration of a photosensitizing agent followed by tissue exposure to a sufficiently powerful laser irradiation in the visible range. When the tumor with accumulated photosensitizer is illuminated by light of the appropriate wavelength, photochemical reactions occur. Most probably, a light induced excitation of the photosensitizer molecules produces a series of molecular energy transfers to ground state oxygen. This last process leads to a generation of singlet molecular oxygen (1O2), highly reactive and cytotoxic species, resulting in cell death.

The singlet oxygen quantum yield (φs) is a key property of a photosensitizing agent. This quantity is defined as the number of molecules of 1O2 generated for each photon absorbed by a photosensitizer. Quantum efficiency is an equivalent term. The production of 1O2 by photosensitization involves four steps: (A) Absorption of light by the photosensitizer; (B) Formation of the photosensitizer triplet state; the quantum yield of this process is the ISC efficiency or triplet yield (φt); (C) Trapping of the triplet state by molecular oxygen within its lifetime; the fraction of trapped triplet states in a given system is designated by Ft; (D) Energy transfer (Et) from the triplet state to molecular oxygen; the probability of this energy transfer is Et; the experimental value of Et is usually unity for those agents in which the fluorescence is not quenched by oxygen. Overall, φs =φt · Ft · Et. Virtually all measurements of φs are scaled to a reference substance. Frequently employed standard values of φs in aqueous media are 0.79 for rose Bengal, 0.52 for methylene blue, and 1.00 for fullerene C60. The published values of φs show considerable variations with the solvent, reaction conditions, and the measurement technique (Redmond and Gamlin 1999).

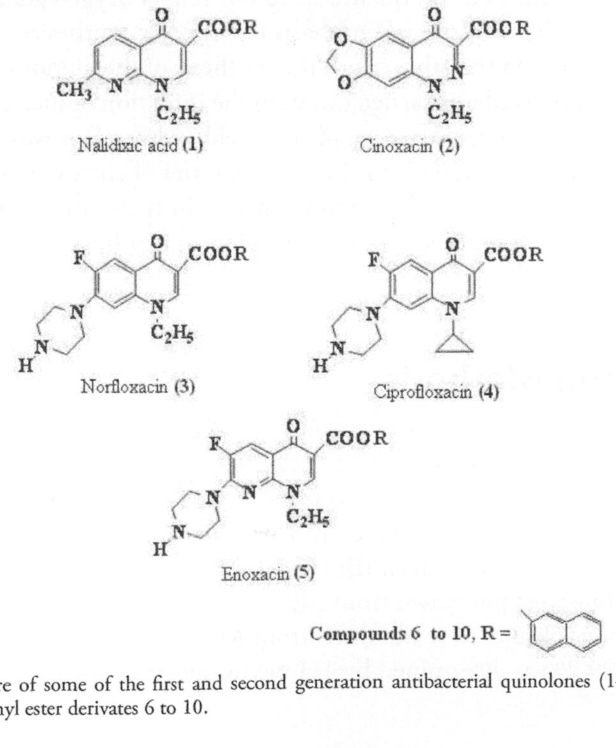

Figure 1. Structure of some of the first and second generation antibacterial quinolones (1–5) and the new synthesized naphthyl ester derivates 6 to 10.

The photophysical and photochemical behavior of photosensitizing drugs under aerobic conditions is particularly relevant to understand the in vivo photobiological effects. In this context, it is somewhat surprising that ofloxacin and rufl oxacin, in spite of their remarkable structural similarity, appear to follow diverging photoreactivity patterns in the presence of oxygen. Thus for example, rufloxacin has a singlet oxygen quantum yield two times higher than that observed for ofloxacin (Sortino et al. 1999; Navaratnam and Claridge 2000).

Expensive methodology or technique used to measure the singlet oxygen yield are given from direct detection of the luminescence produced on relaxation of singlet oxygen (time-resolved or steady-state), calorimetric techniques (photoacoustic calorimetry) and time resolved thermal lensing. It is necessary to perform the appropriate screening for phototoxicity in vitro before introducing drugs and chemicals into clinical therapy. The use of human erythrocytes, lymphocytes and/or neutrophils as cellular systems in our investigation, combined with other in vitro tests employing linoleic acid for lipid photoperoxidation and histidine assay, a sensitive spectrophotometric method to the determination and now for the first time quantified of singlet oxygen, confirmed an important methodology for the study of the phototoxicity of drugs.

The aim of the present study was to establish the histidine assay as a sensitive, fast and inexpensive method to the quantification of singlet oxygen generation by quinolone antibacterials, with special emphasis on the new synthesized naphthyl ester derivative 6 to 10. On the other hand, the synthesis of the quinolone homologous naphthyl ester derived was carried out with the intention of increasing their photostability and their fluorescence quantum yield (whose fact was achieved) and in this way to give to the quinolones better properties of energy transfer when they are subjected to irradiation. This would generate in these compounds, apart from their antibacterials properties, a new behavior and utility like photosensitizer in bacterial media.

Materials and Methods

Chemicals

Nalidixic acid (1), cinoxacin (2), norfl oxacin (3) were purchased from Sigma-Aldrich (St. Louis, MO, USA), while ciprofl oxacin (4) and enoxacin (5) were purchased from Fluka-Riedel-deHaën (Buchs, SG, Schweiz). Histidine and p-nitrosodimethylaniline were purchased from Aldrich (Milwaukee, USA). All analytical or HPLC grade solvents were obtained from Merck (Darmstadt, Germany). Their purity was 99.2% as determined by 1H NMR-spectroscopy (Bruker Aspect

3000, 300 MHz). The spectrophotometrical experiments were recorded with an UV-Vis-Lambda650 spectrophotometer and a Luminescence Spectrometer LS45 Perkin Elmer. The structures of the isolated products were elucidated by ^{1}H NMR and ^{13}C NMR (Brucker Aspect 3000, 300 and 100 MHZ respectively), I.R. (Nicolet DX V 5.07).

Synthesis of Naphthyl Ester Quinolone Derivatives

The synthesis for obtaining the ester compounds were developed with some modifications, taking as example the methodology of the patents Bayer Aktiengesellschaft and Italian Pulitzer S.p.A. (Bayer Aktiengesellschaft Patent 557550, 1987; Pulitzer Italiana S.p.A. Patent 537810, 1984). The esterification of quinolones 1 to 5 with β-naphthol was carried out by making pass a flow of dry and gassy HCL through an equimolar dispersion of the corresponding quinolone and β-naphthol ($3.0 \cdot 10^{-3}$ mol) in CH_2Cl_2 at the reflux temperature during 1 hour. The solid filtrate and washed with dichloromethane were dissolved in cold water and taken to pH 8.9–9.2 (accurately), where a solid Is precipitate In cold overnight.

Irradiation

All processes of irradiation were carried out using a illuminator Cole Palmer 41720-series keeping a distance of 10 cm between the lamp surface and the solution, varying the time periods of exposure at 25 °C under continuous shaking, with a emission maximum in UVA-Vis 320–400 nm (3.3 mW/cm^2, 45.575 Lux/seg) (radiation dose 4.5 J/cm^2) as measured with a model of UVX Digital Radiometer after 1 h continued illumination.

Singlet Oxygen Generation and Quantification

Photosensitized degradation of histidine was measured in the presence of 0.25, 0.50, 1.0, and $1.5 \cdot 10^{-5}$ M solution of compounds 1 to 10 (in etanol/ H_2O 1:10). These solutions were mixed with an equal quantity of L-histidine solution at 0.60 to 0.74 mM in phosphate buffer 0.01 M, pH 7.4. Samples of these mixtures were irradiated with an illuminator Cole Palmer 41720-series keeping a distance of 10 cm between the lamp surface and the solution at 25 °C, with a emission maximum in UV-A-Vis 320–400 nm (3.3 mW/cm2, 45.575 Lux/seg) at time intervals from 45 to 60 min, with the respective controls being protected from light. The concentration of histidine was determined by a colorimetric reaction. The optic density was read on a spectrophotometer at 440 nm against a blank reagent (L-histidine/p-nitrosodimethylaniline/ quinolone derivatives without irradiation) by bleaching of p-nitrosodimethylaniline (Lovell and Sanders 1990; Kraljic and El Mohsni 1978). Rose Bengal, a well known $^1O_2$ sensitizer, was used as a standard for comparison with the compounds 1 to 10 for $^1O_2$ formation, under identical conditions of photolysis. The quantum yield of singlet oxygen generation for Rose Bengal is $\varphi(^1O_2) = 0.76$ (Redmond and Gamlin 1999). This value can be used as a standard to determine a relative quantum yield of the new compounds.

Statistical Treatment of Results

At least three independent experiments were performed except where indicated otherwise. The results of the quantification are expressed as a mean ± S.D. Standard deviation (S.D.) is obtained from 3–4 observations. The level of significance accepted was $p \leq 0.05$. Statistical analyses were performed using t-test.

Results and Discussion

The synthesis of the compounds 6 to 10 were carried out taking as example the procedures of the patented works of Bayer Aktiengesselschaft and Italian Pulitzer S.p.A. with some modifications. Next some physical corrected and spectroscopics data are presented.

The corresponding naphthyl ester (6); 1-ethyl1,4-dihydro-7-methyl-4-oxo-1-,8-naphthyridine-3naphthyl ester (C22H18N2O3, mol wt 358.39), yield: 0.520 g of yellow needles (50.62%), m.p. 127–130 °C. I.R. (KBr): 3205, 1936, 1697, 1616, 1352, 1255, 1220, 744 cm–1.

1H-NMR (300 MHz, CDCl3): δ= ppm 8.86 (d, 1H, J6–5 = 7.30, H-6), 8.65 (m, 1H, H-19), 7.73 (s, 1H, H-10), 7.68 (d, 1H, J18-19 = 9.00, H-18), 7.60 (m, 1H, H-26), 7.37 (m, 1H, H-22), 7.24 (m, 1H, H-24), 7.17 (m, 1H, H-23), 7.11 (m, 1H, H-25), 6.80 (d, 1H, J5-6 = 7.30, H-5), 4.52 (q, 2H, J11-12 = 6.80, H-11), 2.70 (s, 3H, H-13), 1.40 (t, 3H, J12-11 = 6.80, H-12).

13C-NMR (300 MHz, CDCl3): δ= 178.66 (CO-8), 167.28 (C-2), 164.77 (CO-15), 153.00 (C-17), 148.58 (C-4), 148.20 (C-10), 136.05 (C-21), 134.06 (C-6), 129.65 (C-19), 128.74 (C-23), 127.64 (C-26), 126.00 (C-25), 126.31 (C-24), 123.34 (C-5), 122.29 (C-18), 111-22), 109.46 (C-9), 47.44 (C-11), 25.29 (C-13), 15.16 (C-12).

The corresponding naphthyl ester (7); 1-ethyl-1, 4-dihydro-4-oxo-[1,3] dioxolo[4,5-g]cinnoline3-naphthyl ester (C22H16N2O5, mol wt 388.37), yield: 0.490 g of yellow needles (42%), m.p. 143–145 °C. I.R. (KBr): 3282, 1700, 1631, 1469, 1385, 1276, 1242, 743 cm–1.

Compound 6

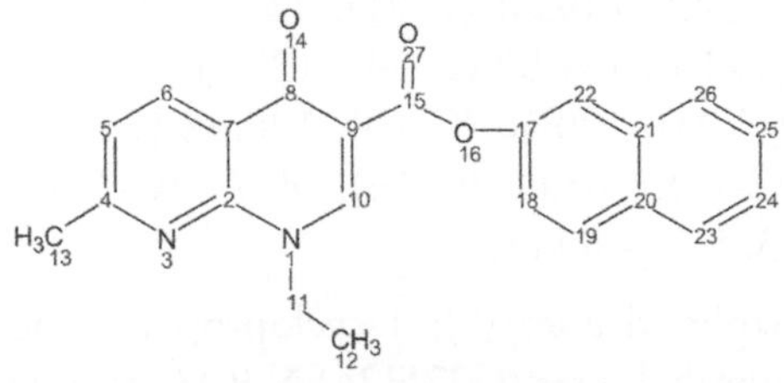

Compound 6

1H-NMR (300 MHz, CDCl3): δ= ppm 7.75 (d, 1H, H-20), 7.67 (d, 1H, H-19), 7.63 (s, 1H, H-23), 7.59 (m, 1H, H-27), 7.37 (m, 1H, H-25), 7.29 (m, 1H, H-24), 7.10 (m, 1H, H-26), 6.50 (s, 1H, H-13), 6.21 (s, 1H, H-6), 5.66 (s, 2H, H-2), 4.56 (q, 2H, J14-15 = 6.88, H-14), 1.50 (t, 3H, J15-14 = 6.88, H-15). 13C-NMR (300 MHz, CDCl3): δ= 169.78 (C-11), 164.71 (C-16), 155.12 (C-5), 153.52 (C-18), 149.14 (C-4), 138.93 (C-7), 134.56 (C-10), 132.84 (C-20), 129.69 (C-21), 128.79 (C-27), 127.67 (C-24), 126.29 (C-26), 123.43 (C-25), 117.81 (C-19), 109.44 (C-23), 103.45 (C-2), 102.20 (C-13), 94.60 (C-6), 53.92 (C-14), 13.74 (C-15).

The corresponding naphthyl ester (8); 1-ethyl-6fluoro-1, 4-dihydro-4-oxo-7-(1-piperazinyl)-3quinolinenaphthyl ester (C26H24FN3O3, mol wt 445.48), yield: 0.430 g of white needles (32%), m.p. 236–237 °C. I.R. (KBr): 3419, 1626, 1489, 1385, 1269, 1030, 931, 825, 741 cm–1. 1H-NMR (300 MHz, CDCl3): δ= ppm 8.65 (s, 1H, H-13), 8.07 (d, 1H, J9-F = 9.20, H-9), 8.00 (m, 1H, H-23), 7.94 (m, 1H, H-22), 7.80 (m, 1H, H-30), 7.40-7.19 (m, 4H, H naphthyl-26, 27, 28, 29), 6.30 (d, 1H, J16-F = 6.20, H-16), 4.31 (c, 2H, J31–32 = 6.88, H-31), 3.36–3.78 (m, 8H, H piperazin-6, 5, 3, 2), 2.29 (s, 1H, H-4), 1.57 (t, 3H, J32-31 = 6.88 H-32).

Compound 7

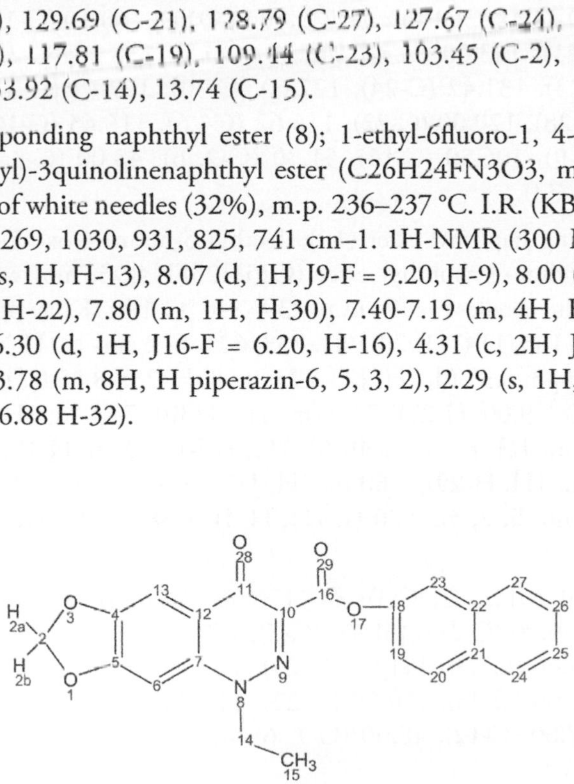

Compound 7

13C-NMR (300 MHz, CDCl3): δ= 176.94 (CO-11), 167.20 (CO-19), 159.40 (C-15), 155.20 (C-8), 149.20 (C-21), 147.77 (C-13), 146.38 (C-7), 137.09 (C-25), 137.00 (C-15), 131.00 (C-24), 127.50 (C-30), 127.52 (C-27), 125.50 (C-29), 125.00 (C-28), 120.50 (C-10), 120.48 (C-22), 118.00 (C-26), 112.94 (C-9), 108.37 (C-12), 105.02 (C-16), 99.75 (C-16), 51.06 (C-2, 6), 49.66 (C-31), 45.84 (C-3, 5), 14.39 (32).

The corresponding naphthyl ester (9); 1-cyclopropyl-6-fluoro-4-oxo-7-piperazin-1-ylquinoline-3-naphthyl ester (C27H24FN3O3, mol wt 457.49), yield: 0.380 g of white needles (31%), m.p. 280–282 °C. I.R. (KBr): 3439, 1624, 1485, 1378, 1183, 1147, 1029, 944, 733, 621, 542 cm–1. 1H-NMR (300 MHz, CDCl3): δ= 8.77 (d, 1H, J9-F = 8.70, H-9), 8.35 (d, 1H, J23-22 = 9.60, H-23), 8.20 (s, 1H, H-13), 8.00 (d, 1H, J22-23 = 9.60, H-22), 7.80 (m, 1H, H-30), 7.60 (d, 1H, J26-22 = 2.60, H26), 7.50 (m, 1H, H-28), 7.40 (m, 1H, H-27), 7.20 (m, 1H, H-29), 6.30 (s, 1H, H-16), 3.30-2.70 (m, 8H, H piperazin-2, 6, 3, 5), 2.50 (s, 1H, H-4), 2.38 (dt, 1H, J31-32 = J31-33 = 5.00, H-31), 2.00 (m 2H, H-32), 1.90 (m, 2H, H-33).

13C-NMR (300 MHz, CDCl3): δ= 175.50 (CO11), 160.10 (C-19), 156.10 (C-8), 150.00 (C-21), 143.00 (C-7), 142.20 (C-13), 138.05 (C-25), 137.00 (C-15), 132.8 (C-23), 131.42 (C-24), 127.60 (C-30), 127.47 (C-27), 126.48 (C-29), 125.50 (C-28), 120.00 (C-22), 116.62 (C-26), 116.45 (C-10), 109.80 (C-12), 107.25 (C-9), 106.50 (C-16), 51.30 (C-2, 6), 45.00 (C-3, 5), 28.45 (C-31), 7.50 (C-32, 33).

The corresponding naphthyl ester (10); 1-ethyl-6-fluoro-4-oxo-7-piperazin-1-yl-[1,8] naphthyridine-3-naphthyl ester (C25H23FN4O3, mol wt 446.47), yield: 0.461 g of white needles (34%), m.p. 239–240 °C. I.R. (KBr): 3439, 1624, 1485, 1378, 1293, 1183, 1147, 1029, 944, 733, 621, 542 cm–1. 1H-NMR (300 MHz, CDCl3): δ= ppm 8.35 (d, 1H, J23-22 = 9.00, H-23), 8.05 (s, 1H, H-11), 7.90 (d, 1H, J22-23 = 9.00, H-22), 7.80 (m, 1H, H30), 7.61 (d, 1H, J26-22 = 2.50, H-26), 7.48 (m, 1H, H-28), 7.40 (d, 1H, J15-F = 9.60, H-15), 7.38 (m, 1H, H-27), 7.12 (m, 1H, H-29), 4.60 (c, 2H, J17-18 = 6.80, H-17), 3.00-2.80 (m, 8H, H piperazin6, 5, 2, 3), 2.20 (s, 1H, H-4), 1.38 (t, 3H, J18-17 = 6.80 H-18).

13C-NMR (300 MHz, CDCl3): δ= 176.00 (CO-13), 166.20 (CO-9), 160.00 (CO-19), 150.00 (C-21), 148.70 (C-7), 147.00 (C-11), 146.50 (C-16), 139.00 (C-25), 133.00 (C-23), 131.05 (C-24), 127.50 (C-30), 127.00 (C-27), 126.40 (C-29), 125.60 (C-28), 120.00 (C-22), 119.70 (C-15), 118.02 (C-26), 112.20 (C-14), 107.60 (C-12), 48.00 (C-2, 6), 47.10 (C-17), 45.88 (C-3, 5), 14.90 (C-18).

Compound 8

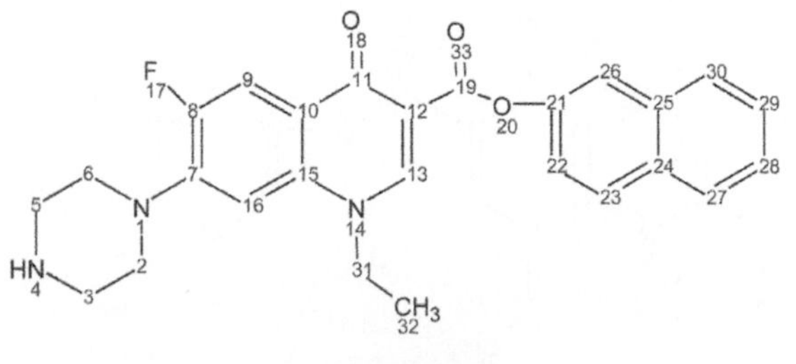

Compound 8

The irradiation conditions described in the experimental section were taken in order to closer resemble the conditions under which their photo-toxicity and the possible phototherapy applications are produced in biological media or in vitro or in vivo cellular systems.

The quantification of singlet oxygen (1O2) was carrying out using the methodology reported by Kraljic in 1978, for its detection (Kraljic and El Mohsni 1978). This methodology is based to the "bleaching" (as secondary reaction) of p-nitrosodimethylaniline (RNO) induced by the selective reaction of singlet oxygen with imidazol derived, in our case we used the histidine like this derived. In the reaction of 1O2 with histidine a trans-annular peroxide takes place as intermediary product, causing the "bleaching" of the group RNO, which can be followed to 440 nm. In absence of RNO the peroxide suffers a rearranges to produce the final products of oxygenation. This methodology can be applied for the quantification of singlet oxygen generated by photosensitizers, since the disappearance of the band of the RNO (to 440 nm) it is a direct measure of the quantity of 1O2 generated by them.

Compound 9

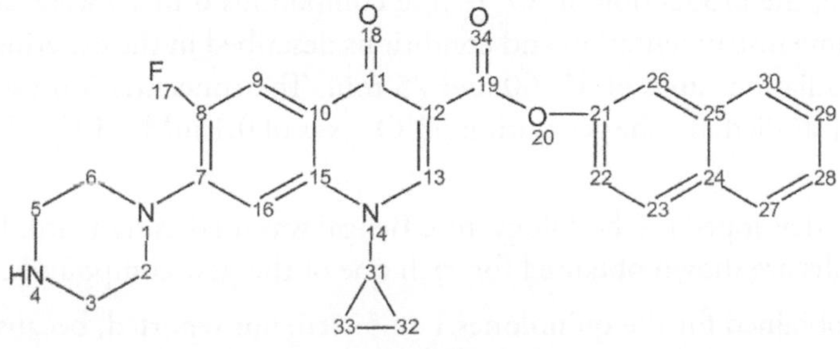

Compound 9

Compound 10

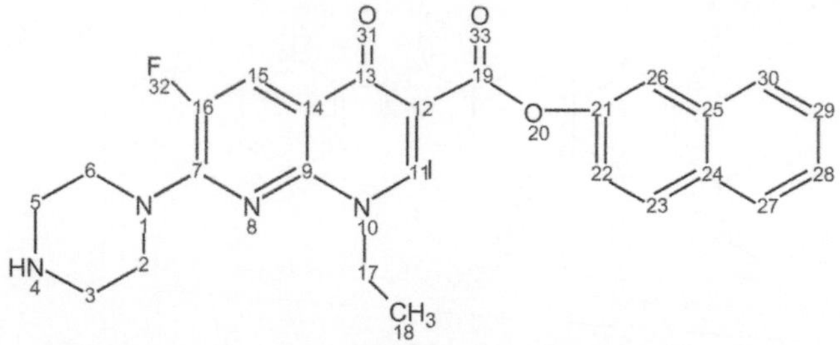

Compound 10

For the quantification based on this methodology we elaborated a calibration curve of the change of optic density as function of the change of p-nitrosodimethylaniline concentration. This curve presents a linear development until to a concentration of 2.00 • 10–5 mol/L (p-nitrosodimethylaniline). For each irradiation time, the variations of optic density were obtained, and by means of the calibration curve of the [p-nitrosodimethylaniline] variations (Fig. 2), they could be related these directly with the quantity of 1O2 taken place by each one of the studied species.

The statistical validation of this linear regression for the validation of the typical errors was performed at the 95% confidential level. The Table 1 shown the figures of merit obtained for the developed of the methodology. The detection limit (DL, 3σ) was calculated for the different irradiation times (45, 60 and 75 min).

Quantification of $^1O_2$ for the Photosensitizers

Kraljic and collaborators report that the generation of $^1O_2$ by photosensitizers is dependent of the irradiation time, as well as of the wave length used in the irradiation. In our case, the production of $^1O_2$ by the compounds 6 to 10 were carried out using the same instrumentation and conditions described in the experimental section with irradiation times of 45, 60 and 75 min. The concentration used for each compound studied, for the generation of $^1O_2$, was of 0,1 mM (H_2O/ethanol, 90:10).

To verify the developed methodology, rose Bengal was used as standard. In the Table 2 the results are shown obtained for each one of the new compounds.

The results obtained for the quinolones 1 to 5 were not reported, because the production of 1O2 was below to the limit of detection of the developed methodology. It is advisable to be carried out by means of the direct measures of detection of the singlet oxygen emission by means of technical of flash photolysis with photomultiplier of germanium for example, although it turns out to be very expensive (Sortino and Scaiano 1999). The formation of singlet oxygen by the photosensitizing mechanism during the photolysis of photosensitizing compounds could be also evidenced by trapping with 2,5-dimethylfuran (GC-mass) (Gollnick and Griesbeck 1983; Costanzo et al. 1989); 1,3-cyclohexadiene-1, 4-diethanoate (HPLC) (Nardello et al. 1997) as 1O2 scavengers, furfuryl alcohol (Haag et al. 1984; Vargas et al. 2001) but these no direct analytical methodology are imprecise and with methodology more complicated that the one developed in this paper on the detection base on histidine. In such a sense we can deduce that most of the phototoxicity taken place by the quinolones 1 to 5 could be produced for the most part by the formation of free radicals, superoxide anion and in smaller grade for singlet oxygen.

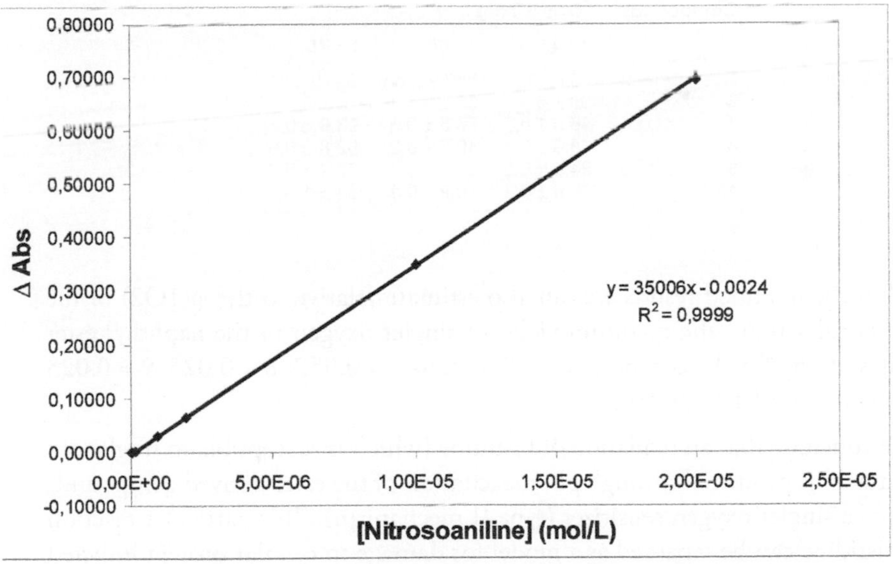

Figure 2. Calibration plot. Date are expressed as means ±S.E.M. Statistically significant ($p \leq 0.05$).

Table 1. Figures of merit.

Regresion line: $\Delta O.D^{a} = a\Delta C^{b} + b$			D.L. (3σ) ($\mu g\ L^{-1}$)		
$a \pm sa$	$b \pm sb$	r^2	$^{c}t = 45$	$t = 60$	$t = 75$
$(350 \pm 1) \times 10^2$	$-0,0024 \pm 0.0009$	0.99993	4,24	1,48	1,29

[a]ΔO.D: Variation of the optic density.
[b]ΔC: Variation of the p-nitrosodimethylaniline concentration.
[c]time of irradiation (t) in minutes.

The ester compounds 6 to 10 are capable of producing singlet oxygen when it is irradiated with UV-A and visible light in the presence of molecular oxygen. This fact can be confirmed by trapping with histidine. We use a simple and sensitive spectrophotometric method for the detection of 1O2 as produced by different sensitizing dyes in neutral air saturated aqueous solutions. The reaction between histidine and 1O2 results in the formation of a trans-annular peroxide. The presence of the latter compound may be detected by bleaching the p-nitrosodimethylaniline at 440 nm. Singlet oxygen alone can not cause the bleaching of the latter compound. No bleaching occurs in the mixture of histidine and p-nitrosodimethylaniline without singlet oxygen (Kraljic and El Mohsni 1978). In order to control the reaction, we observe no measurable loss of the p-nitrosodimethylaniline in the absence of histidine.

Table 2. Generation of $^1O_2$ by the compounds 6–10.

Compounds	**$^1O_2 \pm$ SD (μg L^{-1})**		
	t = 45	**t = 60**	**t = 75**
Rose Bengal	832 ± 1	886,9 ± 0,9	929 ± 2
6	38 ± 4	50 ± 5	70 ± 5
7	56,9 ± 0,1	76,5 ± 0,5	93,9 ± 0,3
8	28,0 ± 0,7	40,0 ± 0,2	52,8 ± 0,9
9	24,9 ± 0,5	27,1 ± 0,7	37,7 ± 0,7
10	55,6 ± 0,3	70,9 ± 0,1	84 ± 1

With the obtained results we can also estimate relative to the φ(1O2) of the Rose Bengal = 0.76, the quantum yields of singlet oxygen to the naphthyl ester derivates. Being for the compound 6 = 0.034; to 7 = 0.052; 8 = 0.025; 9 = 0.023 and to compound 10 = 0.050.

We conclude that an oxidation of histidine (which is susceptible to singlet oxygen attack) is produced through photoexcitation of the ester derived compounds acting as a singlet oxygen sensitizer (type II mechanism). This particular reaction with histidine can be regarded as a model for damage to cellular protein inflicted by photoexcited quinolone antibacterials via formation of singlet oxygen.

On the basis of the results in the present investigation it can be concluded that an analytic method of histidine assay can be used safely to determine the generation of singlet oxygen by drugs.

We have proven that the naphthyl ester derivates of quinolones 1 to 5 produces singlet oxygen under irradiation with visible light. The compounds 7 > 10 > 6 > 8 > 9 (of major to minor generation of singlet oxygen) have the advantages such as sufficient strength to generate under irradiation with visible light 1O2, good hydrophilicity and therefore potential specific affinity for malignant tumors.

These facts are of major significance for the study of its photodynamic action and make these compounds a promising candidate as PDT agents in the medical field. In situ production of the singlet oxygen could be the principle mechanism for tumor destruction in application of photodynamic therapy employing these novel water soluble compounds. Their photobiological properties are deserving of our further investigation.

We don't dare to make any conclusion on a relationship between the structure of the compounds and the generation of singlet oxygen until having a studies it has more than enough theoretical parameters, for example those carried out on phototoxicity of antihyperlipoproteinemic drugs (Aguiar et al. 1995).

References

Aguiar, C., Vargas, F., Canudas, N. and Ruette, F. 1995. Theoretical and experimental studies of intermediate species of photolysis, and phototoxicity of antihyperlipoproteinemic drugs (Fibrates). Molecular Engineering, 4:451–63.

Arata, J., Horio, T., Soejima, R. et al. 1998. Photosensitivity reactions caused by lomefloxacin hydrochloride: a multicenter survey. Antimicrob Agents Chemother, 42:3141–5.

Bayer Aktiengesellschaft Patent 557550, 14/05/1987, Procedure for the obtaining of quinoline carboxilic acid.

Costanzo, L.L., De Guidi, G., Condorelli, G., Cambria, A. and Fama, M. 1989. Molecular mechanism of drug photosensitization II. Photohemolysis sensitized by ketoprofen. Photochem. Photobiol., 50:359–65.

Dougherty, T.J. 2002. Optical methods for Tumor Treatment and Detection: Mechanism and Techniques in Photodynamic Therapy XI. (T.J. Dougherty ed.), Proceedings of SPIE Vol. 4612.

Fasani, E., Profumo, A. and Albini, A. 1998. Structure and medium-dependent photodecomposition of fluoroquinolone antibiotics. Photochem. Photobiol., 68:666–74.

Gollnick, K. and Griesbeck, A. 1983. [4+2]-Cycloaddition of singlet oxygen to 2,5-dimethylfuran: isolation and reactions of the monomeric and dimeric endoperoxides. Angewandte Chemie International Edition in English, 22:726–28.

Haag, W.R., Hoigne, J. and Gassman, A.D. 1984. Singlet oxygen in surface waters part I: furfuryl alcohol as a trapping agent. Chemosphere, 13:631–40.

Kraljic, I. and El Mohsni, S. 1978. A new method for the detection of singlet oxygen in aqueous solutions. Photochem. Photobiol., 28:577–81.

Lovell, W.W. and Sanders, D.J. 1990. Screening test for phototoxins using solutions of simple biochemicals. Toxic in Vitro, 4:318–20.

Martinez, L.J., Li, G. and Chignell, C.F. 1997. Photogeneration of fluoride by the fluoroquinolone antimicrobial agents lomefloxacin and fleroxacin. Photochem. Photobiol., 65:599–602.

Naldi, L., Conforti, A., Venegoni, M. et al. 1999. Cutaneous reactions to drugs. An analysis of spontaneous reports in four Italian regions. Br. J. Clin. Pharmacol., 48:839–46.

Nardello, V., Brault, D., Chavalle, P. and Aubry, J.M. 1997. Measurement of photogenerated singlet oxygen in aqueous solution by specific chemical trapping with sodium 1,3-cyclohexadiene-1,4-diethanoate. J. Photochem. Photobiol. B. Biol., 39:146–155.

Navaratnam, S. and Claridge, J. 2000. Primary photophysical properties of ofloxacin. Photochem. Photobiol., 72:283–90.

Privalov, V.A., Lappa, A.V., Seliverstov, O.V. et al. 2002. Clinical Trials of a New Chlorin Photosensitizer for Photodynamic Therapy of Malignant Tumors, In: Optical methods for Tumor Treatment and Detection: Mechanisms and Techniques in Photodynamic Therapy XI, T.J. Dougherty, Editor, Proceedings of SPIE Vol. 4612, p.178–89. Pulitzer Italiana S.p.A. Patent 537810, 09/11/1984, for the obtaining of piperacinyl-quinoline- benzoylmetil ester.

Redmond, R.W. and Gamlin, J.N. 1999. A complilation of singlet oxygen yields from biologically relevant molecules. Photochem. Photobiol., 70:391–475.

Sauvaigo, S., Douki, T., Odin, F. et al. 2001. Analysis of fluoroquinolonemediated photosensitization of 2′-deoxyguanosine, calf thymus and cellular DNA: determination of Type-I, Type-II and triplet–triplet energy transfer mechanism contribution. Photochem. Photobiol., 73:230–37.

Sortino, S., Marconi, G., Giuffrida, S. et al. 1999. Photophysical properties of rufloxacin in neutral aqueous solution. Photochem. Photobiol., 70:731–36.

Sortino, S. and Scaiano, J.C. 1999. Photogeneration of hydrated electrons, nitrogen-centered radicals and singlet oxygen from Naphazoline: A laser flash photolysis study. Photochem. Photobiol., 70:590–95.

Tokura, Y. 1998. Quinolone photoallergy: photosensitivity dermatitis induced by systemic administration of photohaptenic drugs. J. Dermatol. Sci., 18:1–10.

Vargas, F., Rivas, C. and Machado, R. 1991. Decarboxylation and singlet oxygen production in the photolysis of nalidixic acid. J. Photochem. Photobiol. B. Biol., 11:81–5.

Vargas, F. 1997. Singlet oxygen generation in the photolysis of drugs. En: Modern Topics in Photochemistry and Photobiology. Editor Franklin Vargas, Publicado por Research Signpost, Trivandrum, India, p 27–40.

Vargas, F. and Rivas, C. 1997. Mechanistic studies on phototoxicity induced by antibacterial quinolones. Toxic Substances Mechanisms, 16:81–5.

Vargas, F., Méndez, H., Fuentes, A. et al. 2001. Photosensitizing activity of the thiocolchicoside. Photochemical and in vitro phototoxicity studies. Die Pharmazie, 56:83–8.

Vargas, F., Rivas, C., Díaz, Y. et al. 2003. Photoinduced interaction of antibacterial quinolones with human serum albumin. Toxicology Mechanisms and Methods, 13:221–3.

Vargas, F., Rivas, C. and Fernández, A. 2006. Photodegradation and phototoxicity of antibacterial agents and the new alternatives for the photo-inactivation of microorganisms. Ciencia, 14:210–17.

CITATION

Zoltan T, Vargas F, and Isso C. UV-Vis Spectrophotometrical and Analytical Methodology for the Determination of Singlet Oxygen in New Antibacterials Drugs. Analytical Chemical Insights. 2007; 2: 111–118.

The Application of Resonance Light Scattering Technique for the Determination of Tinidazole in Drugs

Xin Yu Jiang, Xiao Qing Chen, Zheng Dong and Ming Xu

ABSTRACT

A resonance light scattering technique to determine tinidazole in drugs was developed by tetraphenylboronsodium(TPB). Tinidazole was found to bind B(C6H5)- 4 anion and transformed to tinidazole-TPB aggregate which displayed intense resonance scattering light. Effects of factors such as wavelength, acidity, stabilizers, and interferents on the RLS of tinidazole TPB were investigated in detail. The RLS intensity of the tinidazole-TPB suspension was obtained in sulfuric acid solution (pH = 1.44). The resonance scattering light intensity at the maximum RLS peak of 569.5nm was linear to the concentration of tinidazole in the range of 10.0–30.0 μgm L−1 with a detection limit of 5.0μgmL−1. Good results were also obtained with the recovery range of 95.13–106.76%. The method was applied to determine tinidazole in injections and tablets, showing high sensitivity and accuracy compared with the high performance liquid chromatography method (HPLC) according to Chinese Pharmacopoeia.

Introduction

Tinidazole is chemically 1-(2-ethylsulfonyl-ethyl)-2-methyl-5-nitroimidazole (Figure 1). It is active against protozoa and anaerobic bacteria and is used like metronidazole in a range of infections [1]. The drug is reported to hydrolyze quantitatively in alkaline conditions to 2-methyl-5-nitroimidazole and under photolytic conditions, the drug yields intermediate, rearrangement, and degradation products [2].

Resonance light scattering (RLS) is an elastic scattering and occurs when an incident beam in energy is close to an absorption band. Pasternack et al. first established the RLS technique to study the biological macromolecules by means of an ordinary fluorescence spectrometer [3–5]. Due to their high sensitivity, selectivity, and convenience, RLS studies have attracted great interest among researchers [6–10]. RLS has emerged as a very attractive technique that has been used to monitor molecular assemblies and characterize the extended aggregates of chromophores. In recent years, RLS technique has been used to determine pharmaceutical [11, 12] and various biological macromolecules such as nucleic acid [13, 14], protein [15, 16], metal ion [17], and bacteria [18], while the study and determination of tinidazole with RLS technique were not yet reported.

Several analytical methods for tinidazole have been developed so far such as HPLC [19], LC-MS [20], capillary electrophoresis [21], spectrophotometry [22], voltammetry [23], and electrochemical methods [24, 25]. Among these analytical methods, the voltammetry method according to Chinese Pharmacopoeia is popular and regarded relatively reliable for the determination of tinidazole. Although it often provides very accurate results, it suffered from cost time and complexity. HPLC was also used to determine tinidazole in drugs in Chinese Pharmacopoeia with good result but needs tedious pretreatment.

Herein, we report a robust, quick, and simplemethod for the determination of tinidazole in injections and tablets with NaB(C6H5)4 as a probe by RLS technique. The obtained results were almost in agreement with those obtained by the currently used HPLC method according to Chinese Pharmacopoeia.

Experimental

Apparatus

RLS spectra were obtained by synchronous scanning in the wavelength region from 250 to 750nm on a JASCO FP-6500 spectrofluorometer (Tokyo, Japan) using quartz cuvettes (1.0 cm). The width of excitation and emission slits was set at 3.0 nm. HPLC analysis was carried out on an Agilent 1100 HPLC system (USA) equipped with G1314A isocratic pump, a thermostatted column compartment, a variable wavelength UV detector (VWD), and Agilent ChemStation software. The pH measurements were carried out on a PHS-3C exact digital pH meter equipped with Phoenix Ag–AgCl reference electrode (Cole-Palmer Instrument Co., Ill, USA), which was calibrated with standard pH buffer solutions.

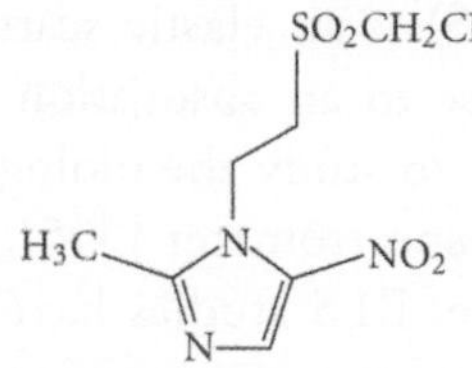

Figure 1. Structure of tinidazole.

Reagents

A working solution of tetraphenylboron sodium (10.0 mgmL^{-1}) was prepared with methanol-water solution (20 : 80, v/v). A stock solution of tinidazole was prepared by dissolving tinidazole (> 99.99%, Sigma) in the doubly distilled water. The working solutions of tinidazole were obtained by diluting the stock solution prior to use. Sulfuric acid solution (0.18mol L^{-1}) was used to control the acidity, while 0.1mol L^{-1} NaCl was used to adjust the ionic strength of the aqueous solutions. All other reagents and solvents were of analytical reagent grade and used without further purification unless otherwise noted. All aqueous solutions were prepared using newly double-distilled water.

Scheme

The composition of precipitate was determined by the Job-Asmus method [26]. The molar ratio tinidazole: TPB was found to be 1 : 1. It is possible that stronger basic secondary amine group in the molecule of tinidazole was transferred to cationic ion and reacted with tetraphenylboron. The precipitation reaction may be as follows.

Standard Procedure

An appropriate aliquot of tinidazole working solution was added to a mixture of 1.0mL of tetraphenyl boron sodium solution (10.0mgmL^{-1}), and 1.0mL sulfuric acid (0.18mol L^{-1}) and diluted to 10mL with water. After standing for five minutes later, the solution was scanned on the fluorophotometer in the region of 250 to 750nm with $\Delta\lambda = 0$ nm. The obtained RLS spectrum was recorded and its intensity was measured at 569.5 nm. The enhanced RLS intensity of tinidazole-TPB system was represented as $\Delta I = I - I_0$ (I and I_0 were the RLS intensities of the system with and without tinidazole). The operations were carried out at room temperature.

Sample

The injections of tinidazole were diluted 100 to 200 folds with pure water. The tablets of tinidazole were dissolved in 500mL pure water and filtered through a 0.45 μm cellulose acetate membrane. A 1.0mL aliquot of the prepared sample solutions was added to a 10 mL volumetric flask instead of tinidazole standard solution.

Results and Discussion

Characteristics of the RLS Spectra

The RLS spectrum of $B(C_6H_5)_4$–Na in sulfuric acid solution (0.018 mol L^{-1}) is shown in Figure 2b. It can be seen that the RLS intensity of $B(C_6H_5)_4$–Na is quite weak in the whole scanning wavelength region. In contrast, upon addition of trace amount of tinidazole to $B(C_6H_5)_4$–Na solution, a remarkably enhanced RLS with a maximum peak at 569.5 nm was observed under the same conditions (Figure 2, c-g). It can be clearly observed that there were two peaks located at 452.0 and 569.5 nm in the RLS spectrum of tinidazole-TPB system. The addition of increasing tinidazole to the $B(C_6H_5)_4$–Na solution leads to the gradual enhancement in RLS intensity, exhibiting a concentration-dependent relationship. The production of RLS and its intensity are correlative with the formation of the aggregate and its particle dimension in solution [3].

As shown in Figures 2a and 2b, when the RLS intensities of tinidazole and NaB(C6H5)4 were considered alone, they were quite weak. It thus can be concluded that $B(C6H5)_4^-$ anion reacted with tinidazole and produced a new-formed compound whose RLS intensity was much higher than that of tinidazole or NaB(C6H5)4 when they existed separately. Moreover, the dimension of tinidazole-TPB particles may be much less than the incident wavelength, and thus the enhanced light-scattering signal occurs under the given conditions. In this way, the resonance light scattering formula [26] could be applicable to the tinidazole-TPB system.

Effects of pH Values in Medium

The newly formed tinidazole-TPB compound may be ascribed to the higher electrostatic attraction between TPB and tinidazole than that of the coexistent sodium ion. Moreover, the RLS is relevant to the dimension of the formed aggregated species. Hence, the pH value may exert certain influence on the attraction strength and the dimension of suspension particles, and thus the RLS production and its intensity. As shown in Figure 3, the RLS intensity of $NaB(C_6H_5)_4$ solution did not change with the variation of pH in range of 1.44–6.44, whereas that of the tinidazole-TPB system presented different traits. The RLS intensity of the tinidazole TPB decreased from pH 1.44 to 6.44. Acidity strongly affected the form of ammonium ion, which reacted with the $B(C_6H_5)_4^-$. A maximum RLS intensity was obtained around pH 1.44 and this value was selected for the subsequent measurements.

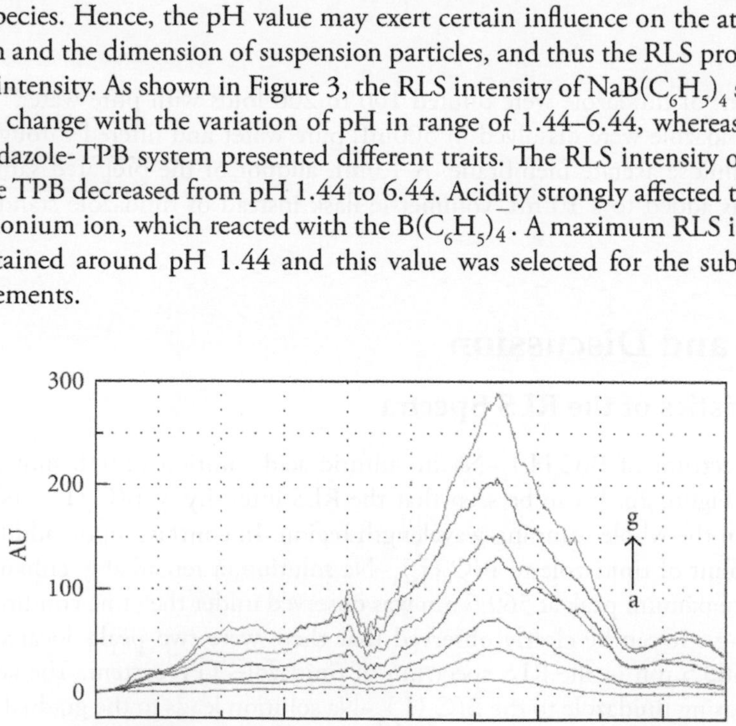

Figure 2. RLS spectra of five tinidazole-TPB systems: (a) tinidazole solution, 50 μgmL^{-1}; (b)-(g) $NaB(C_6H_5)_4$ 1.0mg mL^{-1}, and sulfuric acid solution, 0.018 mol L^{-1}: 0, 10, 15, 20, 25, 30 μgmL^{-1} of tinidazole.

Effect of Ionic Strength

There existed high concentration of sodium chloride (0.9%) in tinidazole samples such as injections. Did the large amounts of Na^+ and Cl^- affect the RLS spectra of tinidazole-TPB system? The Na^+ and Cl^- may interfere with the electrostatic attraction between TPB and tinidazole. Herein, sodium chloride was used to maintain the ionic strength of the solution. The unexpected observation is that both of the RLS intensity of TPB-Na and tinidazole-TPB system hardly changed with the concentration changes of added NaCl (Figure 4). Therefore, the system can be allowed in the solutions with high ionic strength such as injections.

Addition Orders

The effect of addition order on the RLS intensity is listed in Table 1. It was found that the addition orders of reagents have no large effect on the RLS intensity. The proposed assay of tinidazole has a wide pH range.

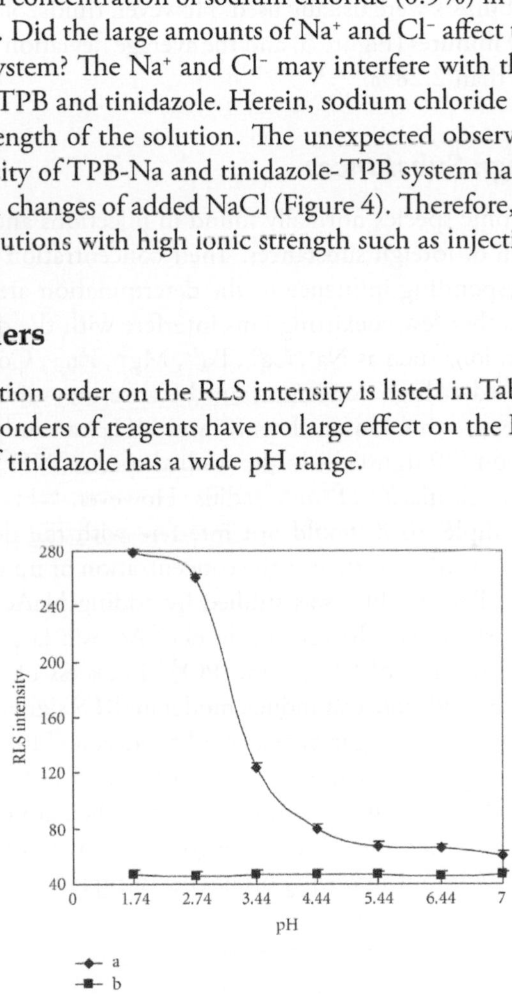

Figure 3. Effect of pH on the RLS intensity. (a) The tinidazole-TPB system suspension: 30 μgmL−1 of tinidazole, and 1mgmL−1 of $NaB(C_6H_5)_4$; (b) 1.0mgmL−1 of $NaB(C_6H_5)_4$.

Stability

The formation process of tinidazole-TPB particles includes three steps: nucleation, crystal growth, and aggregation, which will affect the sizes of the particles directly. Because the size of the particles is one important factor deciding RLS intensity, stabilizer must be used to control the size of particles, prevent the rapid sedimentation of the particles, and improve the reproducibility of RLS intensities of solutions. To improve the reproducibility of RLS intensity of a suspension system, it is crucial to impede the rapid sedimentation of the particles. In this regard, various stabilizers were usually used. However, tinidazole-TPB system is very stable within 20 minutes (Figure 5) and the average deviation of RLS signal was found to be lower than 2.28%.

Tolerance of Foreign Substances

Some cationic and anionic species normally found in injections and tablets were studied by the addition of foreign substances. Their concentration relative to tinidazole and the corresponding influence to the determination are displayed in Table 2. Table 2 shows that few coexisting ions interfere with the determination of tinidazole. Common ions such as Na^+, Ca^{2+}, Ba^{2+}, Mg^{2+}, Zn^{2+}, Co^{2+}, Cu^{2+}, Al^{3+}, and Pb^{2+} can be tolerated at high concentrations because they did not combine with $B(C_6H_5)_4^-$. However, some ions such as K^+ and NH_4^+ can only be tolerated at very low concentration (10 μgmL^{-1}). In the studied species, NH_4^+ and K^+ were affected seriously due to similarity of ionic radius. However, NH_4^+ and K^+ were nearly absent in the sample, so it would not interfere with the determination. The most abundant Na^+ would interfere at the concentration of up to 1000 times than that of tinidazole. Because Na^+ was studied by adding NaAc and Ac^- had more molecular weight than Na^+, the tolerant level of Ac^- was larger. The results demonstrated that the addition of CO_3^{2-} and PO_4^{3-} in excess of 1000 folds in concentration relative to tinidazole can induce moderate RLS signal. This may be due to the formation of extended aggregate around tinidazole-TPB particle cores by the relatively higher negatively charged ions of CO_3^{2-} and PO_4^{3-}. Other studied ions have nearly no effects on the determination when their concentration was the same as or more than tinidazole. Due to the good selectivity of this method, assays can be performed without removing other coexisting ions.

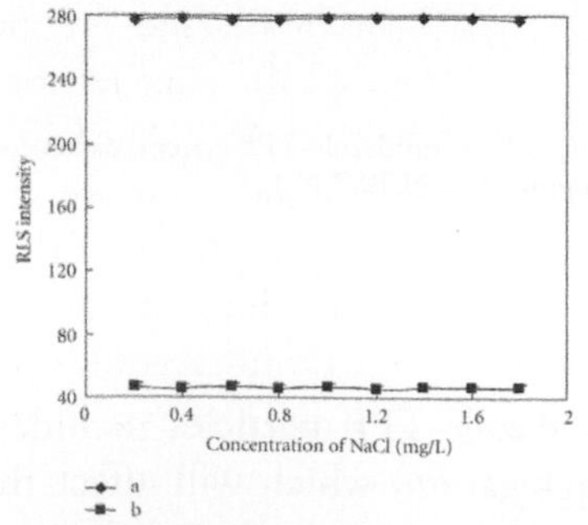

Figure 4. Effect of NaCl concentration on the RLS intensity. (a) The tinidazole-TPB system suspension: 30μgmL^{-1} of tinidazole, $NaB(C_6H_5)_4$ 1mgmL^{-1},and sulfuric acid solution,0.018molL–1; (b) the blank solution (without tinidazole).

Table 1. Effect of adding order.

1	2	3	ΔI(%)
TPB	Tinidazole	Sulfuric acid	100
TPB	Sulfuric acid	Tinidazole	97.19
Sulfuric acid	Tinidazole	TPB	98.36

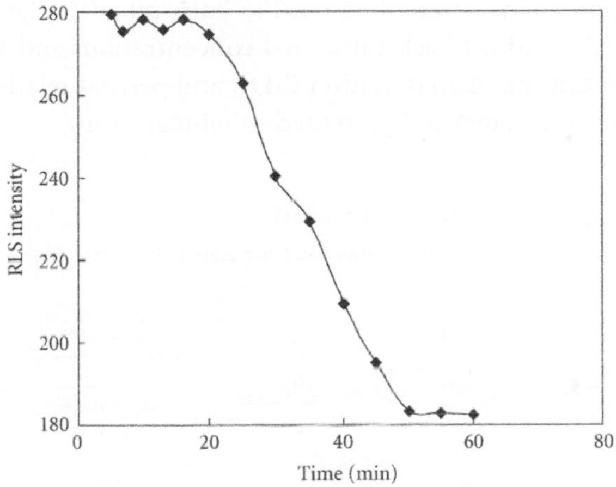

Figure 5. Stability of tinidazole-TPB system: 30 μgmL^{-1} of tinidazole, $NaB(C_6H_5)_4$ 1 mgmL^{-1}, and sulfuric acid solution 0.018 mol L^{-1}.

Analytical Applications

Detection and Quantification Limits

The detection limit was calculated as $s_b + 3s$, where s_b is the average signal of ten blank solutions and s the standard deviation. The quantification limit was calculated as $s_b + 10s$, where s_b is the average RLS signal of ten blank solutions and s the standard deviation. When the RLS intensity at 569.5 nm was selected, the detection limit and quantification limit were calculated to be 5.0 μgmL^{-1} and 10.0 μgmL^{-1}, respectively, indicating high sensitivity of this method for the determination of tinidazole. The sensitivity of the RLS method is prominently higher than that of turbidimetry (results are not presented).

Detected Wavelength and Calibration Curves

From the RLS spectra (Figure 2) of tinidazole-TPB system, three peaks are located at 452.0 and 569.5 nm. The maximum RLS peak is located at 569.5 nm.

Calibration curves were determined for five different concentrations of tinidazole standard solutions under these two wavelengths. Each calibration sample was detected in triplicate. According to the above standard procedure, the calibration curves were obtained by plotting the concentration of tinidazole against the intensity of RLS spectra at 452.0 and 569.5 nm (Figure 6). Table 3 lists the parameters and correlation coefficients of the calibration plots with two wavelengths. The $\Delta I(y)$ and the tinidazole concentrations (x) were fit to the linear function. The results of the regression analysis were then used to back-calculate the concentration results from the ΔI, and the back-calculated concentrations and appropriate summary statistics (mean, standard deviation (SD), and percent relative standard deviation (RSD)) were calculated and presented in tabular form.

Table 2. Effect of interfering ions on the tinidazole determination [(a)].

(a) **The** cationic ions were added in the form of chloride, and the anionic ions were added in the form of sodium.

Foreign ion	Ratio of the concentration (foreign ions/[tinidazole])	Change in RLS (%)	Foreign ion	Ratio of the concentration (foreign ions/[tinidazole])	Change in RLS (%)
Na^+	1000	1.54	NH_4^+	1	4.11
Ca^{2+}	100	1.89	Pb^{2+}	10	3.68
Ba^{2+}	100	2.23	SO_4^{2-}	1000	0.87
Mg^{2+}	100	1.59	NO_3^-	1000	0.94
Zn^{2+}	100	3.42	CO_3^{2-}	1000	4.12
Co^{2+}	100	2.46	SO_3^{2-}	1000	1.53
Cu^{2+}	100	4.15	PO_4^{3-}	1000	3.91
Al^{3+}	100	3.76	Cl^-	1000	0.71
K^+	1	4.28	Ac^-	1000	1.54

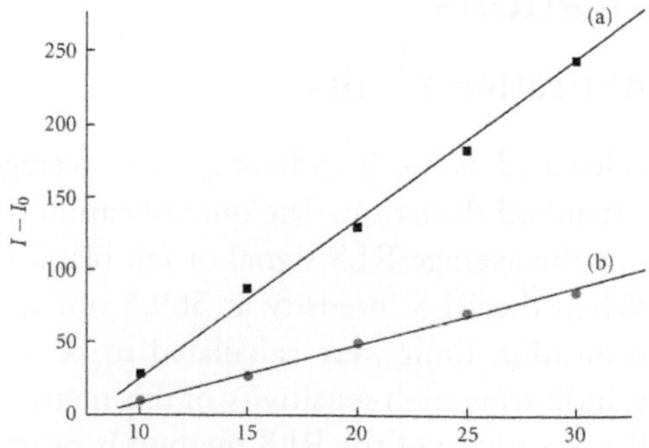

Figure 6. Relationships between RLS intensities of tinidazole-TPB system and the detected wavelengths. ((a) λ = 569.5nm, (b) λ = 452.0nm).

From Table 3, detected wavelength has obvious effect on the linear relationship of this method. To different detected wavelength, the RLS intensity of the system is also different. It offers a wide detected range for different concentration of tinidazole in samples. The lowest detection limit and quantification limit took place at 569.5 nm, because 569.5 nm is the maximum RLS peak.

Table 3. Effect of wavelengths on the linear relationship [(a)].
(a) y is ΔI; x is the tinidazole concentrations (μg mL^{-1}).

Wavelength (nm)	Linear regression equation	Correlation coefficient
452.0	$y = -29.308 + 3.8868x$	0.9958
569.5	$y = -75.356 + 10.472x$	0.9962

[(a)] y is ΔI; x is the tinidazole concentrations (μg mL^{-1}).

Precision

The precision study was comprised of repeatability and reproducibility studies. These were developed in five different samples. The repeatability was established by analyzing the samples five times. The reproducibility was determined by analyzing each sample on three different days over about one month. The repeatability and the reproducibility are < 2.37% and < 3.96%, respectively. These results indicate that the present method can be used for quantitative analyses of tinidazole.

Recovery

To establish the accuracy of the method, this procedure was also performed on tinidazole added to samples. Table 4 shows the recoveries of tinidazole applying this analytical method. From Table 4, good results are obtained with the recovery range of 95.13–106.76%.

Comparison of RLS and HPLC Methods

As shown in Table 5, the proposed method was applied to determine tinidazole concentration in injections and tablets. The attained results were compared with that of HPLC method. From Table 5, it was seen that the RLS results were in agreement with the HPLC method according to Chinese Pharmacopoeia. The average RSD of the RLS method is 0.79%–1.83%, which is slightly lower than that of the HPLC method (1.22%–2.48%), which proved that the RLS assay of tinidazole in drugs was practical.

In this paper, we compared two methods to analyze tinidazole in injections and tablets. These two methods, RLS and HPLC, can give similar results for tinidazole content in drugs (Table 5). However, the operations of RLS and HPLC methods were significantly different. The HPLC method appears to suffer from complexity and cost time, whereas the RLS method described here is robust, cost effective, and simple while retaining sufficient sensitivity. It took more than 10 minutes for an HPLC analysis, but only 1minute for RLS analysis. Second, the RLS analysis was not affected by small variation in temperature, so it could be carried out at room temperature. But the temperature had a significant effect on the HPLC analysis. The HPLC column temperature was set at a fixed temperature. Third, the RLS analysis did not use organic solvent, but toxic acetonitrile is used in the HPLC analysis.

Mechanism Discussion

Light scattering is caused by the presence of fine particles. Because the dimension of tinidazole-TPB particles is much less than the incident wavelength, it should be in accordance with the resonance light scattering formula, which is shown as follows [27]:

$$R(\theta)=\frac{9\pi^2}{2\lambda^4}\left(\frac{n_1^2-n_0^2}{n_1^2+2n_0^2}\right)^2 N_0\nu^2(1+\cos^2\theta), \tag{1}$$

where $R(\theta)$ is the resonance light ratio at the scattering angle θ, which is equal to the ratio of the scattering intensity of incident light $I(\theta)$ at the angle θ to the intensity of incident light I_0; n_1 and n_0 are the refractive indices of solute and medium, respectively; N_0 is the number of particles per unit volume; υ is the volume of the particle; and λ is the wavelength of incident light in the medium.

If c is the concentration of tinidazole-TPB solution, and ρ is the density of each particle, so N0υ is equal to c/ρ. The formula above can be expressed as

$$R(\theta)=\frac{9\pi^2}{2\lambda^4}\left(\frac{n_1^2-n_0^2}{n_1^2+2n_0^2}\right)^2 \frac{c}{\rho}\nu(1+\cos^2\theta). \tag{2}$$

In the experiment, θ is 90°, υ remained nearly constant because the experiment conditions such as acidity and the adding volume of stabilizer and other reagents were kept as identical as possible to obtain the same size particles; n1, n0, λ, and ρ were all constant. According to (2), RLS intensity is proportional to the concentration of tinidazole-TPB suspension (c) or the number of particles in the unit volume (N0). Therefore, tinidazole can be determined based on this theory.

Conclusion

In this contribution, we proposed a resonance light scattering technique to determine tinidazole in drugs. The analytical results showed that our method is rapid, sensitive, selective, and potential to be put into practice. This method may also be a valuable approach for the development of detection of tinidazole in serum.

Acknowledgement

The authors show their appreciation to Hunan Wantwant Hospital for their kindness in providing the samples of tinidazole injections and tablets.

References

1. S. C. Sweetman, *Martindale: The Complete Drug Reference*, Pharmaceutical Press, London, UK, 2002.
2. H. Salomies, "Structure elucidation of the photolysis and hydrolysis products of tinidazole," *Acta Pharmaceutica Nordica*, vol. 3, no. 4, pp. 211–214, 1991.
3. R. F. Pasternack, C. Bustamante, P. J. Collings, A. Giannetto, and E. J. Gibbs, "Porphyrin assemblies on DNA as studied by a resonance light-scattering technique," *Journal of the American Chemical Society*, vol. 115, no. 13, pp. 5393–5399, 1993.
4. R. F. Pasternack and P. J. Collings, "Resonance light scattering: a new technique for studying chromophore aggregation," *Science*, vol. 269, no. 5226, pp. 935–939, 1995.
5. P. J. Collings, E. J. Gibbs, T. E. Starr, et al., "Resonance light scattering and its application in determining the size, shape, and aggregation number for supramolecular assemblies of chromophores," *Journal of Physical Chemistry B*, vol. 103, no. 40, pp. 8474–8481, 1999.
6. Z. De Liu, C. Z. Huang, Y. F. Li, and Y. F. Long, "Enhanced plasmon resonance light scattering signals of colloidal gold resulted from its interactions with organic small molecules using captopril as an example," *Analytica Chimica Acta*, vol. 577, no. 2, pp. 244–249, 2006.
7. Z. Chen, J. Liu, Y. Han, and L. Zhu, "A novel histidine assay using tetraphenylporphyrin manganese (III) chloride as a molecular recognition probe by resonance light scattering technique," *Analytica Chimica Acta*, vol. 570, no. 1, pp. 109–115, 2006.

8. C. Z. Huang, W. Lu, Y. F. Li, and Y. M. Huang, "On the factors affecting the enhanced resonance light scattering signals of the interactions between proteins and multiply negatively charged chromophores using water blue as an example," *Analytica Chimica Acta*, vol. 556, no. 2, pp. 469–475, 2006.
9. Y. J. Long, Y. F. Li, and C. Z. Huang, "A wide dynamic range detection of biopolymer medicines with resonance light scattering and absorption ratiometry," *Analytica Chimica Acta*, vol. 552, no. 1-2, pp. 175–181, 2005.
10. Z. Chen, W. Ding, F. L. Ren, J. Liu, and Y. Liang, "A simple and sensitive assay of nucleic acids based on the enhanced resonance light scattering of zwitterionics," *Analytica Chimica Acta*, vol. 550, no. 1-2, pp. 204–209, 2005.
11. C. Z. Huang and Y. F. Li, "Resonance light scattering technique used for biochemical and pharmaceutical analysis," *Analytica Chimica Acta*, vol. 500, no. 1-2, pp. 105–117, 2003.
12. J. B. Xiao, C. S. Yang, F. L. Ren, X. Y. Jiang, and M. Xu, "Rapid determination of ciprofloxacin lactate in drugs by the Rayleigh light scattering technique," *Measurement Science and Technology*, vol. 18, no. 3, pp. 859–866, 2007.
13. Q. Wei, H. Zhang, B. Du, Y. Li, and X. Zhang, "Sensitive determination of DNA by resonance light scattering with pentamethoxyl red," *Microchimica Acta*, vol. 151, no. 1-2, pp. 59–65, 2005.
14. Z. Chen, W. Ding, Y. Liang, F. L. Ren, Y. Han, and J. Liu, "Determination of nucleic acids based on their resonance light scattering enhancement effect on metalloporphyrin derivatives," *Microchimica Acta*, vol. 150, no. 1, pp. 35–42, 2005.
15. Z. Chen, T. Zhang, F. L. Ren, and W. Ding, "Determination of proteins at nanogram levels based on their resonance light scattering decrease effect on the dibromo-onitrophenylfluorone-sodium lauroyl glutamate system," *Microchimica Acta*, vol. 153, no. 1-2, pp. 65–71, 2006.
16. Y. Zhou, S. She, L. Zhang, and Q. Lu, "Determination of proteins at nanogram levels using the resonance light scattering technique with a novel PVAK nanoparticle," *Microchimica Acta*, vol. 149, no. 1-2, pp. 151–156, 2005.
17. J. B. Xiao, J.W. Chen, F. L. Ren, Y. Y. Chen, and M. Xu, "Highly sensitive detection of trace potassium ion in serum using resonance light scattering technique with sodium tetraphenylboron," to appear in *Microchimica Acta*.
18. J. B. Xiao, J. W. Chen, F. L. Ren, C. S. Yang, and M. Xu, "Use of 3-(4,5-dimethylthiazol-2-yl)-2,5-diphenyl tetrazolium bromide for rapid detection of methicillin-resistant *Staphylococcus aureus* by resonance light scattering," *Analytica Chimica Acta*, vol. 589, no. 2, pp. 186–191, 2007.

19. K. Rajnarayana, M. R. Chaluvadi, V. R. Alapati, S. R. Mada, G. Jayasagar, and D. R. Krishna, "Validated HPLC method for the determination of tinidazole in human serum and its application in a clinical pharmacokinetic study," *Pharmazie*, vol. 57, no. 8, pp. 535–537, 2002.

20. M. Bakshi and S. Singh, "HPLC and LC-MS studies on stress degradation behaviour of tinidazole and development of a validated specific stability-indicating HPLC assay method," *Journal of Pharmaceutical and Biomedical Analysis*, vol. 34, no. 1, pp. 11–18, 2004.

21. L. Zhang, Z. Zhang, and K. Wu, "In vivo and real time determination of ornidazole and tinidazole and pharmacokinetic study by capillary electrophoresis with microdialysis," *Journal of Pharmaceutical and Biomedical Analysis*, vol. 41, no. 4, pp. 1453–1457, 2006.

22. P. Nagaraja, K. R. Sunitha, R. A. Vasantha, and H. S. Yathirajan, "Spectrophotometric determination of metronidazole and tinidazole in pharmaceutical preparations," *Journal of Pharmaceutical and Biomedical Analysis*, vol. 28, no. 3-4, pp. 527–535, 2002.

23. C. Yang, "Voltammetric determination of tinidazole using a glassy carbon electrodemodified with single-wall carbon nanotubes," *Analytical Sciences*, vol. 20, no. 5, pp. 821–824, 2004.

24. A. Z. Abu Zuhri, S. Al-Khalil, R. M. Shubietah, and I. El-Hroub, "Electrochemical study on the determination of tinidazole in tablets," *Journal of Pharmaceutical and Biomedical Analysis*, vol. 21, no. 4, pp. 881–886, 1999.

25. S. A. Özkan, Y. Özkan, and Z. Sentürk, "Electrochemical reduction of metronidazole at activated glassy carbon electrode and its determination in pharmaceutical dosage forms," *Journal of Pharmaceutical and Biomedical Analysis*, vol. 17, no. 2, pp. 299–305, 1998.

26. S. K. Sindhwani and R. P. Singh, "Spectrophotometric determination of osmium using acenaphthenequinonemonoxime," *Microchemical Journal*, vol. 18, no. 6, pp. 627–635, 1973.

27. S. Z. Zhang, F. L. Zhao, K. A. Li, and S. Y. Tong, "A study on the interaction between concanavalin A and glycogen by light scattering technique and its analytical application," *Talanta*, vol. 54, no. 2, pp. 333–342, 2001.

CITATION

Jiang XY, Chen XQ, Dong Z, and Xu M. The Application of Resonance Light Scattering Technique for the Determination of Tinidazole in Drugs. Journal of Automated Methods and Management in Chemistry 2007; 2007: 86857. doi: 10.1155/2007/86857

Enhanced Trace-Fiber Color Discrimination by Electrospray Ionization Mass Spectrometry: A Quantitative and Qualitative Tool for the Analysis of Dyes Extracted from Sub-millimeter Nylon Fibers

Albert A. Tuinman, Linda A. Lewis and Samuel A. Lewis, Sr.

ABSTRACT

The application of electrospray-ionization mass spectrometry (ESI-MS) to trace-fiber color analysis is explored using acidic dyes commonly employed to color nylon-based fibers, as well as extracts from dyed nylon fibers. Qualitative information about constituent dyes and quantitative information about the relative amounts of those dyes present on a single fiber become readily available using this technique. Sample requirements for establishing the color-identity of different samples (i.e., comparative trace-fiber analysis) are shown to be submillimeter. Absolute verification of dye-mixture identity (beyond the comparison of molecular weights derived from ESI-MS) can be obtained by expanding the technique to include tandem mass spectrometry (ESI-MS/MS). For dyes of unknown origin, the ESI-MS/MS analyses may offer insights into the chemical structure of the compound – information not available from chromatographic techniques alone. This research demonstrates that ESI-MS is viable as a sensitive technique for distinguishing dye constituents extracted from a minute amount of trace fiber evidence. A protocol is suggested to establish/refute the proposition that two fibers – one of which is available in minute quantity only – are of the same origin.

Introduction

An important aspect of forensic fiber examinations involves the comparison of dyestuffs used to impart color on or in textile fibers. Information obtained from dyes used to color fibers can provide supporting evidence in forensic casework when comparing two fibers obtained from different locations. To determine that two fibers are of the same origin, it is necessary that they be shown to have the same dye components and that the ratio in which these components are present should be identical. Comparisons of absolute dye concentrations (i.e. nano-grams dye per mm fiber) may not be necessary – or even advisable. Dye intensity may not be distributed uniformly along different fibers from the same coloring batch, or even along the length of a particular fiber. Thus, a forensic evaluation should comprise a qualitative evaluation of dye content and a quantitative determination of the relative amounts in which those dyes are present.

A review of current textile dying techniques found that, in most cases, manufacturers use the same dye constituents in differing ratios to impart different colors to their products. This practice facilitates computer-assisted production control. Most textile dyers use three dyes; a yellow, a red, and a blue to produce the desired effect. Although there are numerous yellow, red, and blue dyes from which to choose, an individual textile manufacturer may use only a small selection to produce the myriad hues in his product line. Thus, the ability to determine dye-constituent ratios, as well as the actual dyes used in a coloring process, is virtually essential to definitively compare dyed fibers.

The predominant techniques currently employed in forensic fiber-color examinations include microspectrophotometry [1, 2] and thin-layer chromatography (TLC) [3]. Microspectrophotometry is the most widely utilized color comparison technique in federal, state, and local forensic laboratories. To the forensic scientist, nondestructive analysis of evidence and application to extremely small sample sizes are the most attractive characteristics of this method. However, the lack of discriminatory power is an inherent limitation. Microspectrophotometry evaluates the spectral characteristics of the composite-dye mixture, but says nothing about the individual dye components. Considerable diagnostic information is therefore left unexamined.

In contrast, TLC is a destructive method in which dyes are extracted from the fiber and subsequently separated and qualitatively compared. The method is relatively insensitive, and may require more fiber for an analysis than the forensic scientist is willing to sacrifice. Furthermore, TLC is generally able to distinguish only the dyes present, but not the ratios in which they are present.

High-performance liquid chromatography (HPLC) [3] is a third analytical method that has been employed to a limited extent in the forensics laboratory for dye analysis. HPLC has been applied to acid-[4], disperse-[5,6], and basic-[7] forensic dye analysis. HPLC offers better separation than TLC, and provides quantitative information. However, HPLC columns and gradient- or isocratic elution systems are specific to only a limited group of similar compounds. The columns required for analysis are generally expensive, and large amounts of solvent may be needed compared to other separation methods. In addition, HPLC fiber length requirements are comparable to the lengths necessary for TLC analysis. Application of modern techniques of micro-HPLC may alleviate one or more of these shortcomings, but we have not found literature references to such application in the field of forensic fiber comparison.

Capillary electrophoresis is another method that shows significant promise as a forensic screening tool for trace-fiber dye analysis [8,9]. In a recent study, the ability to detect acid dyes extracted from nylon fibers between 1 to 3 mm, for dark and light colored fibers respectively, was achieved using large volume stacking with polarity switching [8]. This technique was sensitive to dye constituents as well as manufacturing additives and impurities yielding unique "fingerprint" electropherograms that could be compared to a potential source material. However, quantitative results for this method indicated that the spectral-based detection lacked the sensitivity required for precise comparative measurements.

This paper discusses the application of electrospray ionization mass spectrometry (ESI-MS) to the qualitative and quantitative aspects of comparative fiber-dye analysis. Qualitative identity is primarily established by comparison of the observed masses for each peak in the ESI-MS of each fiber extract. This may be further verified by comparing the tandem mass spectra (ESI-MS/MS) for each of the dye peaks observed in the ESI-MS. Comparison of the relative intensities of the individual dye peaks in the ESI-MS provides a measure of the concentration of each dye in each of the fibers being compared. In the modern forensics laboratory nylon fibers are frequently encountered [10], and this study has accordingly concentrated on such fibers and the dyes (usually acidic) used to color them. It is anticipated however that the methods described below may be equally applicable to other fiber types and to inks.

Experimental Section

Materials

To undertake this study, a supply of real-world nylon fibers, and information regarding the dyes used to color them, was required. Shaw Industries, Inc. (Dalton, GA) supplied numerous colored nylon carpet samples with the associated dyes used to color the samples, and Collins & Aikman (Dalton, GA) provided many colored nylon windings and information regarding the dyes used to color the fibers. Ciba Specialty Chemicals (High Point, NC) provided the dyes used to color those nylon windings. For both carpet- and winding samples, the manufacturers often used identical dye components to produce different colors by varying the ratio of dyes mixed. Only a small selection of the fiber and dye samples kindly provided by these manufacturers were used in this study. They are listed in Table 1.

Table 1. Dyes and Colored Windings Used.

DYES

Lab-Code #	Commercial Name	CI Name
1	Telon Yellow FRL01 200	Proprietary Mix
2	Telon Red 2BN 200	Proprietary Mix
3	Nylanthrene Red CRBS 200	Proprietary Mix
4	Telon Red FRLS 175	Acid Red 337
5	Telon Blue BRL 200	Acid Blue 324
6	Tectilon Blue 4RS 200%	Unknown
7	Tectilon Orange 3G 200%	Acid Orange 156
8	Tectilon Red 2B 200%	Unknown
9	Dye-O = Lan Black RPL 150%	Acid Black 172

WINDINGS

Manufacurer's Code	Manufacurer	Dye Components	Manufacurer's Color Description
9900	Collins & Aikman	6,7,8	Brown
38920	Collins & Aikman	6,7,8	Olive Green
38011	Collins & Aikman	6,7,9	Black

Fiber Extraction

Known lengths of fiber were cut and placed inside glass Wheaton 0.3-mL "v-vials" (1-mL total volume). 400 μL of 4:3(v/v) pyridine/water was added to each vial, and the caps were tightly sealed. The vials were heated at 100 °C for 30 to 35 min. The fibers were immediately removed from the vials, and the vials were placed in a KD-Tube-Evaporator pre-heated to 90-95 °C with air blow-down set on low for 15 min., followed by air at high for 15 to 30 min. The vials were removed, allowed to cool, and solvent (i-propanol/water 4/1 containing 0.1% ammonia) was added to the equivalent of 60 μL per mm original fiber length.

Mass Spectometry

Negative ion electrospray mass spectrometry was conducted on a Quattro-II (Micromass, Manchester, UK) instrument by direct infusion of the solutions at 5 μL/min. Dye samples were prepared for infusion at 500 ppb (~ 1μM) in acetonitrile/water (1/1). Fiber extracts were diluted to the equivalent of 60μL solvent (i-Propanol/Water (4/1) containing 0.1% NH_3) per mm of fiber extracted. Scans were accumulated over two minutes, and averaged to produce one spectrum. Source temperature was maintained at 90 °C, and the nebulizing- and drying-gas flow rates were 30 and 300 L/hr of N_2 respectively. The capillary, and one (nozzle to skimmer) voltages were 2,5 kV and 45 V, respectively. For collision induced dissociation (CID) experiments, the precursor ion was selected in the first analyzer of the triple quadrupole instrument, then allowed to collide with argon (3.4×10^{-3} mBar) in the rf-only collision quadrupole, and the resulting fragments were analyzed in the last quadrupole. The collision energy (or collision energy ramp as appropriate) for each experiment is presented in the relevant text, Figure, or Table.

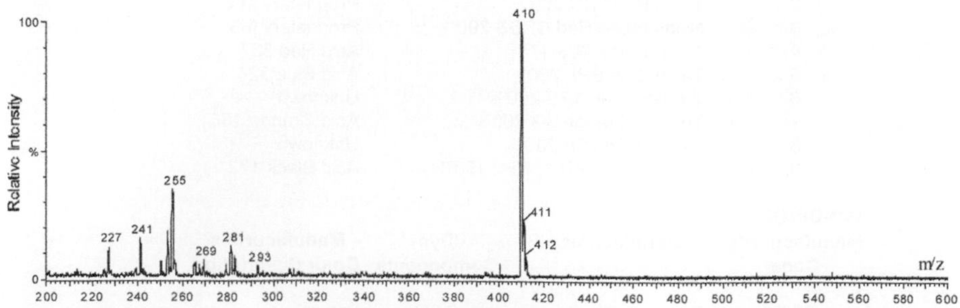

Figure 1. ESI-MS of Dye #4 (500 ppb in CH3CN/H2O 1:1).

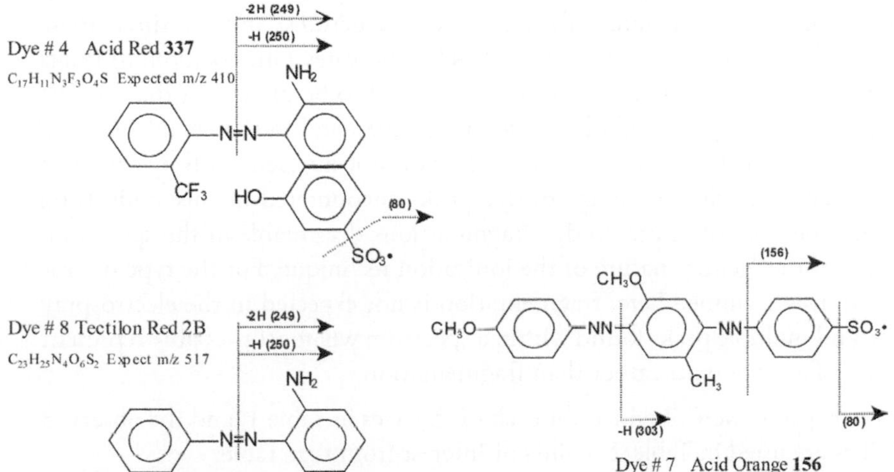

Figure 2. Structures and observed CID fragmentations of the known dyes.

Table 2. Observed Relative ESI Anion Intensities and Charge States for the Dyes Listed in Table 1.

Dye #	Observed Anion m/z Values: 308	408	410	424	427	430	439	444/446	450	462.7	464	467	473.7	481.7	488	489/491	504	517	525	544
1					100^{-1}											$44/34^{-1}$	13^{-1}			
2			100^{-1}									90^{-1}								
3			100^{-1}					$56/22^{-1}$											53^{-1}	
4			100^{-1}																	
5									100^{-1}											
6											37^{-1}									100^{-1}
7							100^{-1}													
8																		100^{-1}		
9	100^{-3}	3^{-1}	3^{-1}	5^{-1}		6^{-2}				9^{-2}			16^{-2}	6^{-2}	2^{-1}					

Observed charge state indicated by superscript; thus,

XX^{-1} = singly charged cluster (^{13}C isotopes @ 1.0 Da)

XX^{-2} = doubly charged cluster (^{13}C isotopes @ 0.5 Da)

XX^{-3} = triply charged cluster (^{13}C isotopes @ 0.3 Da) (requires "high resolution" to observe isotope peaks)

Results and Discussion

ESI Spectra of Individual Dyes

A negative ion electrospray spectrum of dye #4 (500 ppb in acetonitrile/water 1/1) infused into the ESI source is presented in Figure 1, and is representative of all the dye spectra in that: a) The cluster at m/z 410 corresponds to the expected anion for this dye, (see Figure 2 for known dye structures) and the single-mass spacing of the ^{13}C isotope peaks (410, 411, 412) indicates a monoanion. b) Peaks between m/z 200 and 300 are impurities presumed to be present in the solvent. They appear in the spectra of all the individual dyes, and in the "blank" spectra of solvent without the addition of dyes. They do not interfere with dye analysis because they lie outside the range of dye peaks encountered in this study (m/z range 308-544). c) There are no dye-fragment ions discernable in the spectrum, indicative of the "gentle" nature of the ionization technique. For the type of aromatic azo dyes examined here, fragmentation is not expected in the electrospray process, and multiple peaks found within a spectrum will most probably represent a mixture of analyte types rather than fragmentation.

Similar spectra were obtained for each of the dyes in Table 1, and the observed m/z values are listed in Table 2. Points of interest from that Table:

1. In those instances where the chemical structure is known (dyes 4, 7, and 8) the observed mass values correspond to expectation.
2. Several of the dyes (1, 2, 3, 6, and 9) show multiple peaks indicative of mixtures. (Indeed #s 1, 2, and 3 are stated by the manufacturer to be "Proprietary Mixes"; see Table 1.)
3. Dyes #2 and 3 appear to have the same major component (#4) as well as one or two "additives." (The identical structure of the m/z 410 component of #s 2, 3, and 4 is borne out by comparison of their CID spectra; see section "Collision Induced Dissociation..." below).
4. All dyes examined are singly charged (indicated by superscript "-1" in Table 2) indicating monoacidic functionality except #9 (superscripts "-2" and "-3" for two and three charges, respectively).
5. Several of the peaks listed for #9 are closely related and apparently represent the same underlying tri-acidic structure. Thus, triply charged m/z 308.1 represents a trianion with mass 924.3 Da. M/z 462.7 corresponds to the protonated version of the same structure [(924.3+1)/2 = 462.7]. Likewise 473.7 and 481.7 correspond to the sodiated and potasiated structures, respectively. Thus, 308, 462.7, 473,7 and 481.7 do not represent different dyes in the mixture, but rather differently cationized versions of the same tri-acidic dye. However, the singly charged anion at m/z 424, and the doubly charged one at m/z 430 cannot be accommodated similarly, and seem to be different dyes in the mixture. The peaks at 408, 410, and 488 are relatively too small to be used as markers in the ESI-MS fingerprint of dye 9. At somewhat lower signal/noise ratios (S/N) than those attained here, they may become indistinguishable from the background.

6. Each of the dyes examined has at least one peak which is distinct from every other dye in this particular set. If the question were asked "Which dyes are present in a particular fiber extract?", and the choice were limited to this particular set, the answer could be provided by examining the ES-IMS spectrum of the dye mixture without any prior separation into the individual components. Of course, the set of possibilities examined here is extremely limited. If the set were expanded to include the hundreds of acidic dyes used in the coloring of nylon fibers, there would probably be numerous overlaps. In that case, resort to tandem mass spectrometry would provide the needed specificity (see section "Tandem Mass Spectrometry" below).

ESI Spectra from Winding Extracts

The extracts of three windings of dyed nylon threads were examined by ESI-MS. The extracts were generated by the standard procedure (see Experimental) using 50 mm each of windings 38920, 38011, and 9900 (see Table 1), and labeled I, II, and III, respectively. Each extract was diluted to the equivalent of a 1 mm sample dissolved in 60 μL of the solvent used for ESI-MS. A representative example of a windings spectrum is presented in Figure 3. Notable observations from that spectrum apply to the other two which are not separately illustrated: a) The peaks expected for the component dyes of the brown winding (m/z 464 and 544 for dye # 6, m/z 439 for dye #7, and m/z 517 for dye #8) are clearly discernable, as are the "background" peaks (e.g., m/z 253, 255, 281) previously observed for in the spectra of the individual dyes (cf Figure 1). b) A set of peaks apparently representing a homologous series of analytes occurs at m/z 283, 297,... 367. These do not obstruct that region of the spectrum which is most informative for dye analysis (m/z 400-560), therefore they are not considered detrimental to the dye analysis. c) Peaks not attributable to the known dye components are also observed in the "dye region" of the spectrum: viz m/z 417, 451, 487/489. They appear, albeit in varying relative abundances, in all of the winding extracts examined during this study, and may represent degradation products of the nylon fiber generated during the extraction process or compounds which have been added during the dye-fixing process. The "extraneous" peaks which lie within the dye-region of the spectrum have been marked with brackets ( ) in Figure 3, and the recognized dye peaks have been marked with a rectangle.

Reproducibility of ESI-MS Dye-Peak Intensities

The data in Table 3 was gathered by infusing the three winding extracts (I, II, and III) sequentially into the mass spectrometer via a 10 μL Rheodine injection loop.

The 10 μL infusion is equivalent to the extract from 170 μm of the winding. The sequence of three injections was performed four times producing a total of twelve acquired spectra within two hours. Instrumental parameters such as capillary- and cone voltages, source temperature, and nebulizing- and drying gas flowrates were held constant during the full course of these analyses. In Table 3, individual peak intensities have been normalized to the sum of the dye-peak intensities for each spectrum, and then gathered together in groups according to the color of the extracted winding. The average and standard deviation of the observed intensities for each dye component are listed. The calculated standard deviations are certainly adequate to unambiguously differentiate windings I and II (which differ only by the proportions of the dyes) from one another. Whereas winding I ("brown") consists of dyes 6 (blue), 7 (orange), and 8 (red) in the proportion 10/10/3, the proportion for winding II ("olive green") is 16/10/3. It is important to note that the differentiation of these winding extracts, containing identical dye components in somewhat differing ratios, has been accomplished without the need to physically separate the components by chromatography.

Tandem Mass Spectrometry

Molecular ions (or quasi-molecular ions) observed in mass spectra only provide information regarding the mass of the analyte molecule. In the absence of fragmentation there is no structural information. Collision Induced Dissociation (CID) spectra on the other hand can provide significant structural information about the precursor ions. Even if the CID information cannot always be interpreted to draw a definitive structure, the pattern of fragment peaks derived from a particular precursor under defined dissociation conditions does provide a "fingerprint" of that precursor. It is thus particularly useful in establishing that the same analyte mass, observed in separate samples, represent the same structure.

The appearance of a CID spectrum can vary significantly depending on the collision energy (CE) applied. For a "fingerprinting" experiment it is desirable to encompass as many product ions in a single CID spectrum as possible. Figure 4 shows the products of precursor m/z 410 (dye # 4) acquired at four different CEs. At low CE (20 eV; Figure 4a) the precursor ion predominates, and m/z 80 is weak. At higher CE (35 eV; Figure 4d) m/z 80 is of moderate intensity, but m/z 410 has disappeared. In order to incorporate as much information as possible into a singe scan, the CE may be ramped as a function of m/z. Based on the results of Figures 4a-d, a CE-ramp was applied to produce the spectrum displayed in Figure 4e. The ramp ran from 30 eV (at m/z 60) to 20 eV (at m/z 430).

Table 3. Relative intensities, averages, and standard deviations for four repetitions of the ESI-MS spectra of three differently colored windings.

	Dye #	7	9	6	9	9	8	6
	m/z	439	462.5	464	473.7	481.7	517	544
Injection #	**winding**	**% of "total ion current"**						
1	I = "brown"	42.4		10.3			14.0	33.3
4	I = "brown"	44.0		11.4			13.4	31.2
7	I = "brown"	43.9		12.9			12.8	30.5
10	I = "brown"	42.7		14.4			12.0	30.9
	Average	**43.2**		**12.3**			**13.1**	**31.5**
	Stand. Dev.	**0.7**		**1.5**			**0.7**	**1.1**
2	II = "green"	34.7		12.8			12.3	40.1
5	II = "green"	34.2		14.3			11.8	39.6
8	II = "green"	34.1		17.0			10.3	38.6
11	II = "green"	34.3		17.9			11.0	36.8
	Average	**34.4**		**15.5**			**11.4**	**38.8**
	Stand. Dev.	**0.2**		**2.0**			**0.8**	**1.3**
3	III = "black"	19.8	29.4	9.7	15.8	4.0		21.3
6	III = "black"	20.1	29.8	10.2	15.6	3.8		20.5
9	III = "black"	20.4	28.2	12.4	16.0	2.2		20.7
12	III = "black"	21.0	27.2	12.8	14.2	4.1		20.7
	Average	**20.3**	**28.6**	**11.3**	**15.4**	**3.5**		**20.8**
	Stand. Dev.	**0.5**	**1.0**	**1.3**	**0.7**	**0.8**		**0.3**

Table 4. Mass range scanned, collision energy (ramp) employed, and fragments observed for each of the precursor ions enumerated in Table 2.

Precursor m/z	Dye # or Winding #	Scanned m/z Range	Collision Energy Ramp	Observed CID fragments [m/z]
308	9	80-550	19	203 / 308 / 316 / 352 / 408 / 424 / 430
410	2,3,4	60-430	30-20	80 / 168 / 249 / 250 / 410
417	I, II, III	60-437	40-20	83 / 189
427	1	60-447	30	292 / 409 / 427
424	9			Too small for good CID @ 500 ppb of dye #9
430	9			Too small for good CID @ 500 ppb of dye #9
439	7	60-459	40-20	80 / 156 / 303 / 395 / 423 / 439
444/446	3	60-466	30-20	80 / 168 / 249 / 250 / (444/446)
450	5	60-470	35	344 / 386 / 450
451	I, II, III	60-471	35-30	83 / 125 / 207 / 225 / 333 / 433 / 451
462.7	9	60-482	30-20	142 / 172 / 267 / 281 / 352 / 424 / 463
464	6	60-484	35	344 / 400 / 464
473.7	9			Too small for good CID @ 500 ppb of dye #9
467	2	60-487	30-20	80 / 168 / 249 / 250 / 467
481.7	9			Too small for good CID @ 500 ppb of dye #9
487/489	I, II, III	60-510	25	283 / (487/489)
489/491	1	60-510	30-15	94 / 123 / 186
504	1			Too small for good CID @ 500 ppb of dye #1
517	8	60-537	30	80 / 168 / 249 / 250 / 340 / 517
525	3	60-535	30-20	168 / 185 / 249 / 250 / 264 / 340 / 525
544	6	60-564	35	356 / 420 / 480 / 544

The "ideal" CE ramp for any given precursor ion will depend on its fragmentation characteristics. However, it can always be determined empirically. The same sequence of CID experiments described for dye #4 above was implemented for each of the precursor ions enumerated in Table 2, as well as those identified as "extraneous" in Figure 3. The results – indicating the mass range to scan, the CE-ramp to apply over that mass range, and the product ions observed from such a ramped experiment – are listed in Table 4. Thus, each row in Table 4 describes a CID method and the resulting "fingerprint" for one of the dyes under investigation. Within the context of the present study, this information may seem superfluous because each of the dyes already displays a unique pattern in its ESI-MS spectrum (cf Table 2) without the necessity for further fingerprinting. However, if a much larger database is eventually employed incorporating hundreds of dyes, some or many of them will almost certainly display equivalent m/z values in their ESI-MS spectra. CID tables such as Table 4 may then be used as a means of differentiating those isobaric compounds.

Collision Induced Dissociation and Chemical Structure

Of the nine dyes examined in this study the chemical structures of only three (#s 4, 7, and 8 see Figure 2) are known to us, either via information from the manufacturer or from the literature. For dye # 7 the structure and the CID spectrum are readily correlatable as depicted by arrows in Figure 2 showing fragmentation of the most labile bonds. Dyes 4 and 8 possess identically substituted naphthalene moieties bonded by a diazo group to a substituted benzene ring. Not coincidentally they also exhibit the same CID fragmentation pattern (product ions at m/z 80, 168, 249/250; cf Table 4). The 249/250 peaks can be assigned to fragmentation between the nitrogen atoms with concomitant migration of hydrogen(s) by analogy to a similar fragmentation investigated by Brumley et al. [11]. The m/z 168 peak may result from the loss of an HSO_3 radical from 249, but the driving force for such a loss is not clear at this time. The CID spectra of the m/z 410 precursor ions derived from dyes 2, 3, and 4 are identical. This strongly supports the supposition that this m/z 410 component is identical in each of those dyes. The additional components of dyes 2 and 3 (additional to the main component of m/z 410) display the same fragments as those of 4 and 8, implying a similar moiety is present. It is therefore postulated that m/z 467 (of #2) and m/z 444/446 and 525 (of #3) each contain the 2-amino-8-hydroxy-3-sulfonate-2-azo configuration. Furthermore, the ~1/3 ratio of the 444/446 peaks in #3, combined with the 34/36 Da mass difference to #4 strongly implies that 444/446 is identical to the structure of 4, but with a chlorine atom attached to the phenyl ring (ortho- or para position cannot be determined from these spectra).

At this point, it should be re-emphasized that, although it is intellectually pleasing to correlate the observed CID spectrum with the expected fragments of a known structure, it is not necessary to know the chemical structure of a compound in order for its CID spectrum to serve as a fingerprint. Observing two compounds (in this case dye anions) with identical ESI-MS peaks – which also produce identical CID spectra from those peaks – essentially ensures that the compounds are identical or at least very closely related in structure. Rare exceptions could arise e.g., in cases of positional isomerism such as the ortho or para substitution pattern alluded to above for the extra chlorine in dye #3.

Proposed Protocol to Compare Dye Extracts from Carpet Fiber Samples

Proposition

Two fiber samples are believed to be identical in their dye content. Sample "A" is available in large quantity (50 mm or more). Sample "B" is available in small quantity only (one mm can be spared for destructive analysis). They appear identical in color by microscopic examination.

Objective

Prove/disprove the proposition that the dye contents of the two fibers are identical using ESI-MS and CID.

Proposed Procedure

1. Extract a 5 mm long sample of "A" as well as a 1 mm samples of both "A" and "B", and dissolve the extracts in 500, 100, and 100 μL, respectively, of the appropriate solvent mixture. Using a 10 μL loop injector, sequentially infuse samples of the 1 mm extracts and obtain the ESI-MS. If the ESIMS are not essentially identical the proposition is refuted and the experiment is complete. If the ESI-MS spectra are essentially identical in m/z content and relative intensity, proceed to 2.
2. If the ESI-MS of dye ions from 1. above are represented in a "product ion" database (such as Table 4 above) confirm the assignments by running CID experiments under the previously established "ideal" conditions as indicated in the database.
3. If one or more of the major peaks in the ESI-MS are not represented in a "product ion" database, use the 5 mm extract to establish the "ideal"

conditions for obtaining a fingerprint CID of that (those) precursor ions (as described in Section "Tandem Mass Spectrometry").

4. Using the CID conditions established above for each of the major peaks in the ESI-MS, do identical multifunction (one function per dye precursor mass) CID experiments on the 1 mm extracts of "A" and "B". If the CID spectra show identical fragmentation patterns for each pair (one from extract "A", and one from extract "B") of precursor ions, the two fibers can be deemed to be identical.
5. Even if as many as eight precursor ions need to be examined by CID, with an analysis time of 1 min each, and at a flow-rate of 5 μL/min, this will consume only 40 μL of solution. Together with the 10 μL consumed in 1. above this accounts for 50 μL of the original 100, leaving 50 μL for other methodologies, or for a repeat of this one as necessary.
6. Note that application of the above procedure does not require physical separation of the component dyes prior to spectroscopic analysis. Nevertheless, the procedure is "two dimensional" because separation of the ESI-MS ions does occur within the first mass filter of the triple quadrupole mass spectrometer. The use of a tandem mass spectrometer affords the advantages of a two dimensional analysis, without some of the disadvantages associated with physically separating a mixture of compounds into its individual components.

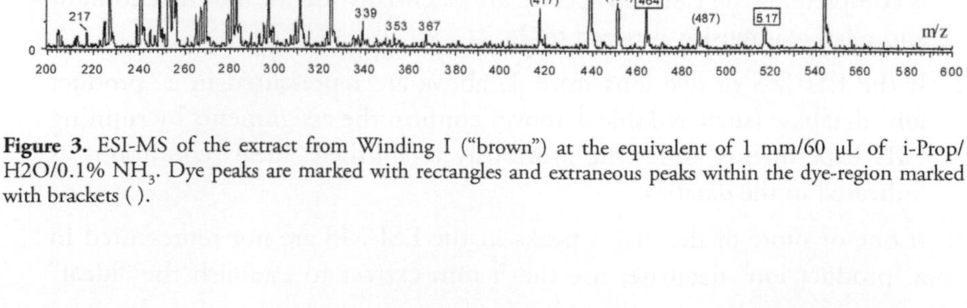

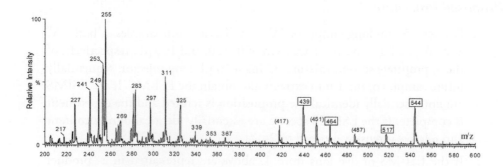

Figure 3. ESI-MS of the extract from Winding I ("brown") at the equivalent of 1 mm/60 μL of i-Prop/H2O/0.1% NH_3. Dye peaks are marked with rectangles and extraneous peaks within the dye-region marked with brackets ( ).

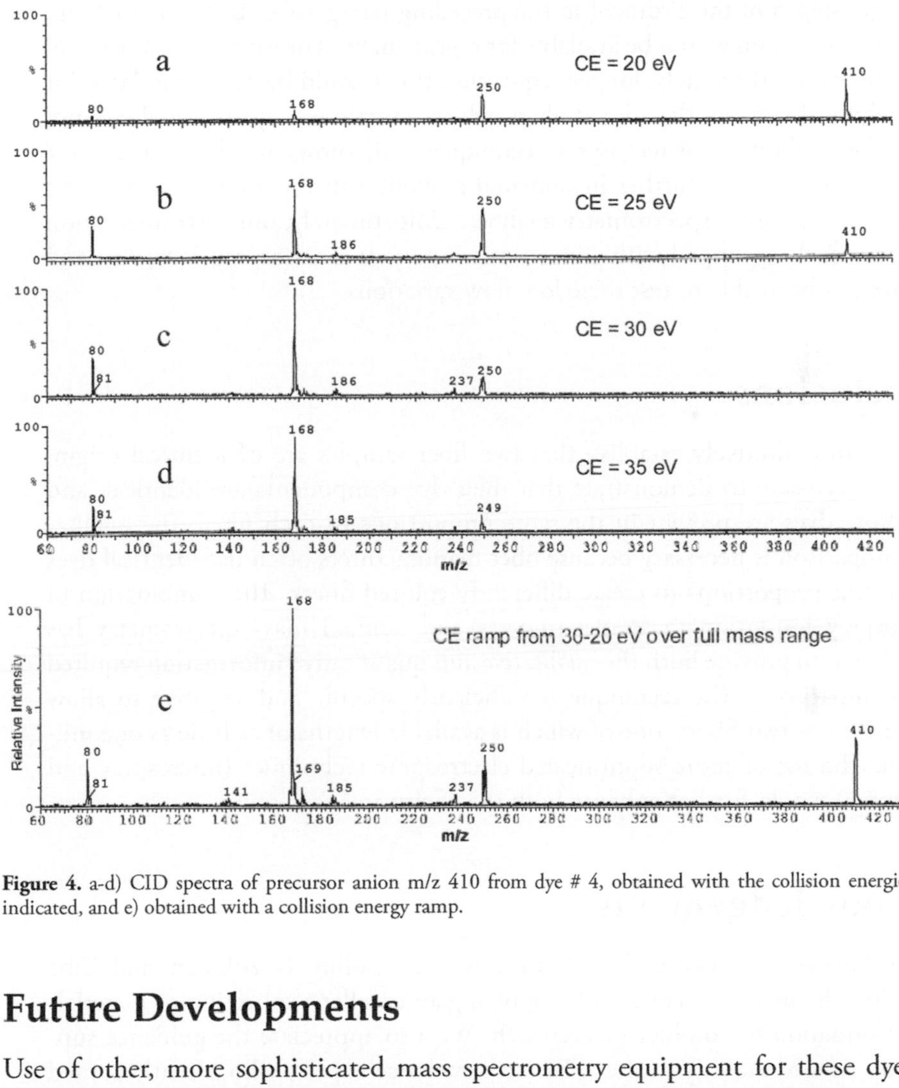

Figure 4. a-d) CID spectra of precursor anion m/z 410 from dye # 4, obtained with the collision energies indicated, and e) obtained with a collision energy ramp.

Future Developments

Use of other, more sophisticated mass spectrometry equipment for these dye-comparison problems is likely to yield even greater specificity and greater sensitivity. The analysis of dye extracts described above (Figure 3 and Table 3) used infusion of a solution at 5 μL/min for a period of two minutes to acquire each spectrum, i.e., 10μL of solution, representing the extract of 170 μm of fiber. Using microspray technology rather than standard electrospray technology would allow slower infusion (< 500 nL/min) without loss of signal intensity and without deterioration of the signal/noise ratio [12]. Thus, in a microspray variation of the Proposed Protocol the total amount of solution needed would be reduced from

50 μL (cf step 5 of the Protocol in the preceding paragraphs) to less than 5 μL. The "extra" solution would be available for repeat analyses or for the application of other methods. Alternately, longer acquisition times could be accommodated for the analysis of very lightly colored fibers, which contain less dye material per mm fiber. The application of nanospray techniques [13] (flowrates down to 20 nL/min) would provide a further incremental reduction in the amount of solution required for the mass spectrometry analyses. Unfortunately, our instrumentation is not currently equipped with either a microspray or a nanospray accessory, and we have not been able to test these low-flow variations.

Conclusions

In order to definitively establish that two fiber samples are of identical origin, it is i.a. necessary to demonstrate that their dye components are identical, and that those dyes are present in the same proportions in each fiber. The qualitative comparison is necessary because fiber manufacturers often use identical dyes in different proportions to create differently colored fibers. The combination of electrospray ionization mass spectrometry and tandem mass spectrometry has been shown to provide both the qualitative and quantitative information required such comparisons. The technique is sufficiently specific and sensitive to allow comparison of two fibers, one of which is available lengths of as little as one millimeter. The use of more sophisticated electrospray techniques (microspray and nanospray) would further enhance both specificity and sensitivity.

Acknowledgements

The authors wish to thank Shaw Industries, Inc., Collins & Aikman, and Ciba Specialty Chemicals for generously supplying an excellent collection of materials and information to conduct this research. We also appreciate the guidance supplied by Ron Menold, a former FBI trace-evidence expert, who recently turned in his labcoat to pursue a career as an FBI agent. This work was funded by the FBI under DOE Project Number 2051-1119-Y1 with BWXT Y-12 L.L.C., U.S. Department of Energy, under contract number DE-AC05-00OR22800.

References

1. Erying, M. B. *Anal. Chim. Acta* 1994, *288*, 25–34.
2. Robertson, J. Forensic Fiber Examination of Fibers; Ellis Horwood: New York, 1992; Chapter 4.

3. Robertson, J.; Forensic Fiber Examination of Fibers; Ellis Horwood: New York, 1992; Chapter 5.

4. Laing, D. K.; Gill, R.; Blacklaws, C.; Bickley, H. M. *J. Chromatogr.* 1988, *442*, 187–208.

5. West, J. C. *J. Chromatogr.* 1981, *208*, 47–54.

6. Wheals, B. B.; White P.C.; Patterson M.D. *J. Chromatogr.* 1985, *350*, 205–215.

7. Griffin, R. M. E.; Kee T. G.; Adams R.W. *J. Chromatogr.* 1988, *445*, 441–448.

8. L. Lewis *et al.*, manuscript in preparation for submission to *Forensic Science.*

9. Xu, X.; Leijenhorst, H.; Van Den Horen, P.; Koeijar, J. D.; Logtenberg, H. *Sci. Justice* 2001, *41*, 93–105.

10. Private communication from Ron Menold, Federal Bureau of Investigation, January 1999.

11. Brumley, W. C.; Brulis, G. M.; Calvey, R. J.; Sphon, J. A. *Biomed. Environm. Mass Spectrom.* 1989, *18*, 394–400.

12. Caprioli, R.M.; Emmett, M. E.; Andren, P. *Proceedings of the 42nd ASMS Conference on Mass Spectrometry and Allied Topics*, Chicago, IL, May 29-June 3, 1994; p754.

13. Wilm, M; Mann, M. *Anal. Chem.* 1996, *68*, 1–8.

CITATION

Tuiman AA, Lewis LA, Lewis SA. Enhanced Trace-Fiber Color Discrimination by Electrospray Ionization Mass Spectrometry: A Quantitative and Qualitative Tool for the Analysis of Dyes Extracted from Sub-millimeter Nylon Fibers. Technical Information Center, Oak Ridge, TN contract DE2002-8-5817. Contracted by the US Government.

The pK_a Distribution of Drugs: Application to Drug Discovery

David T. Manallack

ABSTRACT

The acid-base dissociation constant (pK_a) of a drug is a key physicochemical parameter influencing many biopharmaceutical characteristics. While this has been well established, the overall proportion of non-ionizable and ionizable compounds for drug-like substances is not well known. Even less well known is the overall distribution of acid and base pK_a values. The current study has reviewed the literature with regard to both the proportion of ionizable substances and pK_a distributions. Further to this a set of 582 drugs with associated pK_a data was thoroughly examined to provide a representative set of observations. This was further enhanced by delineating the compounds into CNS and non-CNS drugs to investigate where differences exist. Interestingly, the distribution of pK_a values for single acids differed remarkably between CNS and non-CNS substances with only one CNS compound having an acid pK_a below 6.1. The distribution of basic substances in the CNS set also showed a marked cut off with no compounds having a pK_a above 10.5. The

pK_a distributions of drugs are influenced by two main drivers. The first is related to the nature and frequency of occurrence of the functional groups that are commonly observed in pharmaceuticals and the typical range of pK_a values they span. The other factor concerns the biological targets these compounds are designed to hit. For example, many CNS targets are based on seven transmembrane G protein-coupled receptors (7TM GPCR) which have a key aspartic acid residue known to interact with most ligands. As a consequence, amines are mostly present in the ligands that target 7TM GPCR's and this influences the pK_a profile of drugs containing basic groups. For larger screening collections of compounds, synthetic chemistry and the working practices of the chemists themselves can influence the proportion of ionizable compounds and consequent pK_a distributions. The findings from this study expand on current wisdom in pK_a research and have implications for discovery research with regard to the composition of corporate databases and collections of screening compounds. Rough guidelines have been suggested for the profile of compound collections and will evolve as this research area is expanded.

Keywords: pK_a, dissociation constant, distribution, drugs, absorption, ADME, bioavailability, drug discovery, pharmacokinetics, acids, bases, ampholytes

Introduction

An awareness of the influence of the acid-base dissociation constant, pK_a, on the biopharmaceutical properties of drugs and chemicals has long been established within the pharmaceutical and chemical industry. As the majority of drugs are weak acids and/or bases, knowledge of the dissociation constant in each case helps in understanding the ionic form a molecule will take across a range of pH values. This is particularly important in physiological systems where ionization state will affect the rate at which the compound is able to diffuse across membranes and obstacles such as the blood-brain barrier (BBB). The pK_a of a drug infl uences lipophilicity, solubility, protein binding and permeability which in turn directly affects pharmacokinetic (PK) characteristics such as absorption, distribution, metabolism and excretion (ADME) [1–5]. The well established association between pK_a and PK has also resulted in the requirement for pK_a values to be measured for regulatory compliance (e.g. FDA6). Formulation procedures for optimizing drug delivery also benefit from the determination of the pK_a. Given the importance of this parameter to the drug industry [7], it follows that an ability to estimate or measure [8] the pK_a, together with a knowledge of their distribution, will be of great benefit. This is particularly important when contemplating the large number of compounds that can be considered for screening purposes (e.g. combinatorial libraries, third party compound collections). Ideally, these sets of compounds should be representative of drug-like substances as a whole with regard to the proportion of ionizables and the distribution of the pK_a values themselves.

An estimate of likely ADME characteristics can be obtained using pKa values and various other properties such as molecular weight (MW), partition coefficient (logP), number of hydrogen bond donors (hdon) and acceptors (hacc), and polar surface area (PSA) [9]. The pKa values themselves represent useful pieces of physicochemical information but in isolation they have limited value. From the perspective of designing combinatorial libraries or buying sets of compounds from third party suppliers then it is important to know what the overall profile of a collection should resemble with regard to a range of physicochemical properties. Therefore, in order to complement properties such as MW, logP, hdon, hacc and PSA, information regarding the proportions of acids and bases, and the distribution of pKa values is required. In medicinal chemistry there are many instances where research is infl uenced by rules of thumb. This could be described as a collective wisdom amongst the medicinal chemistry community where the 'rules' have not been fully researched or described. Such might have been the case with the Lipinski study [10] where some of the underlying principles were roughly known and applied prior to their publication. Certainly for pKa distributions, these have not been fully documented in the literature. It is on this basis that the current study has sought to explore the proportions of acids and bases and to detail the distribution of pKa values for a set of general drug-like molecules.

Drug-Likeness

In recent years there have been numerous studies exploring methods to improve the efficiency of the early stages of new medicines research. The aim of all these studies has been to reduce the development time from the initiation of a project through to the selection of a clinical candidate. Much of it has focused on the 'drug-like' or 'lead-like' nature of screening compounds or synthetic candidates [10–14]. The argument raised was that if compounds were selected for optimization that required a considerable number of synthetic cycles to produce novel analogues that address ADMET (T = toxicity) deficiencies then this lengthened the time needed to arrive at a clinical candidate. If, however, the compound was 'drug-like', or perhaps more preferably 'lead-like' [15] from the outset, then it should be easier to arrive at the appropriate biopharmaceutical properties and in a shorter time frame [16]. Such aspirations are based on sound logic and have been implemented within the current practices of the pharmaceutical industry [10,17]. One of the simplest of these procedures is a structure and functional group filter that removes compounds considered unsuitable as hits such as those containing toxic functional groups [18].

Research into drug-like and lead-like concepts has explored a range of ideas looking at structural characteristics and physicochemical properties. These studies have included examinations of molecular frameworks [19,20], molecular properties [12–14, 21, 22] and the prediction of ADME parameters [23] to name but a few. In addition, compounds that target the CNS have also been analyzed to profile their physicochemical characteristics and to predict CNS activity [24–26]. As such, it is becoming entrenched within the medicinal chemistry community to look extremely closely at the characteristics of the molecules they deal with and to work on those known to have suitable properties. Once again, it makes logical sense to operate most of your time in areas where there is a history of successful outcomes and where efficiencies can be garnered.

Our knowledge of the overall proportion of acids, bases and pKa distributions is less understood than other aspects of drug and lead-likeness. For example, statements often describe drugs as 'typically weak acids and/or weak bases.' The proportion of drugs with an ionizable group has been estimated at 95% [27] while an analysis of the 1999 World Drug Index (WDI [28]) showed that only 62.9% of that collection were ionizable between a pH of 2 and 12 [29, 30]. Wells also estimated that 75% of drugs are weak bases, 20% weak acids and the remainder contained non-ionics, ampholytes and alcohols [27]. A breakdown of the WDI set of ionizable compounds showed that two thirds of them had either a single basic group or two basic groups (Figure 1A). The next major group of compounds containing one or two acids made up 14.6% of this set while simple ampholytes with one acid and one base comprised 7.5%. To analyze the WDI database (51,596 compounds) the Chem-X software [31] was used to discriminate acids and bases. The details of which functional groups were used is not easily discernable, however the concept of exploring a pKa range of 2–12 is admirable given that the term ionizable used by Wells may possibly have encompassed a greater proportion of compounds. This may also suggest why only 62.9% of compounds in the WDI were considered of interest compared to Wells figure of 95% [27]. It should be noted (and presumed) that the two sets of drugs considered by the individual authors would have differed. On a smaller set of compounds (n = 53) with known capacity to cross or not cross the BBB it was [32] concluded that "compounds with minimally one charge with a pKa <4 for acids and correspondingly a pKa >10 for bases do not cross the BBB by passive diffusion." The references cited above are among the few that touch on both pKa and the proportion of ionizable compounds within a set of drugs. It may be that dealing with pKa is occasionally troublesome for a number of reasons, e.g. access to measured data is not simple, calculation of large numbers of pKa values is cumbersome and compounds may contain variable numbers of ionizable groups. Consequently the pKa does not lend itself to simple calculation and comparison, such as molecular weight or polar surface area (PSA) might allow.

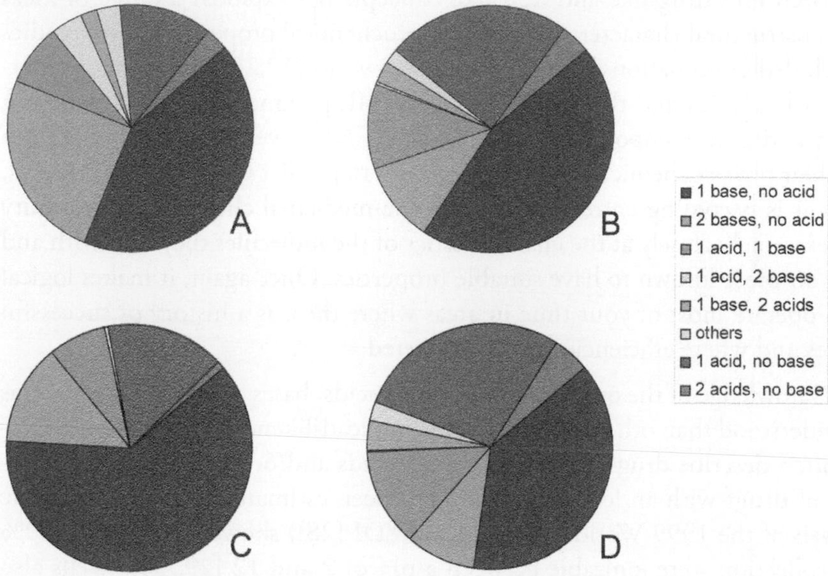

Figure 1. Pie charts showing the distribution of acids and bases from the findings of Comer and Tam [29] outlining the survey conducted by Tim Mitchell [30] on the acid base distribution using the 1999 WDI database, (A); the results from the 582 Williams compound dataset [33], (B); the 174 CNS compound subset, (C); and the 408 non-CNS compound subset, (D). The data associated with these diagrams is given in Tables 1 and 2.

pK_a Data Sources and Analysis

In order to conduct an analysis of the proportion of acids and bases, and pK_a distributions, suitable databases of pK_a values are required. Several sets of pK_a values are available such as PhysProp (Syracuse Research Corporation, North Syracuse, USA), Williams compilation in Foye's textbook [33], the Merck Index [34], Avdeef [35], IUPAC and related compilations [36–41], CRC Handbook [42], Lange's Handbook of Chemistry [43], ACD/labs software and database [44] as well as the general literature. In some cases these data resources do not assign pK_a values to particular functional groups. The Williams set (see Methods for details) [33] used in this current study simply specifies whether the pK_a value is derived from an acid or a base and this feature was an important factor in selecting this dataset for the analysis. Other issues to keep in mind are data quality as these compilations stem from many laboratories. In an ideal world it would be prudent to return to the original study to investigate how the measurements were undertaken and how problems (e.g. apparent pK_a values, decomposition, precipitation, poor UV absorbance, use of cosolvents, complex multi functional compounds) were handled. This perhaps is another reason why the pK_a distribution of drugs has not been described in detail for the analysis of drug-like character.

The goal of these analyses is to provide an indication of the spectrum of pKa values and the proportion of acids and bases within a drug discovery environment. This is with particular regard to drugs that have made it to the marketplace so that this may influence drug discovery processes in general. It could be envisaged that analyses of corporate collections, third party suppliers and combinatorial libraries (real or virtual) are undertaken to determine whether their distributions match that of marketed drugs. Following this, decisions could be made to add to collections where certain classes of compounds or pK_a ranges are underrepresented and to influence synthetic directions. In the simplest sense it may add information regarding the overall composition of compound collections which can be discussed accordingly. Computational tools oriented to looking at ionizable groups as well as tautomer states [45] have recently been established. One example is the ProtoPlex module within Sybyl [46] which can populate a database with alternative tautomers and protomers for each compound. Other workers have also striven to represent compounds in the most appropriate way by considering ionizable groups and tautomers. Kenny and Sadowski [47] described their technique which is able to apply formal charges to selected functional groups. They also emphasized the importance of their work in procedures such as virtual screening. Pospisil and co-workers also showed that tautomer state affected docking scores in virtual screening [45] thus emphasizing the importance of considering pK_a on how we conduct drug discovery. Overall it is clear that the pK_a value(s) of a substance is fundamental to many areas of early and late stage discovery and that knowledge of pK_a distributions will be similarly important to improve how we discover and develop new medicines.

Methods

To explore the proportion of ionizable compounds to non-ionizable compounds the World Health Organization's (WHO) essential medicines list was employed (March 2005 [48]). This represents a list of "minimum medicine needs for a basic health care system," together with a set of complimentary medicines for priority diseases. It may be viewed as a mini-pharmacopoeia, however the makeup of the set will differ somewhat to more extensive lists of drugs. Nevertheless it serves to encompass a range of drug classes for a wide range of medical needs. Compounds were classified into three groups: those with an ionizable group within the pKa range of 2–12 (determined using the ACD/labs software [44]), those without an ionizable group and a miscellaneous set containing proteins, salts and others (e.g. gases, mixtures, polymers, metal complexes, etc). The proportion of ionizable compounds was determined for the entire set and a selected subset that excluded the miscellaneous set.

The list of pKa values compiled by Williams [33] was used as the source of data for the present study. An examination of the list was undertaken and the original set of 599 compounds was reduced to a final set of 582 for analysis. Within this list the source references are given and most of the values come from Hansch in Comprehensive Medicinal Chemistry Volume 649 which is itself a secondary literature compilation. The Williams list [33] was chosen for its assignment of acids and bases, accessibility and representation of a range of compound classes. The initial curation step included removing duplicates (e.g. bupivacaine and levobupivacaine; where the pKa is equivalent) and those compounds without a pKa value. For inclusion the compound was required to have a clinical use (either past or current use) or was considered safe for human consumption or represented an interesting chemotype (e.g. saccharin). Data misplaced in columns was adjusted and where pKa values for acid and base groups had been swapped this was amended. In some cases incorrect values were revised (e.g. tiaprofenic acid) and compounds with non-standard names were excluded where this led to ambiguity of the correct substance.

In addition to this examination, an assessment was made regarding whether the compound was intended for CNS use. In some cases this was not easy to define particularly when the drug has been targeted towards peripheral sites but has CNS side effects. A classic example is the first generation of histamine H1 receptor antagonists that were developed for the treatment of hay fever but often caused drowsiness. Where sedative activity was listed as an indication for the drug then it was annotated as a CNS drug (e.g. trimeprazine). Cocaine, albeit used clinically as a local anaesthetic, has well known CNS effects and was also classified as a CNS substance. In some cases the classification was difficult to assign and, for the most part, the decision was based on the intended uses of the drug.

Table 1. A list of the number of acids and bases in the Williams dataset [33] and associated subsets.

	No acid	1 acid	2 acids	3 acids
Entire dataset				
No base	0	142	22	3
1 base	264	65	16	4
2 bases	61	1	0	0
3 bases	4	0	0	0
CNS subset				
No base	0	27	2	0
1 base	108	13	0	1
2 bases	23	0	0	0
3 bases	0	0	0	0
Non-CNS subset				
No base	0	115	20	3
1 base	156	52	16	3
2 bases	38	1	0	0
3 bases	4	0	0	0

Analysis of the distribution of pKa values was applied to three groups of compounds: those containing a single acid, a single base and ampholytes with 1 acid and 1 base. Histograms for the distributions required binning the compounds into ranges (i.e. $0.5 < X \leq 1.5$, $1.5 < X \leq 2.5$, etc). In each case column heights were expressed as a percentage. Ampholytes (1 acid, 1 base) were also further classified as either ordinary (base pK_a < acid pK_a) or zwitterionic (base pK_a >acid pK_a) compounds. In order to plot and compare the ampholytes the isoelectric point was determined ([acid pK_a + base pK_a)/2] and the values binned in a similar manner to the pK_a values.

Results

(a) Acid and Base Proportions

The proportion of acids and bases in the Williams [33] dataset of 582 compounds was determined by reviewing the pK_a data and summing the number of compounds containing a single base, single acid, and so forth. Table 1 (Entire dataset) shows that almost half the compounds had a single base (45.4%) while single acid compounds made up about a quarter of the total (24.4%). Ampholytes comprised 14.8% of the total of which 65 compounds (11.2%) were considered to be simple ampholytes containing a single acid and base. The other major group was those compounds with two basic groups representing 10.5% of the total. Figure 1B clearly shows the distribution of the 582 compounds demonstrating that over half the compounds are basic in nature (56.5%) [i.e. containing 1, 2 or 3 basic groups without an acidic group].

Splitting the entire list into CNS (n = 174) and non-CNS (n = 408) compounds allowed the construction of pie charts for each of these individual groups. Figure 1C, together with Table 1 (CNS subset) show that the CNS class of compounds is dominated by those containing a single basic group (62.1%). If these compounds are combined with those possessing 2 bases this represents 75.3% of the total. The proportion of compounds containing a single acid was 15.5% while ampholytes (13 compounds) only made up 7.5% of this subset.

The non-CNS group of compounds showed a distribution similar to the entire dataset of 582 compounds and this no doubt was influenced by the large number of compounds that make up this set (n = 408). Figure 1D and Table 1 (Non-CNS) demonstrate that compounds with one or two basic groups now comprise less than half the total (47.5%). The single acids comprised 28.2% and if combined with compounds containing two and three acids these make up about one third of the total. Simple ampholytes on the other hand made up 12.7% of this subset consisting of 52 compounds. Table 2 compares the percentage of compounds containing acids and bases between the Williams lists [33] and the analysis conducted on the WDI [29, 30]. In general the WDI has fewer compounds containing a single acid and a greater number of compounds with two basic groups. The number of compounds with a single basic group was similar between the entire Williams list and the WDI.

The Williams [33] compilation did not, of course, list non-ionizable compounds as its prime interest was in those substances with a pKa value. To estimate the proportion of non-ionizable compounds in a similar manner to the analysis by Comer and Tam [29,30] the WHO essential medicines list was used as a minimum set of therapeutic substances and compounds. The WHO list was consolidated to 301 compounds from their March 2005 edition. Of these, 196 (65.1%) contained an ionizable group with a pKa in the range 2–12. This result is very similar to that obtained by Mitchell of 62.9% [29, 30]. If we remove the miscellaneous compounds (e.g. proteins, salts, mixtures, polymers, gases, etc) from the analysis then we obtain a figure of 77.5% of compounds that contain a relevant ionizable group. This is in contrast to the 95% estimate of Wells [27] and may be a consequence of the small size of the WHO dataset and the inherent limitations for compounds to be included in the list. Alternatively, Wells [27] may have included compounds with ionizable groups outside the pKa range of 2–12.

Table 2. Percentage of acid and base containing compounds in the Williams [33] and WDI datasets [28].

List	1 acid	1 base	2 acids	2 bases	1acid + 1 base	Others
Entire dataset	24.4	45.4	3.8	10.5	11.2	4.8
CNS subset	15.5	62.1	1.1	13.2	7.5	0.6
Non-CNS subset	28.2	38.2	4.9	9.3	12.7	6.6
WDI*	11.6	42.9	3.0	24.6	7.5	10.4

* Data taken from Comer and Tam[29,30]

(b) pK_a Distribution of Single Acid Containing Compounds

From the Williams set [33] single acid containing compounds consisted of 142 substances and a representative sample of these is shown in Figure 2. The distribution of pK_a values is shown in Figure 3 and this also illustrates both the CNS and non-CNS classes. Each column is given as a percentage to allow for the differing sizes of each group. An examination of all 142 acids shows that there is a bimodal distribution with a dip in numbers at a pK_a of around 7.0. Compounds at the lower end of the scale largely contain carboxylic acids while those peaking around a pK_a value of 8.0 contained a large proportion of barbiturates.

Within the CNS class only 27 compounds had a single acid. While this is a low number, the distribution of pKa values was nonetheless very interesting. Figure 3 shows that the majority of acids had a pKa above 7 and only one fell below 6.1 (valproic acid = 4.8).

When the non-CNS class was inspected the bimodal distribution of pKa values was again portrayed showing the dip in frequency close to 7.0. Within this set of 115 compounds those with lower pKa values were predominantly carboxylic acids.

(c) pK_a Distribution of Single Base Containing Compounds

In contrast to the distribution of acids and perhaps as expected, the base pK_a values peaked at a value of 9.0. The majority of compounds had a pK_a value above 6.5 and these compounds typically contained a basic amine group. At the lower end of the pKa scale various functional groups were represented (e.g. nitrogen containing heterocycles). Figure 4 shows a set of representative bases containing various heterocycles and amines. In all, 264 compounds contained a single base making up just under half of the total set analyzed. Figure 5 shows the distribution of base pK_a values ranging in value from 0.1 to 12.3. Once again the CNS and non-CNS classes have been included to allow a comparison of the three groups.

The CNS class (n = 108) showed a clear cut off at the high end of the pKa scale. Indeed, there were no bases with a value above 10.5. Once again the majority of compounds had a pKa above 7 and mostly consisted of amines. The distribution for the non-CNS class closely matched the overall pattern found for the entire dataset with a peak in pKa values at around 9.0. pKa values for the non-CNS compound set (n = 156) ranged from 0.3 to 12.3.

(d) pK_a Distribution of Simple Ampholytes

In order to analyze the distribution of simple ampholytes (i.e. single acid and base) they were first classified as either ordinary or zwitterionic ampholytes and the isoelectric points were calculated. Figure 6 illustrates the range of isoelectric points for both the ordinary and zwitterionic ampholytes. While no clear pattern emerges this may be a reflection of the limited number of compounds (65) available for this analysis. The larger number of ordinary ampholytes at the high end of the scale represent simple phenols with alkylamine side chains (e.g. phenylephrine). If these compounds are left aside, those that remain tend to have isoelectric points between 3.5 and 7.5.

When the CNS and non-CNS drugs were compared interesting differences were observed. For the CNS class there were 13 simple ampholytes which made up only 7.5% of the 174 CNS compound subset. Of these 13 compounds there were six opioids and six benzodiazepines all of which were ordinary ampholytes. In contrast, the non-CNS subset contained 52 ampholytes comprising 20 zwitterions and 32 ordinary ampholytes. No doubt the predominance of ordinary ampholytes in the CNS class reflects the neutral character of these compounds at their isoelectric point where neutrality would favour CNS penetration.

Discussion

Overview of Findings

One concern over the analyses conducted in this study may be the choice of datasets used. This is a problem that plagues any analysis of drug sets that aim to tease out trends in physicochemical characteristics. The set employed should of course be representative of drugs as a whole to enable reasonable conclusions to be drawn. To look at the proportion of ionizables the WHO essential medicines list [48] was used which represents a small pharmacopoeia for priority health care needs. It is overrepresented in certain drug classes (e.g. antibiotics) and lacks a range of medicines which are costly or merely enhance the quality of life (e.g. selective serotonin reuptake inhibitors, HMG-CoA reductase inhibitors, PDE 5 inhibitors, etc.). Nevertheless it is a well thought-out list covering the majority of therapeutic classes. In contrast, the WDI dataset used by Comer and Tam [29, 30] consisted of 51,596 compounds and could be viewed perhaps as a master list of drugs. The WDI, however, includes pesticides, herbicides and compounds that did not reach the market place. Given our desire to be representative of drugs it is not an ideal set and may be considered too encompassing. Our analysis therefore of the proportion of compounds that are ionizable is very dependent on the dataset used and provides results specific to that set. Another option is to examine all the drugs used commercially around the world such as those listed in Martindale [50]. This contains over 5000 drug monographs and an analysis based on this set would be an onerous task. The obvious alternative is to choose a smaller set that has undergone an evolutionary process to select useful therapeutic substances (e.g. through evidence-based therapy), such as the AHFS Drug Handbook [51] (a subject of future research in this laboratory). Until such time that an agreed set of compounds can be selected to determine how many are ionizable the numbers generated here using the WHO list (65.1%) is comparable to the WDI findings of Comer and Tam [29,30] (62.9%) and is far less than the 95% estimate described by Wells [27]. It is not clear which compounds Wells considered or how an 'ionizable compound' was defined. A more interesting analysis might be where strict criteria are used for compounds to be included in a survey. For example, organic compounds of molecular weight <1000 together with a use in mammalian therapy in an oral (or injected) form. For small organic substances this would give a better indication of the proportion of compounds possessing an ionizable group.

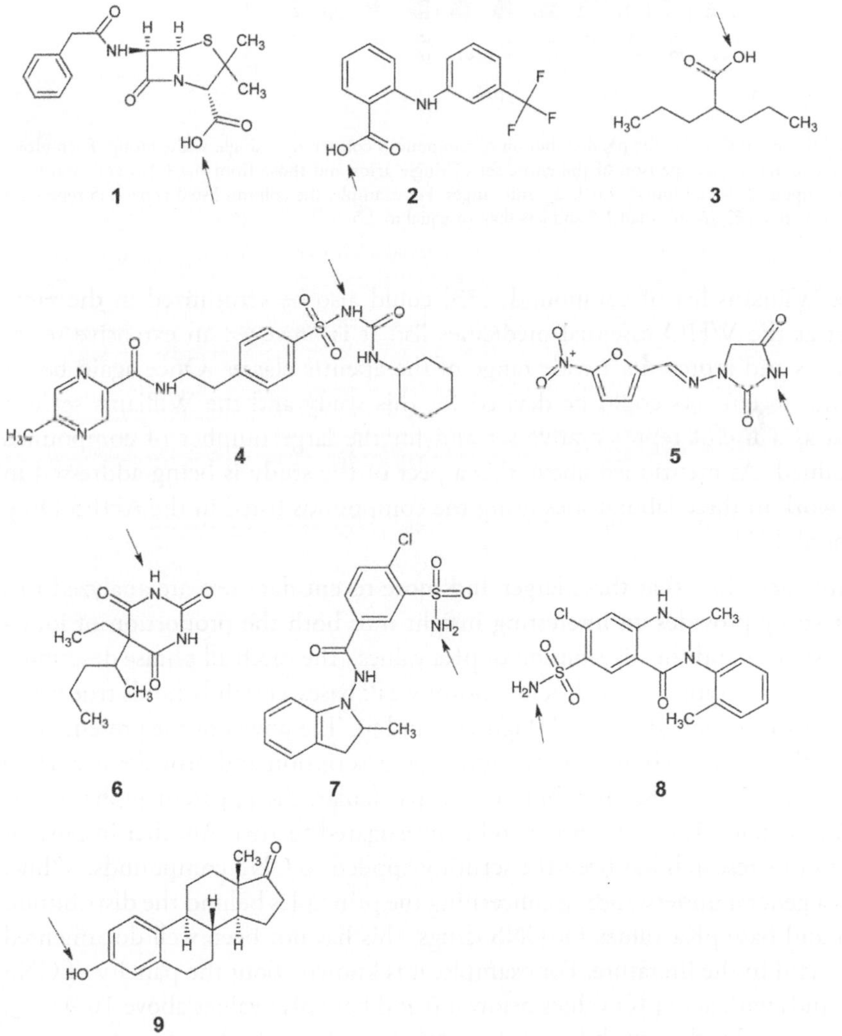

Figure 2. Chart showing nine acids with a range of pK values. In each case the acidic group has been highlighted with an arrow. Penicillin G (1, pK = 2.8), Flufenamic acid (2, pK = 3.9), Valproic acid (3, pK = 4.8), Glipizide (4, pK = 5.9), Nitrofurantoin (5, pK = 7.1), Pentobarbital (6, pK_a = 8.1), Indapamide (7, pK_a = 8.8), Metolazone (8, pK_a = 9.7), Estrone (9, pK_a = 10.8).

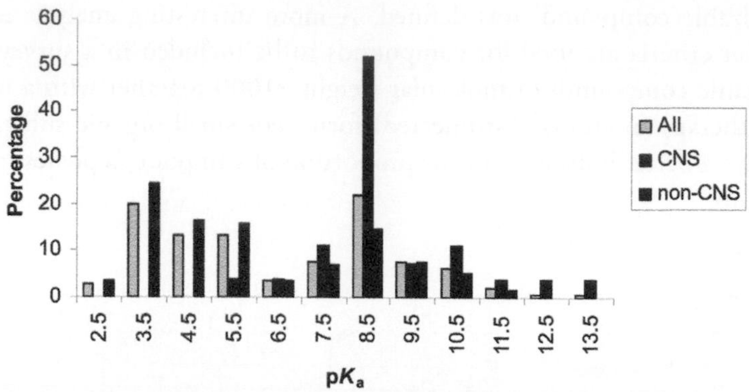

Figure 3. Histogram showing the pK distribution of compounds containing a single acidic group. Each group of columns contains a comparison of the entire set of single acids and those from the CNS and non-CNS subsets. Compounds were binned into 1 log unit ranges. For example, the column listed above 2.5 represents compounds with a pK_a greater than 1.5 and less than or equal to 2.5.

The Williams list of compounds [33] could also be scrutinized in the same manner as the WHO essential medicines list. It is however, an extensive set of substances and represents a wide range of therapeutic classes. Once again better and more recent sets could be devised for this study and the Williams set was selected as a useful representative set and for the large number of compounds it contained. As mentioned above this aspect of the study is being addressed in future work in these laboratories using the compounds listed in the AHFS Drug Handbook [51].

Until such time that these larger and more recent data sets are analyzed this present study provides an interesting insight into both the proportion of ionizable substances and the distribution of pKa values. The catch all phrase describing drugs as mainly 'typically weak acids and/or weak bases' certainly holds true when the pKa distributions are viewed (Figures 3 and 5). The power of the present analysis is to flesh out the bones to this simplistic description and provides a starting point for discussing pKa distributions. In particular, the apparent biphasic distribution of acid pKa values needs to be investigated further. Another important aspect to this research has been the scrutiny applied to CNS compounds. While, there is a general understanding concerning the principles behind the distribution of acid and base pKa values for CNS drugs, this has not been well documented or presented in the literature. For example, it is known about the paucity of CNS compounds with acid pKa values below 4.0 and base pKa values above 10.0 [32]. Also recognized is the sensibility of these values as charged substances do not easily cross the BBB. Acids with pKa values below 4 will be in a charged state over 99% of the time at physiological pH as will bases with a pKa above 10. The cutoff values described by Fischer and coworkers [32] concur with the observations presented here, although only one compound had an acid pKa below 6.1. The important aspect of this present study was to outline the distributions themselves to demonstrate the spectrum of pKa values. Indeed, the overall implication is that this is valuable information when contemplating the properties needed for a drug or sets of screening compounds.

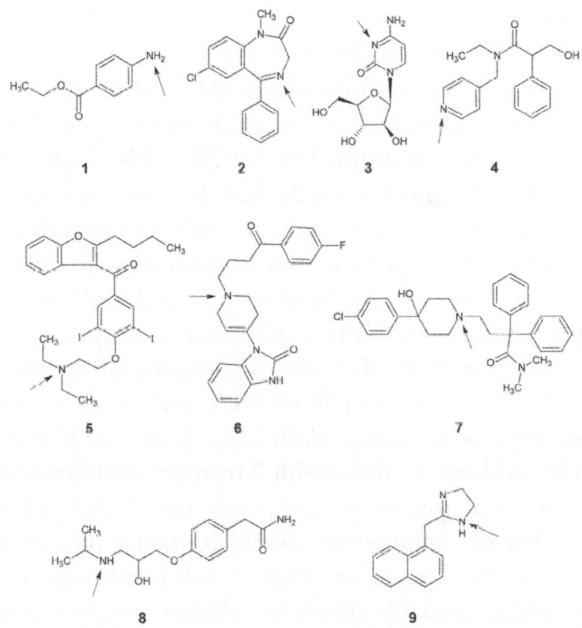

Figure 4. Chart showing nine bases with a range of pK values. In each case the basic group has been highlighted with an arrow. Benzocaine (1, pK = 2.5), Diazepam (2, pK = 3.4), Cytarabine (3, pK = 4.3), Tropicamide (4, pK = 5.3), Amiodarone (5, pK = 6.6), Droperidol (6, pK = 7.6), Loperamide (7, pK_a = 8.6), Atenolol (8, pK_a = 9.6), Naphazoline (9, pK_a = 10.9).

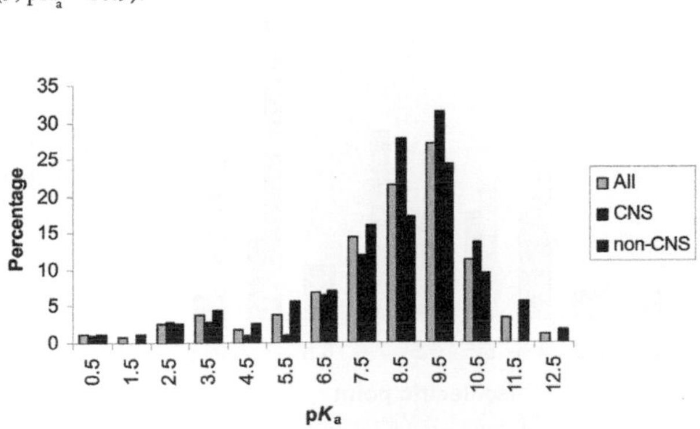

Figure 5. Diagram showing the pK distribution of compounds containing a single basic group. Each group of columns contains a comparison of the entire set of single bases and those from the CNS and non-CNS subsets. Compounds were binned into 1 log unit ranges as per Figure 3.

Application of Findings

The utility of the distributions described here may be applied to third party supplier databases for purchasing decisions regarding screening compounds. Either the ratio of ionizable to neutral compounds could be applied or the pK_a distributions could be used in the selection process. One thing that needs to be borne in mind is that the work described in this study has emerged from an analysis of drugs. Given that current screening efforts are oriented to lead-like molecules [15] then the distributions need to be considered in this light. Certainly an analysis of an ideal screening set of lead-like compounds would yield the appropriate data. In the absence of this we need to look at the guidelines suggested for lead-like character. These follow the criteria outlined here: MW < 350, logP < 3 and affinity approximately 100 nM16. In other words there is scope for chemists to take a small molecule with reasonable activity and enter this into rounds of optimization for activity, selectivity and biopharmaceutical properties. The physicochemical criteria listed above are very simple, however pKa and logD are not considered. Perhaps a simple ratio of ionizable to non-ionizable compounds needs to be suggested (e.g. 3:1, respectively). Furthermore the makeup of the ionizables also needs to be considered by selecting compounds with single acids, single bases and ampholytes, in approximately the ratios outlined in Table 2. More complicated combinations of acids and bases or those with 2 or more acids and bases should be kept to a minimum. These suggestions are purely speculative and are open to debate; suffice to say that the compounds should contain a mix of neutral and ionizables in roughly the ratios seen for drugs as well as allowing chemists the possibility of adding further ionizable groups to enhance activity and biopharmaceutical characteristics as part of the optimization process.

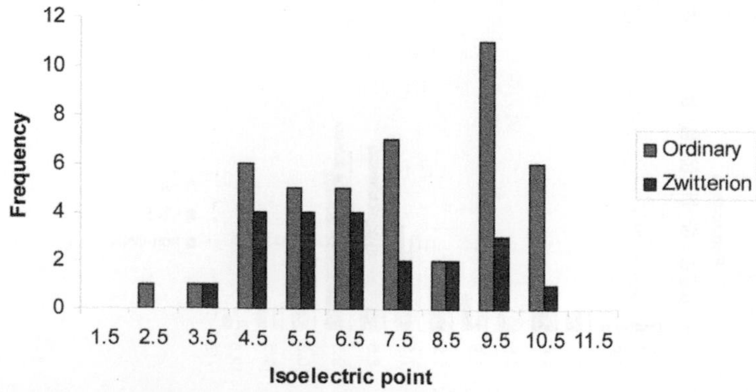

Figure 6. Histogram comparing the isoelectric points of both ordinary and zwitterionic ampholytes. In this case the frequencies of the distributions were shown to reflect the differing number of ordinary ampholytes (44 compounds) and zwitterionic ampholytes (21 compounds). Compounds were binned into 1 log unit ranges as per Figure 3.

Perspectives and Future Directions

Ionizable groups on drug molecules have two principal functions. The first is to modify overall polarity, which in turn controls other physicochemical properties, such as aqueous solubility or hydrophilicity. The second is to provide functional groups that can interact with target macromolecules in specific ways. Organic chemists, on the other hand, do not necessarily consider ionizable groups as first priority groups to include on a novel compound. A chemist, for ease of synthesis may prefer to work with non-polar compounds that are soluble in organic solvents. Another human consideration is the simplicity of the chemistry. Straightforward synthetic schemes will no doubt predominate to reduce the number of steps required. Given that ionizable groups often require protection means that additional synthetic steps are needed and introduces a further level of difficulty. Taking all this together suggests that organic compounds made to date will largely be lacking in ionizable groups. Furthermore, many of the third party suppliers need a large number of new substances for their catalogues which means that a high throughput is required from their chemists. High throughput will be a driver for simpler chemistry and, using the argument above, will result in compounds lacking ionizable groups. Of course, this trend has been identified and is being specifically addressed for compounds with utility in medicinal chemistry. This refers to Lipinski's 10 observations but the historic collections available will certainly be influenced by the (Darwinian) principle of 'simple chemistry wins.'

Medicinal chemists also follow the principles of organic chemistry and prefer to introduce polar (ionizable) groups in the latter stages of a synthesis (e.g. protecting group removal). The last step of a synthesis can also be engineered to be one that can introduce diversity to generate a set of analogues. Third party screening compound suppliers, however, obtain a proportion of their catalogue from organic chemists rather than medicinal chemists. As such it may be that these offerings do not follow the same acid/base/pKa distributions as drugs. Consequently, an examination of acid/base/ pKa distributions will be beneficial to ensure that a suitable mix of compounds is chosen for screening, irrespective of the source.

An overriding question fundamental to this study concerns the pKa distributions themselves. Two separate influences will ultimately shape these findings. The first is chemical in nature concerning the functional groups that comprise the acid and base moieties. If we took the universe of organic compounds (a good representative subset might be the organic compounds contained in the CAS collection) and produced pKa distribution plots then it would be possible to see how drugs compare. It may be that single acid containing compounds don't exhibit a bimodal distribution and that drugs specifically lack groups with pKa values around 7.0. Similar arguments could be directed at basic compounds and that the distributions we observe for drugs are a function of the regularly seen groups used in these compounds. Certainly, toxic functional groups will be very limited in the Williams set [33] and this may also affect the pKa distribution. The second driver for the pKa distributions is biological in nature and is affected by membrane properties and the drug targets themselves. It is known that 7-transmembrane G-protein coupled receptors (7TM GPCR's) have a key aspartic acid residue to recognize the amine group on their endogenous ligands [52]. The need for an amine in drugs that interact with 7TM GPCR's is almost an absolute requirement. If we combine this with the fact that a high percentage of drug targets are 7TM GPCR's [53] then it will follow that amines will be well represented (particularly for CNS compounds) in the Williams set [33]. Our knowledge of pKa distributions for a number of functional groups is quite reasonable but not when these are considered collectively. Presumably the pKa value is a quantity which does not have a smoothly distributed continuum of values, but is necessarily multimodal because of the types of functional groups that exist in organic chemistry. In that sense, it is unlike logP, which has a much more broadly distributed set of values. This is a research area that will no doubt develop as larger populations of compounds are studied.

The task of identifying acids and bases in a database is a readily achievable task. A more difficult procedure is to estimate the pKa values for these compounds. With regard to accuracy we preferably seek to predict within one log unit of the measured value. A variety of computational approaches are available and this topic was reviewed recently by Wan and Ulander [7]. A number of methods are used within the commercial packages (e.g. ACD/Labs44) such as the use of QSAR models based on Hammett analyses. Typically, a molecule is fragmented and the pKa of the functional group is estimated by referring to a database of values with associated QSAR equations. Artificial neural network methods have also been used to estimate pKa and the software available from Simulations Plus is one such example [54]. The ADME Boxes package from Pharma Algorithms [55] also estimates the total number of ionizable groups and predicts the principle pKa values. The other primary method of estimating pKa values is through quantum mechanical techniques. The advantage here is that they can adapt to new chemical classes and do not necessarily need prior examples within the algorithm. In each case, and to differing degrees, estimates can be complicated by conformational flexibility, solvent handling, conjugated systems and a lack of relevant examples. The needs of the pharmaceutical industry are challenging as they regularly explore novel structural scaffolds to enter new patent territory. If the software requires prior examples of a functional group or scaffold then accuracy may be compromised. For the purposes of characterizing a database, speed of calculation is a priority and may take precedence over accuracy. There are many computational hurdles yet to be tackled to provide a chemist friendly, fast and accurate system of estimating pKa values within large databases (100,000's compounds). Among the considerations are problems such as conformational flexibility, internal hydrogen bonding, solvent effects and multiprotic influences [7]. Fortunately, several groups are working on better prediction methods and this will ultimately influence how we undertake research for new medicines.

Conclusion

This study has begun to explore the overall composition of drugs with regard to the proportion of those compounds containing an ionizable group. Within the WHO essential medicines list 65.1% of compounds had an ionizable group with a pK_a in the range 2–12 and this number rises to 77.5% when non drug-like compounds are removed. Other estimates give this number as anywhere between 62.9% [29,30] and 95% [27]. It is certainly clear that this figure is influenced by the collection being studied and how 'ionizable' is defined, and will be the subject of future research from our laboratories.

Analysis of Williams collection of drugs [33] has led to a description of the relative proportions of compounds containing acidic and basic functionality. More importantly, the distribution of pK_a values has been outlined in detail for the first time. Two clear findings emerged upon examination of the distributions particularly when a distinction was made between CNS and non-CNS drugs. Firstly, acid pK_a values for CNS drugs rarely fell below 6.0 and secondly, base pK_a values for CNS drugs were not observed above a value of 10.5. From an ionization viewpoint these observations are entirely reasonable when considering the nature of the BBB and the passage of charged substances across membranes. As such, these observations consolidate current wisdom in the area and open the way for larger collections to be compared to these distributions.

Without doubt pK_a is of paramount importance to the overall characteristics of a drug and has considerable influence on biopharmaceutical properties. Current trends indicate that future research is placing an increased focus on pK_a with the advent of high throughput measurement techniques and improvements to computational prediction software [7]. By taking pK_a into account allows the researcher to begin ADME profiling early in the discovery process. Moreover, with large collections of compounds such as corporate databases, third party supplier offerings and virtual sets of compounds (e.g. virtual combinatorial libraries), the researcher can examine both the proportion of ionizable compounds and with prediction methods can start to look at pK_a distributions. If these differ largely from the observations outlined in the current study then it allows the opportunity to amend synthetic directions or screening compound selections.

The drive to consider the physicochemical properties of drugs to understand biopharmaceutical characteristics began many years ago (e.g.10). This has fundamentally changed how discovery work is undertaken and was oriented to improving the efficiency and productivity of pharmaceutical companies. Likewise, the need to explore pKa will begin to influence how we work. The findings presented here go some way to understanding the distribution of pKa values and further guidelines will evolve as larger datasets are analyzed.

Acknowledgements

The author thanks doctors Richard Prankerd and David Chalmers for their insightful discussions and valuable suggestions.

Note

Since this article was written, Lee et al. (Lee P.H., Ayyampalayam S.N., Carreira L.A., Shalaeva M., Bhattachar S., Coselmon R., Poole S., Gifford E. and Lombardo F. 2007 In Silico Prediction of Ionization Constants of Drugs. Mol., Pharm. 4:498–512.) have described their SPARC program which predicts pK_a values for drug-like compounds. Comparisons of predicted against measured pK_a values for a set of 123 drugs gave a root mean square error of 0.78 log units. The program is also capable of running in batch mode and may be extremely useful for characterizing large data-sets of compounds. Interested readers can also view the software at http://sparc.chem.uga.edu.

References

1. Kerns, E.H. and Di., L. 2004. Physicochemical profiling: overview of the screens. Drug Discov Today: Technologies, 1:343–8.
2. Avdeef, A. 2001. Physicochemical profiling (solubility, permeability and charge state). Curr. Top. Med. Chem., 1:277–351.
3. Xie, X., Steiner, S.H. and Bickel, M.H. 1991. Kinetics of distribution and adipose tissue storage as a function of lipophilicity and chemical structure. II. Benzodiazepines. Drug Metab. Dispos., 19:15–9.
4. Jones, T. and Taylor., G. 1987. Quantitative structure-pharmacokinetic relationships amongst phenothiazine drugs. Proc. - Eur. Congr. Biopharm. Pharmacokinet. 3rd, 2:181–90.

5. Mitani, G.M., Steinberg, I., Lien, E.J., Harrison, E.C. and Elkayam, U. 1987. The pharmacokinetics of antiarrhythmic agents in pregnancy and lactation. Clin. Pharmacokinet, 12:253–91.

6. www.fda.gov.

7. Wan, H. and Ulander, J. 2006. High-throughput pK(a) screening and prediction amenable for ADME profiling. Expert Opin. Drug Metab Toxicol, 2:139–55.

8. Zhou, C., Jin, Y., Kenseth, J.R., Stella, M., Wehmeyer, K.R. et al. 2005. Rapid pKa estimation using vacuum-assisted multiplexed capillary electrophoresis (VAMCE) with ultraviolet detection. J Pharm. Sci., 94:576–89.

9. Hou, T., Wang, J., Zhang, W., Wang, W. and Xu, X. 2006. Recent advances in computational prediction of drug absorption and permeability in drug discovery. Curr. Med. Chem., 13:2653–67.

10. Lipinski, C.A., Lombardo, F., Dominy, B.W. and Feeney, P.J. 1997. Experimental and computational approaches to estimate solubility and permeability in drug discovery and development settings. Adv. Drug Del. Rev. 46:3–26.

11. Teague, S.J., Davis, A.M., Leeson, P.D. and Oprea, T. 1999. The Design of Leadlike Combinatorial Libraries. Angew Chem. Int. Ed. Engl., 38:3743–8.

12. Lajiness, M.S., Vieth, M. and Erickson, J. 2004. Molecular properties that influence oral drug-like behavior. Curr Opin. Drug Discov. Devel., 7:470–7.

13. Proudfoot, J.R. 2002. Drugs, leads, and drug-likeness: an analysis of some recently launched drugs. Bioorg. Med. Chem. Lett., 12:1647–50.

14. Oprea, T.I., Davis, A.M., Teague, S.J. and Leeson, P.D. 2001. Is there a difference between leads and drugs? A historical perspective. J Chem. Inf. Comput. Sci., 41:1308–15.

15. Hann, M.M. and Oprea, T.I. 2004. Pursuing the leadlikeness concept in pharmaceutical research. Curr. Opin. Chem. Biol., 8:255–63.

16. Wunberg, T., Hendrix, M., Hillisch, A., Lobell, M., Meier, H., Schmeck, C., Wild, H. and Hinzen, B. 2006. Improving the hit-to-lead process: data-driven assessment of drug-like and lead-like screening hits. Drug Discov. Today, 11:175–80.

17. Ghose, A.K., Herbertz, T., Salvino, J.M. and Mallamo, J.P. 2006. Knowledge-based chemoinformatic approaches to drug discovery. Drug Discov. Today, 11:1107–14.

18. Hann, M., Hudson, B., Lewell, X., Lifely, R., Miller, L. and Ramsden, N. 1999. Strategic pooling of compounds for high-throughput screening. J Chem Inf Comput. Sci., 39:897–902.

19. Bemis, G.W. and Murcko, M.A. 1996. The properties of known drugs. 1. Molecular frameworks. J Med. Chem., 39:2887–93.
20. Bemis, G.W. and Murcko, M.A. 1999. Properties of known drugs. 2. Side chains. J Med. Chem., 42:5095–9.
21. Ajay, A., Walters, W.P. and Murcko, M.A. 1998. Can we learn to distinguish between "drug-like" and "nondrug-like" molecules? J Med. Chem., 41:3314–24.
22. Sadowski, J. and Kubinyi, H. 1998. A scoring scheme for discriminating between drugs and nondrugs. J Med. Chem., 41:3325–9.
23. Clark, D.E. and Pickett, S.D. 2000. Computational methods for the prediction of 'drug-likeness'. Drug Discov. Today, 5:49–58.
24. Ajay, Bemis, G.W. and Murcko, M.A. 1999. Designing libraries with CNS activity. J Med. Chem., 42:4942–51.
25. Keseru, G.M., Molnar, L. and Greiner, I. 2000. A neural network based virtual high throughput screening test for the prediction of CNS activity. Comb. Chem. High Throughput Screen., 3:535–40.
26. Clark, D.E. 2005. Computational prediction of blood-brain barrier permeation. Annual Reports in Medicinal Chemistry; Elsevier: San Diego, pp 403–15.
27. Wells, J.I. 1998. Pharmaceutical Preformulation; Eills Hoowood Ltd.: London, 25.
28. WDI The World Drug Index is available from Derwent Informatin, London, U.K. www.derwent.com.
29. Comer, J. and Tam, K. 2001. Lipophilicity profiles. Pharmacokinetic Optimization in Drug Research: Biological, Physicochemical and Computational Strategies; Wiley: Zurich. pp 275–304.
30. Mitchell, T. 2006. Personal communication: Cambridge, UK.
31. Nicklaus, M.C., Milne, G.W. and Zaharevitz, D. 1993. Chem-X and CAMBRIDGE. Comparison of computer generated chemical structures with X-ray crystallographic data. J Chem. Inf. Comput. Sci., 33:639–46.
32. Fischer, H., Gottschlich, R. and Seelig, A. 1998. Blood-brain barrier permeation: molecular parameters governing passive diffusion. J Membr. Biol., 165:201–11.
33. Williams, D. A. and Lemke, T. L. 2002. pKa values for some drugs and miscellaneous organic acids and bases. Foye's Principles of Medicinal Chemistry; 5th edition ed.; Lippincott, Williams and Wilkins: Philadelphia. pp 1070–9.
34. Merck. 2006. The Merck Index; 14th edition ed., Merck Publications.

35. Avdeef, A. 2003. Absorption and Drug Development: Solubility, Permeability, and Charge State; Wiley: Hoboken.

36. Kortum, G., Vogel, W. and Andrussow, K. 1961. Dissociation Constants of Organic Acids in Aqueous Solution; Butterworths: London.

37. Sillén, L.G. and Martell, A.E. 1964. Stability Constants of Metal-Ion Complexes, Special Publication 17; Chemical Society: London.

38. Sillén, L.G. and Martell, A.E. 1971. Stability Constants of Metal-Ion Complexes, Special Publication 25; Chemical Society: London.

39. Perrin, D.D. 1965. Dissociation Constants of Organic Bases in Aqueous Solution; Butterworths: London.

40. Serjeant, E.P. and Dempsey, B. 1979. Ionization Constants of Organic Acids in Aqueous Solution; Pergamon: Oxford.

41. Smith, R.M. and Martell, A.E. 1974. Critical Stability Constants, Vols. 1–6; Plenum Press: New York.

42. Lide, D.R. 2006. CRC Handbook of Chemistry and Physics; 87th ed.; Lide, D.R. Ed., CRC Press: Boca Raton, Fla.

43. Dean, J. A. 1999. Lange's Handbook of Chemistry; 15th ed., McGraw-Hill: New York.

44. ACD/Labs ACD/pKa DB version 9.0; 9.0 ed.; Advanced Chemistry Development, Inc. www.acdlabs.com: Toronto.

45. Pospisil, P., Ballmer, P., Scapozza, L., Folkers, G. 2003. Tautomerism in computer-aided drug design. J. Recept. Signal. Transduct Res., 23:361–71.

46. Tripos Sybyl version 7.0, ProtoPlex module; 7.0 ed., Tripos Inc. www. tripos. com: St. Louis.

47. Kenny, P.W. and Sadowski, J. 2005. Structure modification in chemical databases. Methods and Principles in Medicinal Chemistry, Volume 23: Wiley-VCH: Weinheim, pp 271–285.

48. WHO World Health Organisation. 2005. http://www.who.int/medicines/publications/essentialmedicines/en/, http://whqlibdoc.who. int/hq/2005/a87017_eng.pdf.

49. Hansch, C., Sammes, P.G. and Taylor, J.B. 1990. Comprehensive Medicinal Chemistry, Vol 6: Pergamon Press.

50. Sweetman, S. 2006. Martindale: The Complete Drug Reference, 35th Edition; The Pharmaceutical press: London.

51. AHFS. 2003. AHFS Drug Handbook, 2nd Edition; Lippincott Williams & Wilkins: New York.

52. Huang, E. S. 2003. Construction of a sequence motif characteristic of aminergic G protein-coupled receptors. Protein Sci., 12:1360–7.
53. Overington, J. P., Al-Lazikani, B. and Hopkins, A.L. 2006. How many drug targets are there? Nat. Rev. Drug Discov., 5:993–6.
54. Simulations/Plus ADMET Predictor; Simulations Plus Inc. www. simulations-plus.com: Lancaster, CA.
55. Pharma_Algorithms ADME Boxes 3.5; 3.5 ed., Pharma Algorithms www.ap-algorithms.com: Toronto.

CITATION

Manallack DT. The pKa Distribution of Drugs: Application to Drug Discovery. Perspectives in Medicinal Chemisry, 2007; 1: 25–38.

Determination of μmol^{-1} Level of Iron (III) in Natural Waters and Total Iron in Drugs by Flow Injection Spectrophotometry

B. Sahasrabuddhey, S. Mishra, A. Jain and K. K. Verma

ABSTRACT

The equilibrium problems, characterized by recurring end-points, involved in the reaction of iron (III) with iodide make the batch iodometric determination of iron (III) unsuitable. Since the flow injection determination does not require attainment of steady state either for mixing of reagents or for the chemical reaction, the iodometric determination has been accurately and precisely performed using this technique in the present work. This method does *not require any special reagent, including chelating agents or those which are toxic, and has a limit of detection of 0.2 μmol ^{1}(11 μg l^{1}) of iron (Ill). The interference of fluoride has been avoided by adding zirconyl nitrate to the test sample solution, and of copper (II) by complex formation with 2-mercapto-benzoxazole. The method has been applied to determine iron (III) in natural waters, and total iron in drugs.*

Introduction

Iron is an essential trace element involved in normal growth and development. The importance of iron in nutrition has been recognized and, therefore, this element is often added in certain foodstuffs. Reliable speciation of iron (II) and iron (III) is fundamental for the proper characterization of many of the processes in terrestrial and aquatic environments [1], and is also of practical importance in the investigation of corrosion of iron and the treatment of wastewaters from the mining and steel industries [2]. The determination of the oxidation state of iron in a variety of natural water samples has generally been performed by complexation with specific chelating agents followed by spectrophotometry [3-8] or voltammerry [9-11]. To attain adequate sensitivity of the method in dealing with the analysis of iron at μmoll or lower levels, pre-concentration procedures have been reported using the natural polymer Chitin [12], Chelex 100 [13], melamine-formaldehyde resin [14], or by solid phase extraction on C_{18} cartridge [15,16].

When a chelating agent is developed for determination, one has to be aware of the strength of the iron-chelator complex. This is important in order to assess the usefulness of the chelator with respect to naturally occurring chelators in the ambient water samples. If naturally occurring ligands have large stability constants, e.g. EDTA-iron (III) (β= 1025), then the concentrations obtained with chelating methods are only erroneous [8]. Another aspect deals with the potential changes in the oxidation state of iron in the sample due to the changes in the redox potential between iron (II)/iron (III) oxidation states. This change in redox potential is due to the favourable stabilization of one oxidation state over the other when a chelating agent is added to the natural water samples. This is an important aspect in measuring a specific redox state of iron at low ambient levels. Thus, techniques which eliminate the need for using chelating agents are worth re-evaluation.

In the present work, a flow injection spectrophotometric method is described which does not make use of chelating agent or toxic reagents, but is sensitive to μmol levels of iron (III).

Experimental

Instrumentation

Shimadzu LC-5A reciprocating pump (Tokyo) and Ismatec Mini-S 820 peristaltic pump (Zurich) were used for propelling the carrier stream of water and reagents, respectively. Rheodyne 5020 low pressure PTFE 4-way valve (Anachem, Luton) was used for injection of sample solution. Shimadzu SPD-2A variable wavelength dctector (8 μl flow-through cell) and Shimadzu C-R2AX integrator fitted with a printer was used for measurement of peak height. PTFE tubings (Anachem), 0.5 mm i.d., were used for construction of flow lines. All flanged connections were made with plastic nuts and TEFZEL coupler. Home-made T-joints of 0.8 mm i.d. were employed.

The flow-injection manifold used for the analysis of iron (III) is given in figure 1. It consisted of three channels, the first two, mounted on a peristaltic pump, were used for propelling potassium iodide and hydrochloric acid streams. Both iodide and acid were mixed in a delay coil and downstream merged with the third channel used for the water carrier stream. Thc four-way loop injection valve was mounted on thc water carrier stream. Another delay coil was used for the reaction to occur before detection at 360 nm.

Reagents and Standards

Hydrochloric acid, 0.25 mol^{-1} was used as the carrier in the optimized method.

A stock 0.01 mol 1-1 iron (III) solution was prepared as follows. Iron (II) ammonium sulphate, 3.92g, was dissolved in about 100ml of water, mixed dropwise while stirring with 5 ml of concentrated nitric acid, boiled for 5 min, and then cooled to 50-60 ° C. About 3 g of potassiren peroxodisulphate was added portionwise with stirring. The solution was diluted to about 200 ml and again boiled for about 10 min. The cooled solution was treated with ammonia (1:1, concentrated ammonia-water) till slight precipitation of iron (III) hydroxide. The precipitate was redissolved by adding 0.25 mol 1-1 hydrochloric acid, and the solution made up to in a standard flask with the same solvent.

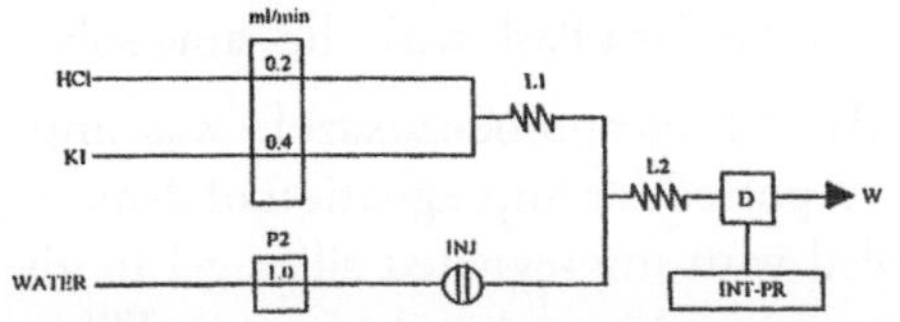

Figure 1. Schematic diagram of the flow injection manifold used for the determination of iron (III). Potassium iodide, 8%, and 0.25M hydrochloricid acid were used as reagents, and deinized distilled water as carrier. P1 = peristaltic pump; P2 = HPLC reciprocating pump; INJ = injection valve (l00-μl loop); L1 = 50 cm mixing coil; L2 = 20 cm reaction coil, D = detector set at 360 nm; INT-PR = integrator and printer; W = waste. All flow lines were made from 0.5mm i.d. PTFE tubing.

For standardization of iron (III) solution [17,18], a 5 ml aliquot of iron (III) solution was treated either with 5 ml of 0.01mol l-1 ascorbic acid or 10ml of 0.01mol l-1 mercaptoacetic acid and swirled for min. About 0.1 g of potassium iodide and ml of 1% starch were added, and the residual amount of ascorbic acid or mercaptoacetic acid from the reaction was evaluated by back titration with 0.01 mol l^{-1} chloramine T, a blue colour was obtained at the end-point. Iron (III) reacts with ascorbic acid or mercaptoacetic acid in a molar ratio of 2:1 and 1:1, respectively. The strength of iron (III) solution determined by two methods agreed within 1%.

The stock solution was sequentially diluted with 0.25mol l-l hydrochloric acid to give test solutions of iron (III).

All other substances used were of high purity, and their solutions were made by dissolving the right amounts in water.

Procedures

Wet Ashing of Iron Tablets

A known number of tablets or contents of capsules were weighed and finely ground. A weight equivalent to a tablet or contents of a capsule was transferred to a 100 ml Kjeldahl flask, mixed with about 20 ml of water and boiled over a small flame till the volume was reduced to about 5 ml. The solution was cooled to room temperature and carefully mixed with 15 ml of concentrated nitric acid and then with 10 ml of concentrated sulphuric acid. The flask was kept at room temperature for about 15 min then heated slowly to boil the contents under a fume-cupboard where copious fumes of nitrogen oxides evolved. The above process was repeated twice by adding both acids, and heating till no more brown fumes evolved. About 50 ml of water along with 5 g of potassium peroxodisulphate was added, and boiling continued till the contents of the flask evaporated to almost dry. The dried matter was boiled with 25 ml of water, and transferred quantitatively to a 250 ml beaker, filtering any insoluble matter. The clear filtrate was treated with 1:1 ammonia-water till slight precipitation of iron (III) hydroxide, the precipitate was redissolved by adding 0.25mol 1^{-1} hydrochloric acid, and finally diluted to 250ml in a standard flask with the same solvent.

Removal of copper (II). 2-Mercaptobenzoxazole was impregnated on silica gel using the procedure as reported for impregnation of 2-mercaptobenzothiazole [19]. The column was filled with impregnated silica gel to give a bed height of about cm which was secured in position by placing one Whatman filter No. 42 disc above and another below the bed. A known volume of drug solution prepared as above was passed through the column under mild suction (1-2 ml min-1). The column was washed with about 2 ml of water. The combined eluent and washings were boiled with 2 ml of concentrated nitric acid, cooled and diluted to a known volume.

Results and Discussion

Though iodimetric titration of iron (III), which involves the reaction of iron (III) with acidified iodide and titration of the liberated iodine with standard thiosulphate, has been recommended in British Pharmacopoeia, 1968, the most frequent [20] difficulties are encountered due to recurring end-points, as the reaction is very slow near the stoichiometric end-point. Extraction of the liberated iodine into chloroform or carbon tetrachloride has been recommended to push the reaction to completion [20]. However, such titrations involving two immiscible phases are always cumbersome and the endpoint is often carried over.

Flow injection analysis does not require the attainment of a steady state for mixing of reactants or for the chemical reaction, and the signals, though transient, are highly reproducible. This unique feature of the flow injection technique was tested for the analysis of iron (III) by its reaction with iodide, and the conditions have been optimized for achieving maximum sensitivity.

Study of Flow Injection Variables

The flow injection manifold, as given in figure 1, was used when the acid and potassium iodide reagents streams were each maintained at 0.2 ml min $^{-1}$, carrier flow rate was ml min^{-1}, loop size was 75 μl, mixing coil and reaction coil were, respectively, 50 and 25 cm. The effect of all the variables was evaluated. There was an optimum peak height observed when 0.25 mol l^{-1} hydrochloric acid was used. The peak height increased when up to 8% potassium iodide was used, then remained practically constant at higher concentration. In the optimized procedure, an 8% potassium iodide was used. Optimum flow rates were: carrier stream ml min^{-1}, acid stream 0.2mlmin-and potassium iodide stream 0.4 ml min^{-1}.

A 50cm coil was necessary for complete mixing of potassium iodide and hydrochloric acid mixed reagents streams. A 20 cm reaction coil was found to give optimum peak height. The signal was optimum when a 100 μl sample was introduced into the flow system.

Validation of the Analytical Procedures

The flow injection analysis of iron (III) was validated by injecting known amounts of standards in the flow system. Calibration graphs were constructed on two concentration ranges of 0.01-0.1 mmol^{-1} and 1-10 μmol 1^{-1} iron (III). The analytical characteristics of calibration graphs are given in table 1. The limit of detection has been found to be .0.2 μmoll^{-1} (11 μg 1^{-1}) iron (III) [S/N = 3; RSD = 3.5%] which compares favourably with that of diverse procedures for iron determination (table 2), and is an important feature of this method since no special reagent is necessary.

Masking of Flouride in Natural Waters

Fluoride forms a stable complex with iron (III), hexafluoroferrat (III), which does not liberate iodine on reaction with iodide. Thus, the interference of fluoride was avoided by adding zirconyl nitrate to the test sample solution when a still more stable complex zirconyl fluoride is formed leaving iron (III) free.

Table 1. Analytical characteristics of the flow injection spectrophotometric determination of iron (III); the absorbance was recorded at 360nm with AFS 0.32. Six calibration points of five replicates each were sampled.

Range	*Slope**	*Intercept**	*r*
0.01–0.1 mmol l^{-1}	12 938.3 IU.l $mmol^{-1}$	−37.31 IU	0.9996
1–10 μmol l^{-1}	11.92 IU.l $μmol^{-1}$	−5.29 IU	0.9992

*IU = integrator arbitrary units.

Table 2. Comparison of diverse methods for iron determination (without preconcentration).

Technique	*Reagent*	*Limit of detection, μmol l^{-1}*	*Ref.*
Spectrophotometry	Iodide	0.2	This work
	1,10-Phenanthroline	0.36	1
	1,10-Phenanthroline	0.2	3,5
	Ferrozine	0.1	15
	Ferrozine	0.2	8
	1-Amino-4-hydroxyanthraquinone	20	22
Spectrofluorimetry	5-(4-methylphenylazo)-8-Aminoquinoline	0.18	23
Potentiometry	Solochrome Violet RS	0.25	24
	1,10-Phenanthroline	0.54	25
Flame AAS		0.2	26

Table 3. Determination of iron (III) and total iron in the presence of copper (II) involving sorption of copper (II) with 2-mercaptobenzoxazole.

		% Found after extraction of copper (II)			
Iron (III) μmol l^{-1} taken	*Copper (II) μmol l^{-1}*	*Iron (III)*	*%RSD (n = 5)*	*Total iron*	*%RSD (n = 5)*
5	2	96.1	1.8	99.9	
	5	97.2			
	10	98.4	2.9	100.1	2.1
10	2			98.7	
	10	97.1			
	20	98.5	2.2	99.6	2.9
20	2	96.2			
	20	98.2		99.2	
	40	98.8	2.5	100.5	2.2
40	2	95.8	2.3		
	40	97.0	2.8	98.9	
	80	98.2		101.2	2.8

Removal of Copper (II) from Drug Formulations

Sorption of copper (II) by complexation with thiol reagents loaded on to silica gel has been documented [19]. Since all general thiols also reduce iron (III), attempts were made to complex copper (II) without significant reduction of iron (III). Reagents which were tried include 2-mercaptobenzothiazole, -benzimidazole and-benzoxazole, and triphenylmethanethiol. Triphenylmethanethiol showed minimal reduction of iron (less than 1%) but was too sluggish in the sorption of copper (II). 2-Mercaptobenzoxazole was found to be the best reagent among those tested. Results are given in table 3 for the determination of iron (III) and total iron in the presence of copper. There was a 4-5% reduction of iron (III) on the sorbent, however, this was less appreciable when the copper (II)/iron (III) ratio was more than 1, ostensibly due to greater reactivity of the sorbent towards copper (II). Any iron (II) formed during sorption was reoxidized before analysis of total iron in drugs.

Application of the Method to Natural Waters

The present flow injection method was applied to determine iron (III) present in certain natural waters. The samples were filtered through a 0.45 gm membrane filter, the filtrate was acidified with 0.5ml of concentrated hydrochloric acid per 500ml of sample, treated with 1% (w/v) zirconyl nitrate to mask fluoride, if also present, and analysed. The method was further validated by standard addition method. The results obtained for the addition of known amounts of iron (III) to natural samples gave an average recovery of 102% with a standard deviation of 5.4% (table 4). The concentration of copper (II) in natural waters is usually low enough so as not to interfere in the determination or iron (III) [21].

Application of the Method to the Determination of Total Iron in Drugs

The present method has been applied to determine total iron in certain pharmaceutical preparations. The results obtained have been compared with those found by a previously checked procedure [18] (table 5). The present method has been found to be rapid and precise.

Table 4. Flow injection spectrophotometric determination of iron (III) in natural waters.

Water sample	*Iron (III) added μmol l⁻¹*	*Iron (III) found μmol l⁻¹*	*Recovery (%)**	*%RSD (n = 6)*
Well water	0	15.0		4.5
	10.0	26.4	114	5.0
Tap water No. 1	0	2.32		6.1
	5.0	7.20	98	5.9
Tap water No. 2	0	3.89		7.1
	5.0	9.00	102	5.6
Ground water No. 1	0	4.05		5.9
	10.0	13.87	98	5.3
Ground water No. 2	0	3.75		7.0
	5	8.55	96	6.5
Ground water No. 3	0	70.4		4.3
	50.0	125.0	109	4.6

* The results are the average of six determinations.

Table 5. Results of the flow injection spectrophotometric determination of total iron in drugs.

	Total iron, mg, per tablet or capsule			
Drug	*Label claimed*	*Found by present method**	*% RSD*	*Found by comparison method [18]†*
Fesovit capsule‡	54.0	48.4	5.6	50.0
Autrin capsule§	113.9	109.3	6.2	105.0
Haematinic tablet¶	25.0	21.2	5.8	20.4

* The results are the average of five determinations.

† The results are the average of three determinations.

‡ Also contains ascorbic acid (50 mg), thiamine mononitrate (2 mg), nicotinamide (15 mg), pyridoxine hydrochloride (1 mg), pantothenic acid (calcium salt) (2.5 mg) and amaranth.

§Also contains ascorbic acid (150 mg), cyanocobalamine (15 μg) and folic acid (1.5 mg).

¶ Also contains ascorbic acid (35 mg), calcium citrate (500 mg), vitamin B2 (2 mg), vitamin B6 (1 mg), folic acid (0.3 mg) and vitamin E (1 μg). Each tablet was mixed with 5 mg of copper (II).

Interferences

Chelating methods are usually severely interfered by fluoride, EDTA and phosphate. Many of these agents either do not interfere in the iodide method or can be masked with suitable reagents.

Chlorine and chlorine dioxide are other oxidizing agents that may be present in drinking water along with iron (III). Their interference can be removed by purging with nitrogen (for chlorine dioxide) or treatment with sodium oxalate (for chlorine) before analysis for iron (III). Nitrite can be decomposed with ammonium chloride in feebly acidic medium.

Conclusions

The sensitivity of the present method involving oxidation of iodide is comparable to those which use chelating agents, e.g. 1,10-phenanthroline or ferrozine, or based on AAS. Flow injection circumvents the equilibrium problems of the reaction which are commonly observed in batch methods.

The present method does not make use of any special reagent. It is simple and rapid, and suitable for sensitive determination of iron (III) in environmental waters, and total iron in drugs. The method has potential for its application to other samples, e.g. foodstuffs.

References

1. Pehkonen, S., 1995, Analyst, 120, 2655.
2. Clark, L. J., 1962, Anal. Chem., 34, 348.
3. Chalk, S.J. and Tyson, J. E, 1994, Anal. Chem., 66, 660.
4. Hase, U. and Yoshimura, K., 1992, Analyst, 117, 1501.
5. Liu, R. M., Liu, D.-J. and Sun, A.-L., 1992, Analyst, 117, 1767.
6. Kang, S.W., Sakai, T., Ohno, N. and IDA, K., 1992, Anal. Chim. Acta, 261, 197.
7. Y., Z., Zhuang, G. S., Brown, E R. and Duce, R. A., 1992, Anal. Chem., 64, 2826.
8. Pascual-Reguera, M. I., Ortega-Carmona, I. and Molina-Diaz, A., 1997, Talanta, 44, 1793.
9. Florence, T. M., 1992, Analyst, 117, 551.
10. Wang, L. Z., Ma, C. S., Zhang, X. L. and Wang, J. G., 1994, Anal. Lett., 27, 1165.
11. Diaz, M. E. V., Sanchez, J. C.J., Mochon, M. C. and Perez, A. G., 1994, Analyst, 119, 1571.
12. Hoshi, S., Yamada, M., Inque, S. and Matsubara, M., 1989, Talanta, 36, 606.
13. Rubi, E., Jimenez, M. S., Bauza De Mirabo, F., Forteza, R. and Cerda, V., 1997, Talanta, 4, 553.
14. Filik, H., Ozturk, B. D., Dogutan, M., Gumus, G. and Apak, R., 1997, Talanta, 44, 877.
15. King, D. W., Lin, J. and Kester, D. R., 1991, Anal. Chim. Acta, 247, 125.

16. Nickson, R. A., Hill, S. J. and Worsfold, E J., 1995, Anal. Proc., 32, 387.
17. Verma, K. K., 1979, Talanta, 26, 277.
18. Verma, K. K., Gupta, A. K. and Bose, S., 1981, Farmaco Ed. Pr., 36, 23.
19. Terada, K., Matsumoto, K. and Kimura, H., 1983, Anal. Chim. Acta, 1983, 153, 237.
20. Garratt, D. C., 1964, The Quantitative Analysis of Drugs (London: Chapman & Hall), p. 350.
21. Yokota, E and ABE, S., 1997, Anal. Commun., 34, 111.
22. Abu-Bakr, M. S., Sedaira, H. and Hashem, E. Y., 1994, Talanta, 41, 1669.
23. Yan, G. E, Shi, G. R. and Liu, Y. M., Anal. Chim. Acta, 1992, 264, 121.
24. Janger, R., Renman, L. and Stefansdottir, S. H., 1993, Anal. Chim. Acta, 281, 305.
25. Hassan, S. S. M. and Marzouk, S. A. M., 1994, Talanta, 41, 891.
26. Cresser, M. S., 1994, Flame Spectrometry in Environmental Chemical Analysis: A Practical Guide (Cambridge: The Royal Society of Chemistry), p. 85.

CITATION

Sahasrabuddhey B, Mishra S, Jain A, and Verma KK. Determination of µmol-1 Level of Iron (III) in Natural Waters and Total Iron in Drugs by Flow Injection Spectrophotometry. Journal of Automated Methods and Management in Chemistry, 1999; 21(1): 11–15. doi: 10.1155/S1463924699000024.

Design and Synthesis of a Noncentrosymmetric Dipyrromethene Dye

M. B. Meinhard, P. A. Cahill, T. C. Kowalczyk and K. D. Singer

ABSTRACT

Semiempirical methods were applied to the design of a new second order nonlinear optical (NLO) dye through noncentrosymmetric modifications to the symmetric dipyrromethene boron difluoride chromophore. Computational evaluations of candidate structures suggested that a synthetically accessible methoxyindole modification would have second order NLO properties. This new dye consists of 4 fused rings, is soluble in polar organic solvents and has a large molar extinction coefficient (86 x 10^3). Its measured hyperpolarizability, β, is -44x 10^{-30} esu at 1367 nm. The methoxyindole therefore induces moderate asymmetry to the chromophore.

Introduction

The large potential market for electrooptic materials and devices for high speed data transfer, either as part of telecommunications or CATV networks or within computer backplanes, has prompted efforts towards the synthesis, processing and evaluation of new organic nonlinear optical (NLO) materials. Such devices would operate through the linear electrooptic effect that requires noncentrosymmetry on both molecular and macroscopic scales. Stability of the noncentrosymmetric poled state is one of many requirements placed on these materials; other requirements include suitable and stable refractive indices for core and cladding, thermal stability as high as 350 °C, electrooptic coefficients greater than 30 pm/V, suitable electrical resistivity in both the core and cladding for efficient poling, and excellent optical transparency (losses < 0.3 dB/cm) at operating wavelengths. Simultaneous attainment of all these parameters has proved extremely difficult. Because organic NLO dyes are often the limiting factor in the ultimate thermal stability of a NLO material, a promising approach to improved materials is through the synthesis of new dyes.

Dye Design

Our approach is based on Marder et al.'s observations that a maximum in β, the first hyperpolarizability, occurs near the zero-bond alternation, or cyanine limit, in noncentrosymmetric chromophores (I). Therefore the nonlinearity of organic dyes may be maximized through asymmetric modifications to known, centrosymmetric cyanine or cyanine-like dyes (Figure 1). This approach might also lead to dyes with narrow electronic absorption spectra and correspondingly lower absorption losses to the red of the principal charge-transfer absorption which gives rise to the nonlinear optical effect. Poling of cationic dyes such as the cyanines is problematic; therefore, charge neutralization must first be addressed.

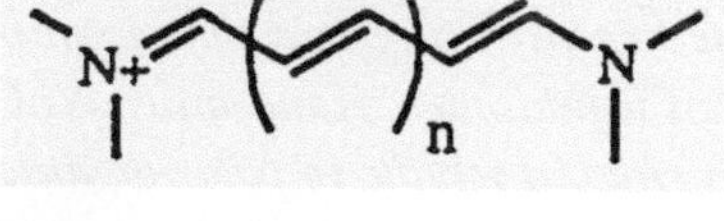

Figure 1. Cyanine dye structure.

An internally charge compensated class of cyanine dyes are the dipyrromethene difluoroborates shown in Figure 2. (2) Highly fluorescent symmetric dipyrromethenes are commercially available as biological probes (3) and laser dyes (4). Derivatives that are soluble in either organic (R2 = alkyl) or aqueous media

(R2 = sulfonate) are known. In addition, the dipyrromethenes are among the most photochemically stable dyes. (5) High fluorescence quantum yields have been reported over a wide spectral range. (3) Therefore, this chromophore is a good starting point for the synthesis of second order NLO dyes by asymmetric modification.

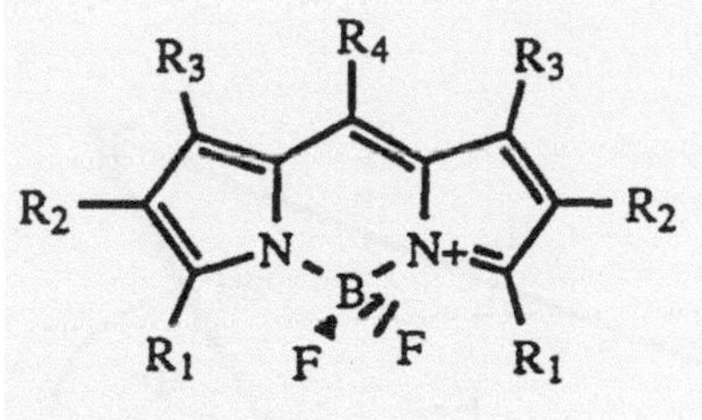

Figure 2. Dipyrromethene difluoroborate dye structure.

The type and location of donor and acceptor groups on this chromophore that will maximize second order NLO properties is not obvious, however, because the ends of the cyanine chromophore are coordinated to the boron atom and therefore not available for direct modification. A noncentrosymmetric substitution pattern must provide for both a charge transfer transition that is related to the strong absorption in the symmetric molecule and a ground state dipole moment which is substantially parallel to this transition. Candidate structures were therefore evaluated computationally with MOPAC using the AM 1 basis for geometry optimization. Spectroscopic INDO/S methods with configuration interaction (ZINDO) were used for electronic spectra estimation (6).

Direct calculation of the first hyperpolarizability of the candidate molecules with these semicmpirical methods was considered, but a lack of closely related model compounds that could be used to verify such methods were not available. The two-state model, however, provides a means of relating molecular hyperpolarizability to information readily obtainable from spectral calculations and was used to estimate the magnitudes of the hyperpolarizability of candidate molecules (7).

$$\beta(-2\omega;\omega,\omega) \cong \frac{3e^2}{2hm} \frac{\omega_{eg} f \Delta\mu}{(\omega_{eg}{}^2 - \omega^2)(\omega_{eg}{}^2 - 4\omega^2)}. \quad (1)$$

In this expression ωeg the frequency of the electronic transition, f is the oscillator strength, and Δμ is the transition dipole moment. A figure of merit composed of the ground and excited state dipole moments, (μg and μe, respectively), oscillator strengths (f) and cos φ, the projection of the transitlon dipole moment onto ps was then used to order and rationally select a synthetic target molecule.

$$FOM = f \bullet \mu \bullet \cos\phi \bullet \lambda^2 \quad (2)$$

Such a figure of merit tacitly assumes that the lowest energy charge transfer absorption gives rise to the majority of the NLO effect, or alternately, that the two-state molecule is a good approximation for this class of dyes. This is probably a valid assumption because the lowest energy electronic transition in dipyrromethenes (and cyanines in general) is well separated from other transitions.

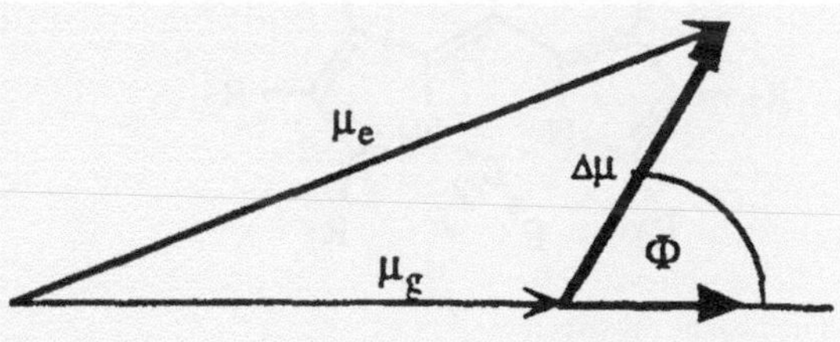

Figure 3. Projection of the change in dipole moment onto the ground state dipole.

Computational Results and Discussion

Direct substitution of the dipyrromethene framework (Figure 2) was attempted first, but substitutions of donor/acceptor pairs at R_1 or R_2 did not couple strongly with the symmetric chromophore's electronic structure, i.e., the HOMO and LUMO coefficients at these carbons are small. After other similar substitution patterns led to the same weak coupling, a variation based on benzannelation at R_1 and R_2 to give the indole-pyrrole-methene chromophore shown in Figure 4 suggested stronger coupling.

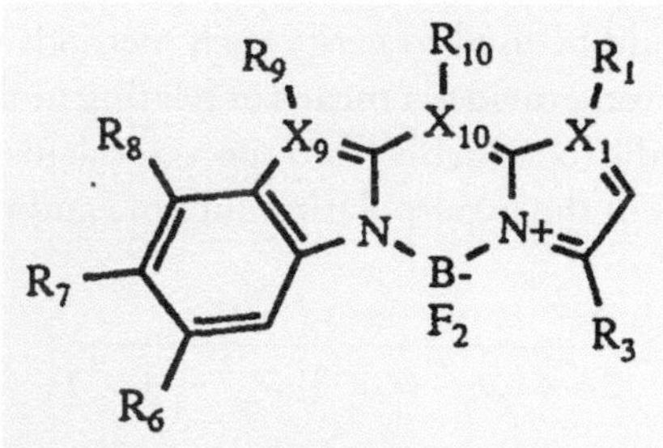

Figure 4. Structure of an indole-pyrrole-methene difluoroborate chromophore.

Ground and excited state dipoles (μ), oscillator strengths (f) and absorption-maxima (λ) were extracted from INDO/S calculations for the series of molecules listed in Table 1. The projection angle φ was calculated from the dot product of the ground and excited state dipoles as depicted in Figure 3. Results of ZINDO and figure of merit calculations from these systematic structural variation of the dipyrromethene framework are given in Table 2.

Table 1. Candidate Indole-Pyrrole-Difluoroborates

Compound	X9	X1	X10	R9	R3	R6	R7	R1	R10
A	C	C	C	Me	Me	MeO	H	Me	H
B	C	C	C	Ph	Me	MeO	H	Me	H
C	C	C	C	Ph	Ph	MeO	H	Me	H
D	C	C	C	Me	Me	MeO	H	Ph	H
E	C	C	C	Me	Me	Me2N	H	Me	H
F	C	C	C	Me	Me	MeO	MeO	Me	H
G	C	C	C	Me	Me	MeO	MeO	Ph	H
H	N	C	C	-	H	MeO	H	H	H
I	N	C	C	-	H	MeO	H	Me	H
J	N	C	C	-	H	MeO	H	Ph	H
K	C	N	C	Me	Me	MeO	H	-	H
L	C	N	C	Me	Me	MeO	MeO	-	H
M	C	N	C	Me	Me	MeO	MeO	-	Me
N	C	N	C	Ph	Ph	MeO	MeO	-	H
O	C	C	N	Me	Me	MeO	H	H	-
P	C	C	N	Me	Me	MeO	H	Me	-
Q	C	C	N	Me	Me	Me2N	H	Me	-
R	C	C	N	Me	Me	MeO	H	MeO	-
S	N	C	N	-	H	MeO	H	Me	-
T	C	N	N	Me	Me	MeO	H	-	-
U	C	N	N	Me	Me	MeO	H	-	-
V	C	N	N	Me	Me	MeO	MeO	-	-
W	N	N	N	-	H	MeO	H	-	-

Table 2. Figure of Merit (FOM) Computations Summary. A C.I. level of 14 was used in these INDO/S calculations.

Compound	μ_g	$\Delta\mu_e$	$\Delta\mu$	λ	f	$\cos\phi$	ϕ	FOM
A	7.73	10.6	16.7	502	1.06	-0.643	130	1.33e+06
B	7.44	11.2	8.21	515	1.14	0.677	47.4	1.52e+06
C	7.37	11.5	8.96	532	1.16	0.623	51.5	1.51e+06
D	7.48	10.9	9.10	509	1.09	0.560	55.9	1.18e+06
E	5.83	11.0	10.3	522	1.10	0.380	67.7	6.61e+05
F	8.96	10.7	9.09	510	1.05	0.583	54.3	1.43e+06
G	8.83	10.8	9.74	513	1.08	0.525	58.3	1.32e+06
H	2.63	11.2	10.5	537	1.10	0.375	68.0	3.15e+05
I	3.24	11.3	12.1	535	1.12	-0.118	96.7	1.22e+05
J	3.34	11.5	12.6	541	1.16	-0.187	101	2.12e+05
K	9.37	9.57	18.9	516	0.835	-0.985	170	2.05e+06
L	10.0	9.75	19.6	534	0.838	-0.959	164	2.29e+06
M	9.97	9.50	3.77	519	0.818	0.926	22.2	2.04e+06
N	10.3	10.5	3.08	552	0.940	0.956	17.0	2.81e+06
O	6.81	11.2	5.91	548	1.07	0.894	26.6	1.95e+06
P	6.56	11.2	17.3	550	1.07	-0.887	152	1.88e+06
Q	3.81	11.7	14.8	570	1.14	-0.739	138	1.04e+06
R	3.71	11.6	14.6	544	1.16	-0.778	141	9.91e+05
S	2.27	12.1	14.2	590	1.16	-0.913	156	8.39e+05
T	10.7	9.46	20.1	578	0.730	-0.990	172	2.57e+06
U	10.9	9.32	2.09	575	0.712	0.990	7.92	2.54e+06
V	10.9	9.57	20.5	600	0.720	-0.999	178	2.83e+06
W	8.50	10.6	18.4	641	0.830	-0.842	147	2.44e+06

The calculations summarized above suggest that benzannelation induces non-centrosymmetry into the chromophore and that the methoxy substituent additionally polarizes the ground state. Phenyl substituents lead to greater bathochromic shifts than methyl, but the effects are small. Such substitutions may separately impart synthetic advantages in directing electrophilic additions or in blocking sites otherwise available for side reactions.

Further examination of Table 2 permits several generalizations to be made concerning structure-property relationships within various substitution patterns and the degree and location of heteroatom (nitrogen) incorporation into the dipyrromethene chromophore. Compounds A-G are modifications of an "all-carbon" frame. These compounds have very similar μg (7.3-8.8 D), λmax (500-530 nm), and good oscillator strengths. As a group these compounds have less than optimal projections of the electronic transitions as reflected by φ in the range of 47-67°. All these compounds include a MeO donor, except for E, which includes a Me2N donor. None of these chromophores include an electron acceptor substituent such as nitro or nitrile because the calculations indicated that the bridging methene (X10) acts as a strong acceptor in this unusual chromophore.

Compounds H-J, where X1 = N, have the lowest FOM for the series. The resulting structure is an indole-imidazole combination in which electron donor are present at both ends of the chromophore. This is consistent with the low calculated values for μg (2.6-3.3 D) and φ's of 67-100°(transition dipole nearly perpendicular to the ground state dipole).

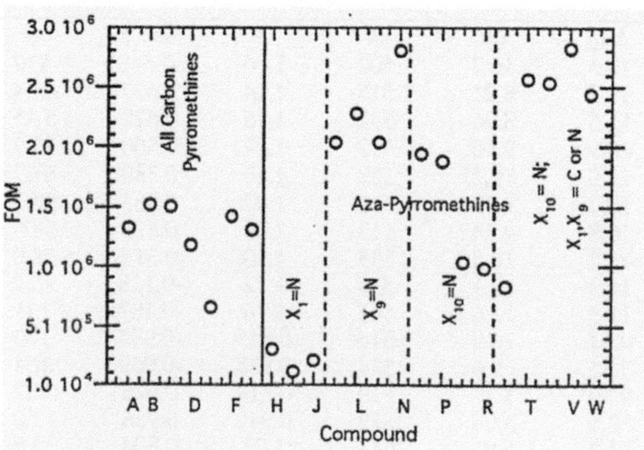

Figure 5. Graph of computational results. The highest Figures of Merit are calculated for benzimidazole (X_9 = N) donors.

Series K-N, X9 = N, are composed of a benzimidazole-pyrrole framework and show the largest figures of merit of the chromophores linked by a carbon at X10. This group has large μg (9.3-10.2 D) and φ's within 20° of parallel with μg

Absorption maxima are on the order of 530 nm. Strong oscillator strengths are also observed. Series O-R, X10 = N, demonstrated low to moderate μg, excellent f and λ in the range of 550 nm. These compounds have variable FOM performance and have moderate μg and mid-range φ values.

Compounds S-W represent a series of di- and tri-aza frameworks with widely varying FOMs. For compound S, μg was exceptionally low (2.2 D) in contrast to diaza compounds T-V with μg values 10.7-10.9 D. This latter group showed the longest absorption maxima, 56-600 nm, with modest values for f (0.71). Excellent projections were noted for these compounds, particularly compound V with a projection within 2° of parallel. The synthesis of these compounds has not yet been examined.

Experimental Results and Discussion

Compound A (Scheme I) was chosen for synthesis on the basis of the calculations of the asymmetry combined with synthetic accessibility. The experimental program consisted of a multistep synthesis and a hyperpolarizability measurement of the final chromophores. Detailed experimental procedures will be available separately, but all compounds were fully characterized by standard spectroscopic methods. Purity was additionally established on the final chromophore by HPLC. Observed absorption maxima were shifted by about 10% to the red of the values predicted from INDO/S computations.

Syntheses

Compound A was chosen as the "all carbon" chromophore for synthesis,in part due to synthetic accessibility as shown in Scheme I, below. The methoxyindole itself is prepared is several steps by literature methods (8-10). Condensation of this indole with a pyrrole aldehyde leads to the desired product which is purified by chromatography in low to moderate yields. Interestingly, symmetric products are also formed and must result by transfer of the aldehyde from the pyrrole to the indole and subsequent reaction between identical ring systems.

Scheme I.

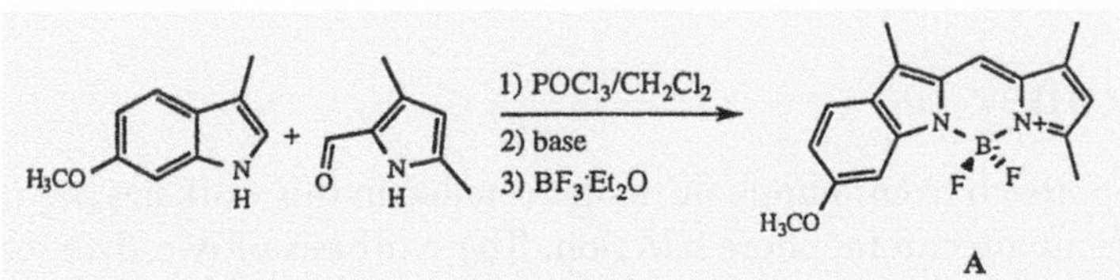

The diphenyl analog to A was also obtained using the 2-formylindole and 2,4-diphenylpyrrole as shown in Scheme II. Symmetrical products were again observed and indicate transfer of the aldehyde functional group between the indole and pyrrole. Reaction product mixtures were generally complex. Yields ranged from 1550% in the final condensation step.

Scheme II.

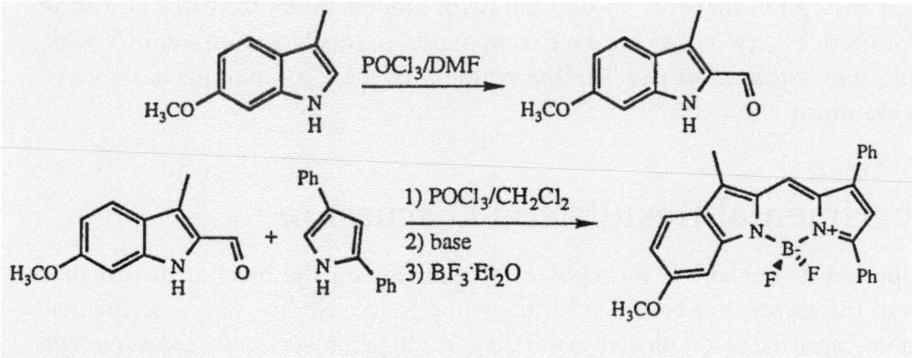

Compound A's absorbance maximum (λmax) is 555 nm in benzene with a molar absorptivity (ε) of 86 x 103 L/mol-cm. The large molar absorptivity indicates that the oscillator strength of the initial cyanine chromophore is not lost in the asymmetric system. The absorption maxima are only slightly solvatochromic.

Hyperpolarizability

The hyperpolarizability of the indolepyrromethene chromophore A was measured by EFISH techniques in dioxane. The value suggests only a slight deviation from centrosymmetry. The data is summarized in Table 3.

Polarizability(α) 4.9 x 10-23esu

Dipole Moment(μ) 7.3 x 10-18

Hyperpolarizability, 1367nm(β1367) -44x 10-30

Hyperpolarizability, extrapolated to zero frequency(β0) -12x 10-30

Conclusions

The rational approach to chromophore design detailed in this work has been demonstrated to facilitate chromophore selection. The synthesis of two dyes selected on the basis of semiempirical calculations has been completed. The magnitude of first hyperpolarizability of the methoxyindole-pyrrole-methene indicates that only slight asymmetry has been induced into the symmetric dipyrromethene chromophore. Stronger donors and acceptors will be needed to maximize nonlinear optical coefficients in this promising series of dyes.

Acknowledgements

This research was performed, in part, at Sandia National Laboratories and was supported by the U.S. Department of Energy under contract DE-AC04-94-AL85000. Helpful discussions with D. R. Wheeler and C. C. Henderson (SNL) and A. J. Beuhler (Amoco Chemical Co., present address: Motorola) are acknowledged. This project could not have been completed without the help of R. Steppel et al. (Exciton) who scaled up the synthesis of the indole-pyrrole dye.

References

1. Marder, S. R.; Gorman, C. B., Meyers, F.; Perry, J. W.; Bourhill, G.; Bredas, J.-L.; Pierce, B. M. *Science* 1994, 265, 632.
2. For a recent review see Boyer, J. H.; Haag, A. M.; Sathyamoorthi, G.; Soong, M.-L.; Thangaraj, K. *Heteroatom Chem.* 1993, 4, 39.
3. Molecular Probes, Eugene, Oregon.
4. Exciton, Dayton, Ohio.
5. Hermes, R. E.; Allik, T. H.; Chandra, S.; Hutchinson, J. A. *Appl. Phys. Lett.* 1993, 63, 877.
6. LaChapelle, M.; Belledte, M., Poulin, M.; Godbout, N.; LeGrand, F.; HCroux, A.; Brisse, F.; Durocher, G. *J. Phys. Chem.* 1991, 95, 9764.
7. Oudar, J. L. and Chemla, D. S. *J. Chem. Phys.* 1977, 66, 2664.
8. Fleming, I., Woolias, M. *J. C. S. Perkins Trans. I* 1979, 827.
9. Silverstein, R. M., Ryskiewicz, E. E. and Willard, C. *Org. Syn. Coll. Vol. III*, 831.
10. Deady, L. W. *Tetrahedron*, 1967, 23, 3505.

CITATION
Meinhard MB, Cahill PA, Kowalczyk TC and Singer KD. Design and Synthesis of a Noncentrosymmetric Dipyrromethene Dye. http://www.osti.gov/scitech/servlets/purl/10110701. US Governnment publication.

The Function of TiO_2 with Respect to Sensitizer Stability in Nanocrystalline Dye Solar Cells

A. Barkschat, T. Moehl, B. Macht and H. Tributsch

ABSTRACT

Dyes of characteristically different composition have been tested with respect to long-term stability in operating standardized dye sensitized cells during a time period of up to 3600 hours. Selective solar illumination, the use of graded filters, and imaging of photocurrents revealed that degradation is linked to the density of photocurrent passed. Photoelectrochemical degradation was observed with all sensitizers investigated. Sensitization was less efficient and sensitizers were less photostable with nanostructured ZnO compared to nanostructured TiO_2. The best performance was confirmed for cis-Ru^{II} $(dcbpyH_2)_2(NCS)_2$ on TiO_2. However, it was 7–10 times less stable under other identical conditions on ZnO. Stability is favored by carboxylate *anchoring and metal-centred electron transfer. In presence of TiO_2, it is enhanced by formation of a stabilizing charge-transfer complex between oxidized Ru dye and back-bonding interfacial Ti^{3+} states. This is considered to be the main reason for the ongoing use of expensive Ru complexes in combination with TiO_2. The local surface chemistry of the nanocrystalline TiO_2 turned out to be a crucial factor for sensitizer stability and requires further investigation.*

Introduction

The sensitization of emulsions of semiconductor particles has played a major role in the development of photography [1, 2]. Photoelectrochemical studies on the mechanisms of spectral sensitization of semiconductors were performed in the late sixties [3–5]. Shortly thereafter, it was proposed to use dye sensitization to drive a new kind of solar cell [6]. The sensitizing Chlorophyll molecules were extracted from spinach to demonstrate solar energy conversion in analogy to primary photosynthesis [5, 6] (this, and nor, as later claimed [7] the so called Grätzel cell, proposed 20 years later, was the first demonstration of bioanalogue solar energy conversion using dye molecules). Although, in the subsequent years, much scientific information on the sensitization process became available and reasonably efficient dye sensitized solar cells were demonstrated [8–14] on the basis of sintered oxide ceramics, it became more and more evident that the stability of organic sensitizers in such cells was a major problem. Nevertheless, in 1980, Matsumura et al. showed that on the basis of zinc oxide ceramics made intentionally porous with addition of aluminium, an efficiency for conversion of light into electrical energy of 2.5% could be achieved (illumination occurred only in the absorption region of the dye) [14]. Up to that time, dyes from very different groups of chemicals and many different large gap semiconductors had been tested for the sensitization phenomenon. In 1980, Dare-Edwards et al. tested Ru-based dyes by attaching them with two ester linkages to TiO2 [15]. At that time, with respect to efficiency, no peculiar advantage of attached $Ru(bipy)_2(bpca)$ on a TiO_2 large gap semiconductor was observed. This was probably due to the specific experimental conditions involved. Some deterioration of current efficiency after many hours of illumination of the sensitized electrode was specifically mentioned. Interestingly, it was a very similar system which became the focus of attention 10 years later when high-efficiency, long-term stable nanocrystalline dye sensitized solar cells were proposed [16]. They used nanostructured TiO_2 material sensitized via chemically attached Ru-based dyes in combination with an I^-/I^{3-} redox electrolyte. With such systems, solar energy conversion efficiencies of 7–10% were demonstrated with at least several cm^2 large cells (tiny cells reached 11-12%). A reasonable short-term stability was obvious when the cells were tightly sealed but a long-term stability of 20 years was claimed on the basis of rather complicated experimental and theoretical arguments [17]. Involvement of the excited Ru-sensitizer in photochemistry was excluded due to a fast (femtosecond) injection rate observed for the electron into TiO_2 [18].

A fast regeneration of the oxidized Ru-sensitizer, on the other hand, was deduced from the presence of a high (0.5 M) concentration of iodide in the electrolyte. Regeneration rates sufficient for stabilization of solar cells for over 108 cycles corresponding to 20 years lifetime were calculated [17]. In the course of intensive research which followed in the nineties, complications with respect to electrolyte chemistry, photochemistry, purity of sensitizers, as well as with respect to sealing and iodine chemistry were noted [19]. The best performing cells saw their sensitizer change from a three nuclear Ru-complex [16] to the mononuclear cis-RuII(dcbpyH2)2(NCS)2 complex (here called N3) and the electrolyte solutions were replaced by less volatile and more stable ones. But after 15 years of research on Ru-based dye solar cells, and after efforts by many groups, reasonably stable dye solar cells have still not found their way into production. In fact, several companies have failed to produce reasonably stable commercial cells, with larger panels still degrading too fast (e.g., on the main building of CSIRO-Energy in Newcastle, Australia).

Since long-term stability of dye sensitized cells is a crucial issue for economic feasibility, long-term studies of dye sensitized cells have been initiated in increasing numbers [20–27] to take advantage of the high energy-conversion efficiency observed [28, 29]. However, a major problem turned out to be the sensitivity of the cells towards electrolyte loss via evaporation and the instability of the electrolyte. The chance of a dye sensitized solar cell to break down because of secondary phenomena not related to photochemical stability was very high. To overcome this problem, photocurrent imaging techniques were applied to visualize the effect of selective illumination of dye sensitized solar cells [30]. Only part of a dye sensitized solar cell was illuminated through a mask with simulated solar light (sulphur lamp) and the development of photocurrents were followed over a longer period of time by measuring spatially resolved images of the photocurrent. In these experiments, a He/Ne-laser spot was used for scanned illumination. In this way, clear differences between illuminated and nonilluminated areas could be observed [19]. This observation confirmed previous experiments with total reflection FTIR-studies performed on working dye sensitized solar cells. They indicated photo-induced generation of oxidation products of the Ru-sensitizer [31]. Significant quantitative differences between the performances of dye sensitized solar cells from different groups were observed. However, they behaved in a similar way qualitatively with respect to photo-induced degradation relative to stability in nonilluminated areas. A theoretical analysis of these preliminary studies indicated that the branching ratio, the ratio of regeneration rate of the oxidized sensitizer to the rate of product formation, is a crucial factor for stability of dye sensitized solar cells. It was estimated that it may be one order of magnitude lower than required for 20 years stability [32].

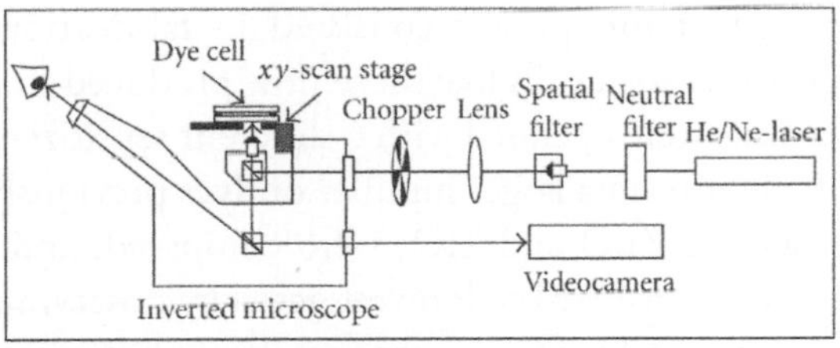

Figure 1. Scheme of the photoelectrochemical set up employing an optical scanning microscope (SMSC).

Some publications related the problem of dye cell degradation to the presence of water in the electrolyte [33] and other influences of the electrolyte [17, 31, 34] to higher temperatures [35] and to UV-light incidence [35]. In a more recent paper, the degradation mechanisms in a dye-sensitized solar cell were further investigated using UV/VIS- and IR-spectroscopy [36]. These results indicate a loss of the isothiocyanato-ligand of the TiO2-adsorbed dye in the presence of air, at higher temperature or in the presence of water in the electrolyte. This effect is accelerated by illumination. It was found that this leads to a decreased, blue-shifted N3-dye absorption [36].

It is therefore necessary to investigate the problem of stability further, especially with respect to the origin of dye degradation. The question should specifically be answered is why, after so many years of intensive research on nanostructured dye solar cells, TiO2 and expensive Ru-based sensitizers remained the by far preferred components of dye solar cells.

The aim of the present contribution is first to investigate different dyes, besides of the mainly used Ru-complex (N3), to study chemical preconditions for improved stability and, further, to find out whether the excited or the oxidized sensitizer is a key problem in inducing photochemical side-reactions. A key question will be the role which TiO2 and its surface chemistry play for sensitizer photodegradation. For this reason, ZnO will be studied as a substrate in parallel. It should be investigated to what extent the chemical quality and surface morphology of nano-TiO2 do have an influence on the survival of oxidized sensitizer molecules.

Photo-induced degradation, which is addressed here, is just one of several problems which can lead to degradation of dye sensitized cells. They have been discussed in a parallel publication [19] and include irreversible changes of the redox electrolyte, Li insertion into the oxide, loss of electrolyte through evaporation, and contamination from the atmosphere through the sealing material.

Experimental

The experimental strategy of this project consisted in fabrication of standard nanocrystalline dye sensitized solar cells like those now produced in many laboratories [19, 37, 38]. The cells were operated with 6 different sensitizers, which were selected as the most efficient from a larger number of dyes previously used. Both, cells made of nanocrystalline ZnO and TiO_2 were compared, and the time-dependent performance of dye sensitized cells investigated by observing their photoelectrochemical performance and studying space resolved photocurrent images in order to distinguish the performance of illuminated and nonilluminated areas.

While screening a larger number of dyes (e.g., Ni-phtalocyanine, fluoresceine, rhodamine B, methylene blue, bromothymol blue, zinkone) for their function in standardized nanocrystalline dye sensitized cells, the work concentrated on the 6 best performing complexes. They were:

(i) bis(tetrabutylammonium) cis-bis(isothiocyanato) bis(2, 2′-bipyridine-4-carboxylic acid, 4′-carboxylate) ruthenium(II) or cis-RuII(dcbpyH2)2(NCS)2, called N3 or Ru535 (obtained from Solaronics Inc., Aubonne, Switzerland),

(ii) porphyrine,

(iii) pyrogallol red,

(iv) sodium salt of copper-chlorophylline,

(v) di-(2, 2′-dipyridylmethylene)[(2, 2′-dipyridylmethylene)malonato] ruthenium(II)dihexafluorophosphate • 2HCL,

(vi) di-(2, 2′-dipyridyl) [(2, 2′-dipyridylmethylene)nomooctadecyl malonato] ruthenium(II).

With the exception of porphyrine when chloroform was used, the dyes were dissolved in ethanol to give 0.3 mM stock solutions which were then applied to the nanocrystalline oxides.

Dye Sensitized Cells

Standard dye sensitized solar cells were prepared as described in the literature [19] and TiO_2-layers were prepared by hydrolysis of Ti(IV)isopropylate [39, 40].

For the preparation of TiO2 layers, 1 mL 65%-HNO3 was added to 120 mL of distilled water, and, for the next 45 minutes, a solution of 10 mL Ti(IV) isopropylate in 10 mL isopropanol was added under constant stirring in an argon atmosphere. A white colloidal precipitation was formed, and the isopropanol was removed by distillation. When the solvent left was only water, the suspension was kept under reflux for 8 hours. The suspension was then heated to 230°C while being stirred in an autoclave for 12 hours. After removing the water by vacuum distillation, 40% polyethylenglycol 20000 (40% relative to the content of TiO2 (11–20%)) was added. This mixture was then applied onto the glass by pulling a glass rod over the surface. The thickness of the layer was kept constant by the use of a mask, which was fixed on the glass. The thickness of the TiO2 nanolayer was adjusted to 5 μm, which is smaller than the optimum for high efficiency (10–15 μm) but adequate for 2–4% efficient experimental cells (excess dye should not be acting as a buffer). After drying for 15 minutes, the layers were tempered at 450°C for 30 minutes (cells of Figures 3(b), 3(c), 4(b), 6, 7).

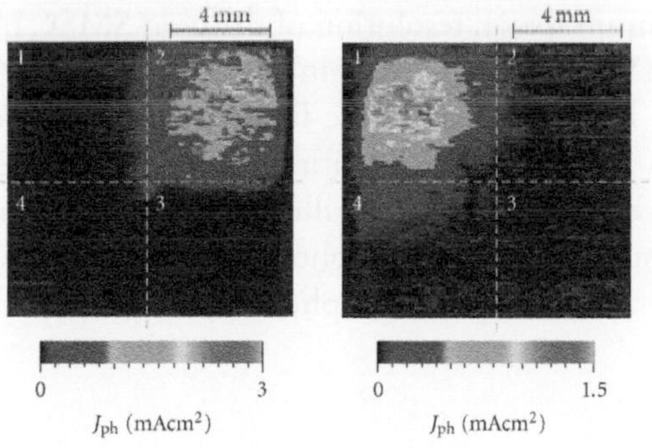

Figure 2. Photocurrent images of dye sensitized solar cells (nanocrystalline zinc oxide) subdivided into four sections for combinatorial testing (clock-wise from top left): (a) neutral red, N3-dye, rhodamine B, methylene blue; (b) N3-dye, neutral red, methylene blue, bromothymol blue. It is seen that only the N3 dye shows significant sensitization.

For ZnO-layers, commercial powders (Alfa, Aldrich) were used to prepare a suspension with the addition of acetyl acetone and water (1:10) and Triton X100 as a detergent.

The suspensions were again applied to the FTO-glass by pulling a glass rod over a mask, which was fixed on the glass. Afterwards, the samples were tempered at 450°C for 30 minutes (cells of Figures 2(a), 2(b), 3(a)).

Additionally, commercially available nanocrystalline TiO2-layers from the Institute for Applied Photovoltaics, Gelsenkirchen (INAP) were used. (cells of Figures 4(a), 4(c), 4(d), 8).

After tempering, the layers were dipped while still hot (approx. 80°C) into a 0.2 mM or 0.3 mM solution of the dye. The second glass with the platinum back contact was fixed above the layer and the cells were sealed with Surlyn 1702 (DuPont). The whole cell was then put into electrolyte solution in an exsiccator. When vacuum was applied, the cell is filled with electrolyte. The electrolyte contained 0.5 M LiI, 50 mM I2 and 0.2 M tert.-butylpyridine in acetonitrile.

Measurements

Dye sensitized solar cells were characterized by photocurrent voltage measurements before exposing them to solar simulated light (sulphur lamp) under short-circuit conditions. A scanning laser spot technique was used to produce images of photocurrents as described in Chaparro et al. [41]. The scanning microscope for semiconductor characterization (SMSC) (see Figure 1) is basically an inverted microscope provided with an x-y scan-stage. The laser spot was focused by the microscope lens onto the dye sensitized solar cell, which was connected to a lock-in amplifier. The maximum spatial resolution of the used SMSC build up at HMI Berlin is 1 µm with a laser spot of 2 µm in diameter. In these experiments, a He/Ne-laser was used for illumination, the light intensity was adjusted to 100 $mWcm^{-2}$. In order to study photodegradation, only well-defined surface areas (using masks) of the dye sensitized cell were illuminated with simulated solar light (solar simulator with sulphur lamp). The photocurrent images were recorded at intervals to investigate the development of photodegradation.

Results

Because quantitative reproducibility of dye sensitized cells turned out to be difficult (efficiency, electrolyte stability, durability of sealing, and long-term behavior varied from cell to cell) combinatorial approaches to dye sensitized properties of different compounds seemed to be appropriate. Figure 2 shows the photocurrent images of two dye sensitized cells, each of which has been separated into four sections dyed with different sensitizers. Three of the sensitizing substances (the Ru-complex N3, neutral red and methylene blue) were identical in order to test the reproducibility. The additional sensitizers were rhodamine B in Figure 2(a) and bromothymol blue in Figure 2(b). As seen in Figures 2(a) and 2(b), only the Ru-complex N3 shows significant sensitized properties in the presence of the same iodide/triiodide-acetonitrile electrolyte. A larger number of potential candidates for sensitized were tested (e.g., Ni-phtalocyanine, fluoresceine, rhodamine B, methylene blue, bromothymol blue, zinkone) but only five additional compounds listed in the experimental section proved to be reasonably efficient for long-term testing. Among these were two additional Ru-complexes and the compounds porphyrine, pyrogallol red, and copper chlorophylline. Other compounds, such as rhodamine B or methylene blue, which had been demonstrated to sensitize ZnO reasonably well under different conditions, turned out not to be sufficiently effective in a nanocrystalline environment in the presence of an iodide/triiodide electrolyte in acetonitrile.

The six sensitizers mentioned were tested under comparable conditions in solar cells which had a nanocrystalline oxide layer of 4 μm only. This is significantly smaller than in high-efficiency dye sensitized cells (8–16 μm) so that the solar cell efficiency reached was correspondingly smaller. In the case of the N3 Ru-complex, the solar cell efficiency reached was between 2 and 3%. The same sensitized cell with ZnO yielded an efficiency between 0.3% and 0.45%. Solar cells fabricated with the above mentioned six different sensitizers were exposed to simulated solar light for periods of up to 30 days, and in the case of the Ru-N3 dye and nanocrystalline TiO2 substrate for periods of up to 280 days. Photochemical degradation was observed in all cases. It was clearly more pronounced in the case of ZnO, even though simultaneously lower solar cell efficiencies were reached.

Figure 3(a) shows the integral photocurrent density-voltage characteristics of a nanocrystalline zinc oxide cell sensitized with the N3 ruthenium complex at the beginning of the experiment and after 20 days of continuous exposure to simulated solar light.

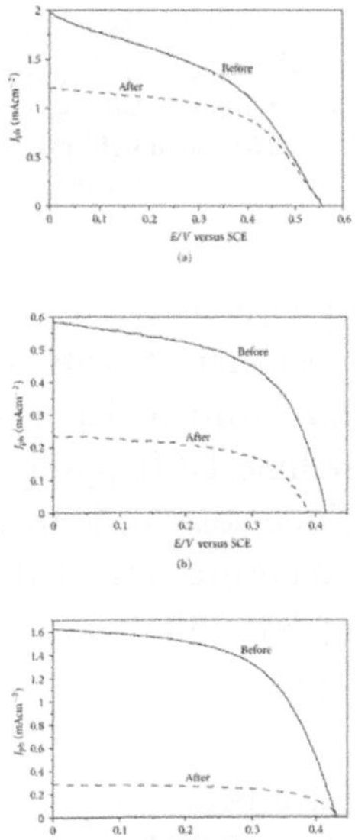

Figure 3. Power output characteristic (photocurrent voltage dependencies) of dye sensitized cells: (a) nanocrystalline ZnO, N3 complex before and after 20 days of simulated solar illumination; (b) nanocrystalline TiO_2, pyrogallol red before and after 11 days of simulated solar illumination; (c) nanocrystalline TiO_2, copper chlorophylline before and after 30 days of simulated solar illumination.

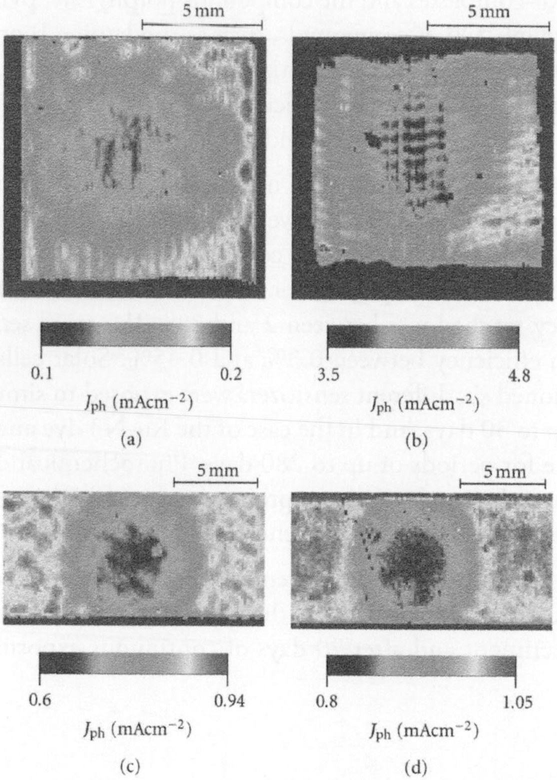

Figure 4. Photodegradation in photocurrent images of four regularly operating nanocrystalline TiO2 solar cells. Sensitization with N3-dye and illumination with simulated solar light in a round central areas only. Illumination periods: (a) 1008 hours; (b) 336 hours; (c) 1248 hours; (d) 1248 hours.

During this period the integral photocurrent density decreased by 35%. It is realized that basically only the photocurrent density decreased and not the photovoltage. Figures 3(b) and 3(c) show similar degradation curves of the power output characteristics of a nanocrystalline TiO2 pyrogallol red cell and of a nanocrystalline TiO2 copper chlorophylline solar cell. In the first case, Figure 3(b), the period of simulated solar light exposure was 11 days, in the second case the cell was exposed to solar light for 30 days. All investigated dyes showed a gradual photochemical degradation when exposed to simulated solar light for a prolonged period.

Among the many dyes which have been screened, only a few approach the properties of Ru-N3, as shown in Figure 2. Ru-N3 turned out to be the most favorable sensitizer, all other dyes exhibited lower efficiency and stronger degradation. Lower efficiencies and stronger degradation were also observed with nanocrystalline zinc oxide solar cells. Comparing different dyes, it was found that certain anchor groups like carboxyl or sulphone groups improve the sensitization properties of compounds. The interaction is apparently less efficient or of different nature with zinc oxide nanoparticles.

Special attention was dedicated to the degradation properties of the N3 ruthenium complex on nanocrystalline TiO2. Here, a long-term experiment, in which sensitized cells were exposed to simulated solar light during a period of 155 days, turned out to be especially informative. The cells were illuminated through a mask with a round opening in order to distinguish between illuminated and nonilluminated areas. During such long-term experiments, the instability of the iodide/triiodide electrolyte turned out to be a major problem. In some cases, depending on small variations in the preparation, the electrolyte was bleached so that the iodine largely disappeared from the electrolyte. This problem was investigated and may be understood in relation to the potential-pH diagram of iodine. Depending on traces of water which leak into the organic electrolyte through the seal from the outside, and due to the presence of tert.-butylpyridine, a slightly alcaline pH can be developed, which allows oxidation of iodide to iodate (IO3-). Since this iodate could not be detected spectroscopically in the electrolyte, it may have been adsorbed on the nanocrystalline TiO2-layer. Whatever the detailed mechanism of this process, it causes the disappearance of iodine and therefore the electron transport through the electrolyte became extremely limited.

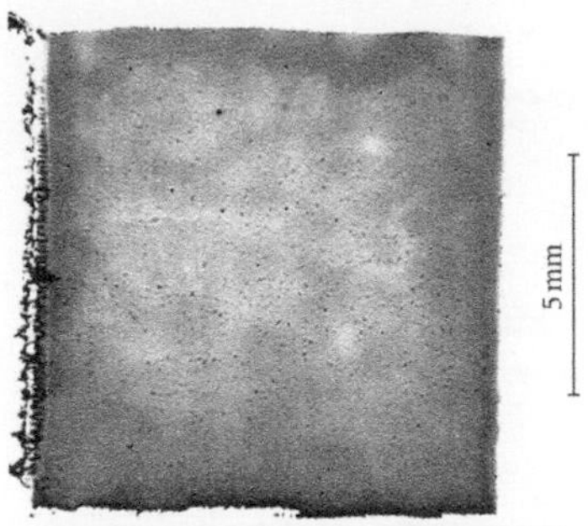

Figure 5. Video image of the cell in Figure 4(a), obtained under light transmission and after contrast amplification.

In the above mentioned long-term experiment, four dye sensitized cells showed bleaching of the electrolyte, but were continuously operated under short-circuit conditions with a very small current flow. This means, since the photocurrent largely collapsed, that the bleached cells had a power output characteristic with low fill factor. At the end of the long-term experiment, all dye sensitized cells, both those that maintained good power output characteristics and those which were bleached and showed very unfavorable power output characteristics were investigated using the scanning laser technique to obtain photocurrent images. During these measurements, only a very small photocurrent generated by the laser spot (laser light intensity not exceeding solar light intensity) had to flow through the dye sensitized cell. Such a small current could also be maintained by those cells which had a bleached, iodine deficient electrolyte.

Figure 4 shows four dye sensitized cells (N3 ruthenium complex on nanocrystalline TiO2) which operated with unbleached electrolyte (Figures 4(a)–4(d)). For comparison, four cells in which the electrolyte was bleached during long-term (3600 hours) illumination are shown in Figure 6. In the first case (Figure 4), photodegradation of the ruthenium complex within the round areas illuminated with simulated solar light can clearly be recognized while in the second case there is no evidence of a difference between illuminated and nonilluminated areas of the dye sensitized solar cell. The difference between the cells in Figures 4 and 6 is simply that in the first case photocurrent could be passed in an uninhibited way, while in the second case photocurrent passage was severely hindered due to lack of iodine. This shows that photons and thermal energy alone are not the primary cause of the degradation if electrons are allowed to react back. Because the mask for forming the light patterns was black and placed directly onto the cell, thermal energy was also generated outside the circular opening. The graded filter also absorbed light to generate thermal energy, so that temperature gradients across nonhomogenously illuminated solar cell surfaces were expected to be low.

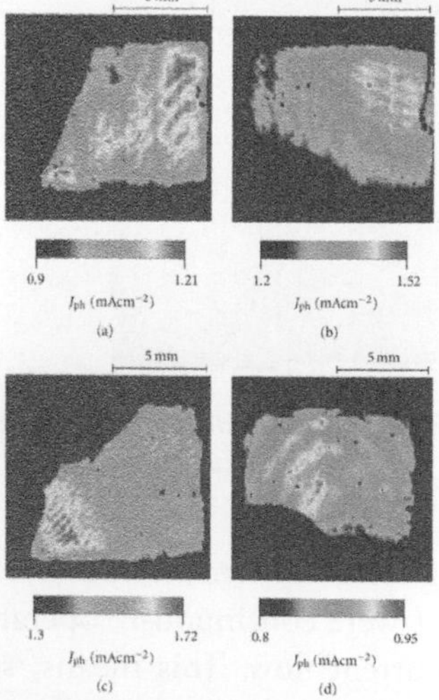

Figure 6. Photodegradation is not detected in photocurrent images of nanocrystalline TiO_2 solar cells with bleached electrolyte, which causes significantly decreased photocurrent densities during illumination. All cells are sensitized with N3 dye and were illuminated with simulated solar light in round central areas only. The illumination period was 150 days (3600 hours). The cells were at the end not completely wetted with electrolyte.

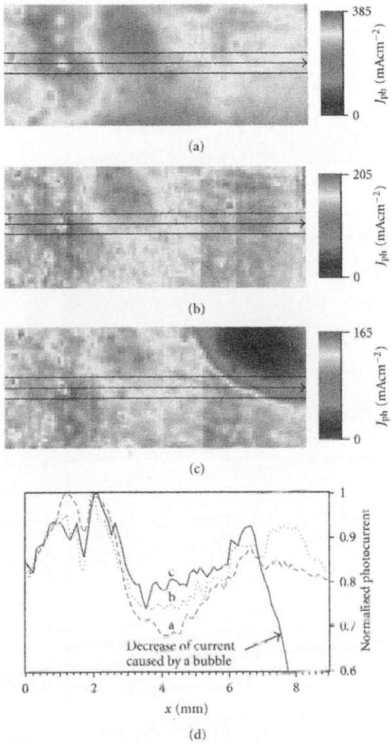

Figure 7. Photocurrent profiles at the same position across a dye sensitized solar cell with inhomogeneous photocurrent distribution. Measurements were done after different periods of illumination with simulated solar light. The photocurrent profiles smooth out during long-term photodegradation, because photodegradation is faster where large photocurrents are flowing. (Experiment performed with copper-chlorophylline on TiO_2): (a) before and (b) after 21 days, and (c) after another 93 days of illumination with simulated solar light.

The significant 15–30% photocurrent decrease observed during the long-term experiments (Figure 4) raises the question as to the optical visibility of the illuminated and degraded areas. Optical transmission was indeed changed.

As Figure 5 shows, after contrast amplification, the transmission pattern of the solar cell matches essentially the photocurrent pattern (Figure 4(a)). This video image (Figure 5) was obtained under light transmission, after cleaving of the cell, from only the front-half which consisted of ITO-glass and a TiO2-layer with adsorbed dye. An influence of the electrolyte or back-contact platinum layer is therefore excluded. However, with the naked eye the transmission change has only occasionally been observed clearly. This means it stayed within a 1-2% limit, which leads to the conclusion that the deactivated sensitizer still absorbs light in the visible spectral region. The degraded complex is still present and has only moderately changed its absorption spectrum in the visible spectral region with respect to the sensitizer.

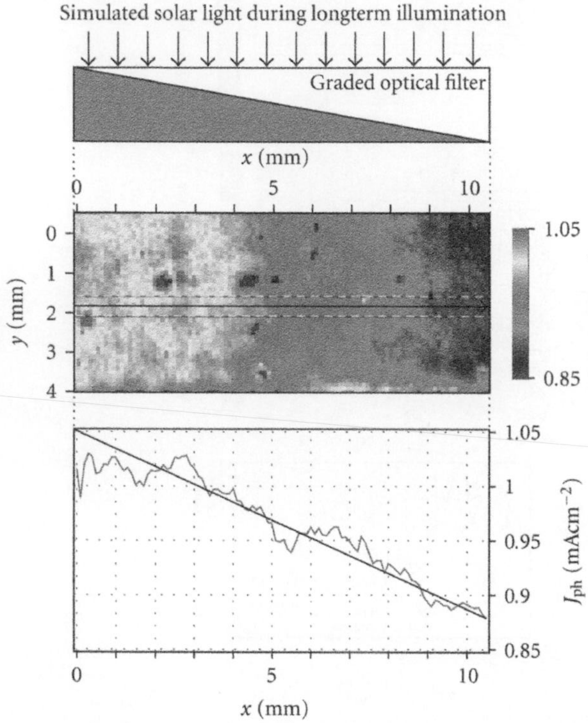

Figure 8. Nanocrystalline TiO_2-N3 sensitized solar cell with graded filter placed on its surface after 1245 hours of exposure to simulated solar light. It is seen that photodegradation, measured as spatially resolved photocurrent output, is proportional to solar light intensity and thus the photocurrent passed.

These results show that photodegradation of the sensitizer must be related to the generation of sensitization photocurrents through the cell. In the case that these photocurrents cannot be generated, which causes the effective rapid recycling of the electrons, photodegradation is suppressed. Therefore, it can be concluded that only the oxidized state of the sensitizer is critical for photodegradation.

In order to test this interesting observation, which may also be relevant for other sensitizers, two additional experiments were performed. In the first, dye sensitized cells with very inhomogeneous photocurrent distributions were investigated. If a higher photocurrent density also means a higher rate of photodegradation, a profile through such an inhomogeneous cell should smooth out during a prolonged illumination. This means the photocurrent in high-photocurrent-density regions should decrease faster than in lower-photocurrent-density regions. Such an experiment is demonstrated in Figure 7 showing the normalized photocurrent-profile through an inhomogeneous dye sensitized cell versus cell operation time. It is indeed recognized that the profiles smooth out with illumination time.

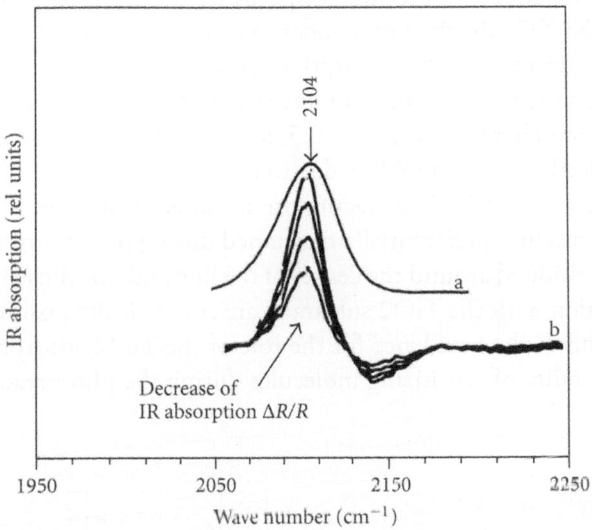

Figure 9. Shape of ex situ IR-CN stretch-line (2104 cm^{-1}) of N3 sensitizer (a) compared to lineshape characteristic for photodegradation loss after passage of 1.4, 5.4, 9.5, and 14 mC cm^{-2} of photoelectrical charge through the dye sensitized cell, (b) measured in total reflection FTIR-experiments.

To confirm the conclusion that higher photocurrent flux means more degradation, a graded filter was placed onto an N3 dye sensitized cell for a long-term study. The cell with originally homogeneous photocurrent density distribution was exposed to simulated solar light attenuated by the graded optical absorption filter. Figure 8 shows that the spatially resolved photocurrent density measured after exposure of the cell to 1248 hours of simulated sunlight actually has a linear degradation profile. The laser-induced photocurrent decreases indeed proportional to the light intensity and thus to the magnitude of the photocurrent generated at the respective location. The higher the photocurrent density is, the higher the rate of degradation will be.

Total reflection infrared spectroscopy on functioning dye sensitized solar cells performed in our laboratory have demonstrated the occurrence of products parallel to photocurrent flow [31, 42]. The integrated infrared signal disappears when the current density goes to 0 and increases with increasing photocurrent density. These results confirm that the generation of oxidation products is related to the extraction of electrons from the sensitizer's environment. Depending, however, on the specific adsorption sites, sensitizer molecules may be involved in quantitatively different reaction behavior. The loss peaks in the in situ spectrum at 1988 cm−1 and 2104 cm−1, with widths of 29 cm−1 and 26 cm−1, respectively, showed a somewhat smaller width than in the ex-situ spectrum where the width was 33 cm−1. This is shown in Figure 9 for the 2104 cm−1 line in the

CN-vibration region. It shows the line shape from an exsitu spectrum of cis-RuII(dcbpyH2)2(NCS)2 adsorbed on nanocrystals within a 1 μm thin TiO2 layer. It also shows the decrease of this absorption (plotted, however, in the opposite direction to permit a better comparison of line shapes) from an in situ experiment after passing the photocharge of 1.4, 5.4, 9.5, and 14 mCcm−2 through the dye sensitized cell. A clearly narrower IR line due to the consumption of the sensitizer is realized, indicating that a limited fraction of adsorbed molecules is selectively turned over. The molecules preferentially consumed during photocurrent flow are those, which are positioned around the center of the line and not those which, due to stronger interaction with the TiO2 substrate, are energetically positioned away from the center. This is clear evidence for the role of the TiO2 adsorption site in determining the stability of sensitizing molecules during the photoreaction.

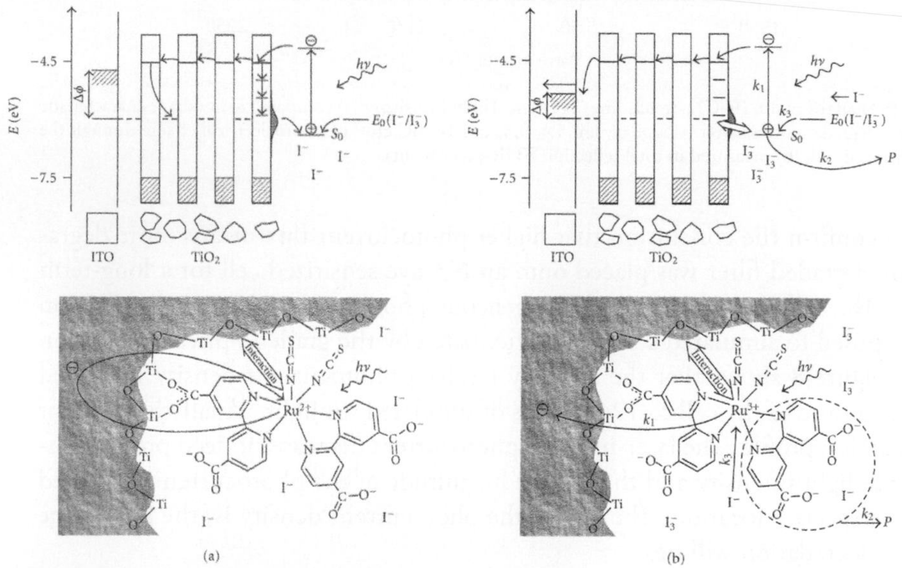

Figure 10. Compares the N3 ruthenium complex, both, in the state of rapid electron recovery (Figure 10(a)) and in the state of electron extraction via a reasonably high photocurrent flow (Figure 10(b)). While present in the TiO_2 interphase, the injected electron may convert Ti^{4+} into Ti^{3+} enabling a back bonding reaction facilitating a temporary Ti^{3+}(ligand)Ru^{3+} charge transfer complex.

Discussion

The main aim of this study was to identify physical-chemical factors relevant for long-term stability or degradation of sensitizers of nanocrystalline ZnO and TiO_2 solar cells to determine the influence of the adsorption site. The focus was on light-induced degradation. An analytical formula describing the nature of various cell parameters, and rate constants which may influence the photocurrent voltage characteristic, had been presented in an earlier publication [32]. The photodegradation of sensitizer molecules is just one possible mechanism of degradation. Other processes involve the kinetics of the rectifying front TiO_2/fluorinated tin oxide contact (it is very critical for the performance and long-term stability of the dye sensitized solar cell), insertion of alkali atoms into TiO_2, irreversible changes of the redox electrolyte, loss of electrolyte, and contamination of the electrolyte from the atmosphere [19].

All sensitizers investigated in the present study, including the Ru-complex (N3), which presently is considered to be sufficiently stable for long-term (20 years) operation in dye sensitized cells, were found to degrade to varying extents. The N3 Ruthenium sensitizer turned out to be the most stable one, but it nevertheless photodegrades (Figure 4) and it degrades proportional to the photocurrent passed through the dye solar cell (Figure 8). This means that photodegradation will be the more pronounced the higher the cell efficiency.

Since this is a somewhat perturbing conclusion and since there have been persistent claims that the Ru-dye does not photodegrade and supports 108 turnovers, let us first discuss our results in relation to other recent studies on dye solar cell performance and stability [20–29]. They, for example, claim significant progress by using alternative electrolytes and improved sealing techniques (e.g., [20]). In this latter work, 0.6 M hexylmethyl imidazolium iodide was added to the acetonitrile electrolyte and a glass frit sealing was used. In stead of exposing the solar cell as it is to the solar simulator, the cell was first exposed to 80°C in the dark (!) for 1400 hours, and subsequently exposed to light for 1700 hours at only 25°C. The claim then was that the dye cell showed a photodegradation of less than 5% during 1700 hours. In fact, the entire procedure (thermal treatment plus subsequent illumination at 25°C) has decreased the solar cell efficiency by 20%, as seen from [20, Figure 10].

In reality, however, strong sunshine can heat a dye solar cell to 80°C, and the most damaging degradation process is the periodically daily heating and cooling of the dye solar cell under solar illumination.

For comparison, our solar cells, the selective degradation of which is depicted in Figures 4(c) and 4(d) decreased their photocurrents during 1248 hours of illumination and solar simulator determined higher temperature by approximately 10%. It has also to be considered that photodegradation was found to be proportional to its efficiency and thus to the photocurrent density passed (compare Figure 8). In addition, we are confronting these results with nearly Redox inactive solar cells, which show practically no degradation after 3600 hours (Figure 6). Our stability results therefore appear to be entirely realistic and contradict possible claims, that more recent cell configurations have solved the stability problem.

Since the most stable dye cell in [20] only had an efficiency of 2%, our results also claim that a corresponding 8% solar cell will, under illumination, degrade 4 times faster, if a linear extrapolation is applicable. It will also degrade faster when illumination is not performed at 25°C, but at an illumination up to 60 degree higher temperature, which can be reached in intensive sunlight.

In our work, it could be confirmed that anchoring groups of the dye support the sensitization effect but that the nanocrystalline ZnO interface is a less favorable substrate for electron injection and long-term stability of Ru-dyes under the investigated conditions. It has, for example, been suggested [43] that the N3 ruthenium complex tends to aggregate by reacting with zinc ions dissolved from the ZnO interface. The oxide surface is electronically and chemically significantly different for TiO2 and ZnO, the latter of which additionally exposes polar surfaces terminating with Zn and O, respectively. In our studies, degradation of this ruthenium complex is significantly faster on ZnO than on TiO2. This degradation is faster by a factor of 7 to 10 on ZnO when the same iodide/triiodide electrolyte in acetonitrile is used.

One reason why metal centered complexes (ruthenium complexes, copper chlorophylline complex) are more stable than sensitizers without transition metal centers may be due to coordination chemistry. The electron transfer occurs metal centred and involves states which are not essentially weakened by electron transfer. In addition, some coordinative interaction between iodide and oxidized metal centered sensitizers can occur. This allows a quite selective electron transfer which does not excessively weaken other bonding states. Nevertheless, as shown in the literature [44], a ruthenium complex can undergo approximately 143 transitions involving different ligand configurations. Depending on the type of transition involved in the sensitization reaction, the oxidized product may have variable survival lifetimes because they involve different portions of the ligand environment. Figure 9 shows that not all sensitizer molecules at different adsorption sites of the TiO2 environment are equally affected. This means that the reaction kinetics of molecules is dependent on the type of adsorption sites. Molecules which do not inject electrons or molecules which are immediately regenerated will not participate in the generation of the in situ infrared signal. Others, due to their peculiar adsorption position, may show preferred tendency towards product formation, that is, they may show a large k2 rate constant and thus contribute to a decrease of life time (1) of dye sensitized cells, formulated below. The interface of ZnO apparently provides more unfavorable reaction possibilities. Alternatively, it may also be that the N3 dye has an especially favorable affinity to TiO2. From Figure 9 it may be tentatively deduced that the weakly interacting N3 molecules and not the strongly interacting (large deviation from mean position of IR band) are preferentially lost to photodegradation. It may be that these molecules, after losing the injected electron, tend to irreversibly react, involving the chemistry of the interface. Recent HPLC-chromatic studies combined with UV/VIS-spectroscopy of N3-dye samples extracted from long-term illuminated solar cells have detected a variety of products which may have formed this way [45]. In this case, the interface of the substrate should be critical, which is indeed supported by the much higher degradation rate of N3 on nanocrystalline ZnO. Such an effect of the oxide interface on the rate k2 of product formation could also explain why at the beginning of a long-term experiment the rate of photodegradation, as reflected in the branching coefficient kbr from relation (1), appears to be larger in some experiments before leveling off [30]. The reason may be that photo-oxidized molecules at selected surface sites, which provide an adequate surface environment, will more easily find reaction possibilities towards irreversible products.

Interestingly, however, as seen from Figure 6, degradation ceases (at least for the time window observed) when photocurrent flow through the cell is suppressed while all other parameters are maintained. This is a very important observation which has to be discussed in some detail. It explains why accelerated dye sensitized solar cell tests under high-illumination intensity, but under open-circuit conditions (no photocurrent flow), yielded no evident degradation and turnover numbers for electrons of the order of 108. This would be necessary for a solar cell operation of 20 years [17]. Complementarily, experiments in Figure 4 clearly show that when photocurrent flows, the stability of the sensitizer is reduced. This is in agreement with the formula for the half-lifetime of a dye sensitized solar cell which has earlier been proposed [32]:

$$T_{1/2}^{0} \sim \frac{S_0}{j_{ph0}} k_{br} \sim \frac{S_0}{j_{ph0}} \bullet \frac{k_3 I^-}{k_2} \tag{1}$$

with the branching coefficient $k_{br} = k_3 [I^-]/k_2$ for the assumed simplest sensitizer reaction, S_0 = initial sensitizer concentration, j_{ph0} = initial photocurrent density, P = product for the reaction sequence

$$\begin{gathered} h\nu + S \xrightarrow{g} S^* \\ S^* \xrightarrow{k_1} S^+ + e^- \\ S^+ \xrightarrow{k_2} P \\ S^+ + I^- \xrightarrow{k_3} S + \frac{1}{2} I_2 \end{gathered} \tag{2}$$

Relation (1) shows that the half lifetime of the dye sensitized cell is proportional to the initial concentration S0 of the sensitizer in the nanostructure and inversely proportional to the initial photocurrent density. The perturbing fact is that the lifetime of the cell will be the shorter the higher its efficiency (and thus the photocurrent passing through it). This is clearly demonstrated by the graded filter experiment of Figure 8. The most successful long-term experiment of an optimized dye solar cell in [20] (1700 hours of solar illumination at only 25°C yielding less than 5% of degradation) was achieved with a cell of only 2% efficiency. A higher efficiency (photocurrent drawn) will correspondingly accelerate photodegradation. Depending on the temperature dependence of k2 and k3 in (2), the life time of the solar cell will be correspondingly temperature-dependent.

Additionally, it has to be considered that the rate of product formation, k2, will be influenced by the chemistry of the oxide/electrolyte interface (compare Figures 3(a) and 4, and 9), so that k2 = k2 (oxide interface). Considering relation (1), it has also to be pointed out that the amount S0 of sensitizer in the nano-oxide layer may codetermine the initial degradation behavior of a cell. If the absorber layer is sufficiently thick and well be covered with dye molecules, excess sensitizer molecules may gradually and temporarily replace oxidized ones as sensitizers.

As discussed elsewhere [32], this branching coefficient may actually still be one order of magnitude too small to guarantee a 20 year lifetime of an efficient dye sensitized solar cell based on the N3 complex. But it turned out to be already surprisingly large for Ru-dyes, when compared with other sensitizers. However, the important finding that the sensitizer stability depends on the photocurrent density, and thus on the solar cell efficiency, explains, why further research on sensitizer stability is urgently needed. The instability problems are expected be the larger the higher dye solar cell efficiency, which is a critical factor for commercialization.

More recent systematic stability studies on dye sensitized cells performed [46] also show a clear dependence of degradation on the nature of the electrolyte (within 2 months approximately 40% in acetonitrile, 20% in propionitrile, 25% in methoxypropionitrile, and 75% in methoxyacetonitrile). But another study claims stability over 1000 hours at high efficiency [47] using a nonvolatile unspecified electrolyte. The experiments in this case were performed at 60°C during illumination, or at 80°C in the dark only. Real problems are to be expected and are actually found when the cell is allowed to heat to 80°C and beyond during periodic daylight illumination, where complex instability problems arise [48]. Only one of the problems involved concerns, in this case, the dye.

If the simplified relation (1) is applied, the expected rate of dye degradation is inversely proportional to the half life time and thus

$$k_{\text{degr}}^{0} \sim \frac{k_2 j_{ph0}}{S_0 k_3 [I^-]}. \tag{3}$$

This means it is determined, besides on the iodide concentration, by the reaction rates k2 and k3, both of which should depend on the nature of the electrolyte as well as on the interfacial chemistry of the sensitizer adsorbed to oxide nanoparticles, and, of course, also on the temperature. It also follows that a negligible photocurrent will not induce significant degradation, as observed by comparing Figures 4 with 6. This, of course, explains why low-efficient dye solar cells yielded an apparent good long-term stability [17, 20]. But in order to understand reported good long-term stability at high current density [46], we have to assume a high rate of regeneration k2 and a low rate of product formation k3. Both rates may depend on interfacial properties of the nano-oxides. In addition, a surplus of sensitizer S0 in solar cells (S0 in (3)) with a thick (15–20 μm) TiO2 layer should theoretically be of advantage. However, it was found that, simultaneously, regeneration of S+ by I– is faced with transport problems, which lead to enhanced degradation with increasing layer TiO2 thickness [49].

A puzzling result from our studies, apart from the fact that initial sensitizer degradation is faster than the later ongoing one, is that the same N3 dye is from 7 to 10 times less stable on ZnO compared with TiO2 and also appears to depend on TiO2 preparation. This means, for example, that the oxidation product of N3 generated on TiO2 (Figure 10(b)) is different in its behavior from that formed on ZnO. This must indeed be concluded since, in both cases, electron transfer to the TiO2 occurs in the pico to femtosecond range via the same carboxyl bridges, and since the possibly slightly different bond strength of which should only have a small influence on the life time of the large oxidized N3 molecule. Second, the initially faster degradation rate of N3 on TiO2, and the difference in N3 degradation on differently prepared and inhomogeneous TiO2 layers indicate that the specific surface chemistry and surface morphology of nano-TiO2 should be critical. Therefore, it had to be investigated how the oxidized N3 species (Figure 10(b)) could differently interact with the TiO2 and ZnO, respectively, to which an electron has previously been donated.

What this interaction could involve has been studied in a parallel paper [50]. Photocapacitive techniques (oxide layers exposed to a gas phase studied in capacitive setup) were applied to study sensitized TiO2 and ZnO layers in contact with a controlled gas phase. When oxygen was removed from the gas phase, formation of Ti3+ states could be spectroscopically observed within the energy gap. When oxygen was removed in presence of the N3 sensitized TiO2, it was found that the N3 dye forms a charge transfer complex with Ti3+ (generated by oxygen extraction from TiO2 into the gas phase) in the TiO2 nanomaterial surface. This charge transfer complex formation is accompanied by a significant (0.4 eV) broadening of the N3 absorption band.

This phenomenon of a charge transfer complex (Figure 10(b)) turned out to be entirely reversible and disappeared when oxygen was added again to the gas phase bordering TiO2, so that Ti3+ states were reconverted into Ti4+ states. This remarkable effect was only observed with the N3 dye and not observed with ZnO nor was it observed with metal cation-free dyes (Pyrogallol, Carmine) studied on the same TiO2.

Our interpretation for the exceptional stability of N3 (and related Ru-based dyes) on TiO2 is therefore tentatively the following: when excited N3 is transferring an electron to TiO2 (Figure 10(b)), Ti4+ may be locally reduced to Ti3+, which now, via its d-state, can get involved in back bonding with the oxidized N3, so that a Ti3+-ligand-Ru3+ charge transfer complex is formed. This is a kind of microscopic feedback reaction. Macroscopically, an autocatalytic (feedback) process leads to selforganization and to a local export of entropy. This creates local order. This may also happen on a microscopic, macromolecular level. The feedback coupling (backbonding of the Ti3+ d-state with the oxidized N3 dye) may stabilize the oxidized N3 species, which temporarily forms a complex with (reduced) TiO2. This could explain the longer survival of N3 sensitizer molecules on TiO2 compared to ZnO in an entirely consistent way. As this phenomenon of a backbonding of N3 and of N3 stabilization had been predicted [51] in 2004 long before the relevant experiments were made [50], this adds credibility to the here discussed charge transfer hypothesis, which is clearly different from the traditional approach to understand electron injection from Ru(dcbpy)2X2 with (X2 = 2 SCN, 2 CN or dcbpy) into TiO2 [52]. It occurs within less than 100 femtoseconds, and sensitization efficiency was found to be dependent on the redox potential of the sensitizer. The latter observation [52] could also be consistent with the here discussed model of a charge transfer complex.

We may talk of a light-induced charge transfer complex formation of N3 on TiO2 nanoparticles. Since TiO2 nanomaterials require a temperature treatment for favorable sensitizer adsorption it may be that the TiO2 surface has to be especially activated or optimized with respect to the charge-transfer interaction proposed. There may consequently be more favorable and less favorable nano-TiO2 substrates.

On the basis of these new viewpoints on the stability of the dye solar cell, it is interesting to reconsider the development history of dye solar cells. The discovery of the improved efficiency and stability of dye solar cells using N3 and related Ru-based sensitizers [16] was essentially a chance discovery, based on a favorable combination of Ru sensitizer and with TiO2. Before, dye solar cell experiments did not encourage major efforts, because the sensitizers were not sufficiently stable on sintered ZnO substrates [14] and earlier studies of Ru-dyes on TiO2 [15] did not use the most favorable electrolytes. It also explains the comparatively flat learning curve [51] for the dye solar cell development, resulting from not recognizing the essential research targets. It equally explains why, after 15 years of intensive research on the Grätzel cell [16], with much effort in direction of cheaper dyes, expensive ruthenium-based complexes (approx. 1000$ per gram) are still the preferred sensitizers.

Now the crucial role of TiO2 chemistry should be highlighted: the most essential difference between sensitization of ZnO and of TiO2 will be a higher life time of the oxidized N3 complex on TiO2 due to the formation of the proposed light-induced Ti3+(ligand)Ru3+ charge transfer complex [50]. While optimizing the expensive Ru-dye further for still higher stability is a possible strategy, a preferable initiative would be to learn to get cheaper metal containing dyes reasonably charge transfer stabilized while reacting on TiO2 surfaces. Can suitable alternative charge transfer complexes be designed to work with cheaper dyes containing more abundant transition metals?

For answering this question the here proposed mechanism will have to be studied theoretically and experimentally in greater detail. Since the nano-TiO2 surface is critical for the formation of the charge transfer complex and for the rate of reaction (k2) for formation of the oxidized sensitizer, it needs increased attention. Specifically, it should also be investigated, whether doping with transition metal ions, which can engage in d-state backbonding with the sensitizer, can increase sensitizer stability.

Interestingly, the here demonstrated influence of the nature of the absorption sites on the photochemical degradation of sensitizers has already been recognized in an early study, 35 years ago, on ZnO single crystals sensitized with chlorophyll [5]. From the degradation curve of chlorophyll sensitized electrodes, it was deduced that sensitizer molecules with widely different stabilities were active on the oxide surface.

In the future, the electronic properties of TiO2 nanoparticle surfaces will have to be optimized for increased sensitizer stability and specific new sensitizer complexes should be tailored if advantage should be taken of the potentially low costs projected for dye solar cells.

References

1. H. Meier, "Sensitization of electrical effects in solids," Journal of Physical Chemistry, vol. 69, no. 3, 719 pages, 1965.
2. H. Meier, Spectral Sensitization, The Focal Press, London, UK, 1967.

3. H. Gerischer, M. Michel-Beyerle, E. Rebentrost, and H. Tributsch, "Sensitization of charge injection into semiconductors with large band gap," Electrochimica Acta, vol. 13, no. 6, 1509 pages, 1968.
4. H. Tributsch and H. Gerischer, "The use of semiconductor electrodes in the study of photochemical reactions," Berichte der Bunsen-Gesellschaft für Physikalische Chemie, vol. 73, 850 pages, 1969.
5. H. Tributsch and M. Calvin, "Electrochemistry of excited molecules: photoelectrochemical reactions of chlorophylls," Photochemistry and Photobiology, vol. 14, no. 2, 95 pages, 1971.
6. H. Tributsch, "Reaction of excited chlorophyll molecules at electrodes and in photosynthesis," Photochemistry and Photobiology, vol. 16, no. 4, 261 pages, 1972.
7. M. Grätzel, "Light and energy, dye sensitized solar cells mimic natural photosynthesis," in Proceedings of "Solar Energy and Artificial Photosynthesis," Satellite Meeting of the SEB 14th International Congress on Photosynthesis (PS '07), The Royal Society, London, UK, July 2007.
8. H. Tsubomura, M. Matsumura, Y. Nomura, and T. Amamya, "Dye sensitised zinc oxide: aqueous electrolyte: platinum photocell," Nature, vol. 261, 402 pages, 1976.
9. T. Watanabe, T. Miyasaka, A. Fujishima, and K. Honda, "Photoelectrochemical study on chlorophyll monolayer electrodes," Chemistry Letters, vol. 7, no. 4, 443 pages, 1978, (Japan).
10. N. Alonso-Vante, M. Beley, P. Chartier, and V. Ern, "Dye sensitization of ceramic semiconducting electrodes for photoelectrochemical conversion," Revue de Physique Appliquée, vol. 16, 5 pages, 1981.
11. M. Nakao, K. Itoh, and K. Honda, "Effect of donor density of semiconductor on spectralsensitization photocurrent," Denki Kagaku Oyobi Butsuri Kagaku, vol. 52, 378 pages, 1984.
12. M. T. Spitler and B. A. Parkinson, "Efficient infrared dye sensitization of van der Waals surfaces of semiconductor electrodes," Langmuir, vol. 2, no. 5, 549 pages, 1986.
13. K. Itoh, M. Nakao, and K. Honda, "Preparation of ZnO thin-film transparent electrodes and their application to electrochemical spectral sensitization," Denki Kagaku Oyobi Butsuri Kagaku, vol. 52, 382 pages, 1984.
14. M. Matsumura, S. Matsudaira, H. Tsubomura, M. Takata, and H. Yanagida, "Dye sensitization and surface structures of semiconductor electrodes,"

Industrial & Engineering Chemistry. Product Research and Development, vol. 19, no. 3, 4157 pages, 1980.

15. M. P. Dare-Edwards, J. B. Goodenough, A. Andrew, K. R. Seddon, and R. D. Wright, "Sensitisation of semiconducting electrodes with ruthenium-based dyes," Faraday Discussions of the Chemical Society, vol. 70, 285 pages, 1981.
16. B. O'Regan and M. Grätzel, "A low-cost, high-efficiency solar cell based on dye-sensitized colloidal TiO_2 films," Nature, vol. 353, 373 pages, 1991.
17. O. Kohle, M. Grätzel, A. F. Meyer, and T. B. Meyer, "The photovoltaic stability of, bis(isothiocyanato)rlutheniurn(II)-bis-2, 2'bipyridine-4, 4'-dicarboxylic acid and related sensitizers," Advanced Materials, vol. 9, no. 11, 904 pages, 1997.
18. R. Eichberger and F. Willig, "Ultrafast electron injection from excited dye molecules into semiconductor electrodes," Chemical Physics, vol. 141, no. 1, 159 pages, 1990.
19. B. Macht, M. Turrion, A. Barkschat, P. Salvador, K. Ellmer, and H. Tributsch, "Patterns of efficiency and degradation in dye sensitization solar cells measured with imaging techniques," Solar Energy Materials and Solar Cells, vol. 73, no. 2, 163 pages, 2002.
20. J. Sastrawan, J. Beier, U. Belledin, et al., "New interdigital design for large area dye solar modules using a lead-free glass frit sealing," Progress in Photovoltaics: Research and Applications, vol. 14, no. 8, 697 pages, 2006.
21. T. Toyoda, T. Sano, J. Nakashima, et al., "Outdoor performance of large scale DSC modules," Journal of Photochemistry and Photobiology A, vol. 164, no. 1–3, 203 pages, 2004.
22. P. Wang, C. Klein, R. Humphrey-Baker, M. Zakeeruddin, and M. Grätzel, "Stable ≥ 8% efficient nanocrystalline dye-sensitized solar cell based on an electrolyte of low volatility," Applied Physics Letters, vol. 86, no. 12, Article ID 123508, 3 pages, 2005.
23. P. Wang, S. M. Zakeeruddin, J. E. Moser, M. K. Nazeeruddin, T. Sekiguchi, and M. Grätzel, "A stable quasi-solid-state dye-sensitized solar cell with an amphiphilic ruthenium sensitizer and polymer gel electrolyte," Nature Materials, vol. 2, 402 pages, 2003.
24. G. E. Tulloch, "Light and energy—dye solar cells for the 21st century," Journal of Photochemistry and Photobiology A, vol. 164, no. 1–3, 205 pages, 2004.
25. K. Okada, H. Matsui, T. Kawashima, T. Ezure, and N. Tanabe, "100 mm × 100 mm large-sized dye sensitized solar cells," Journal of Photochemistry and Photobiology A, vol. 164, no. 1–3, 193 pages, 2004.

26. M. Spath, P. M. Sommerling, J. A. M. van Roosmalen, et al., "Reproducible manufacturing of dye-sensitized solar cells on a semi-automated baseline," Progress in Photovoltaics: Research and Applications, vol. 11, no. 3, 207 pages, 2003.

27. S. Dai, K. Wang, J. Weng, et al., "Design of DSC panel with efficiency more than 6%," Solar Energy Materials and Solar Cells, vol. 85, no. 3, 447 pages, 2005.

28. Y. Chiba, A. Islam, R. Comiya, N. Koide, and L. Han, "Conversion efficiency of 10.8% by a dye-sensitized solar cell using a TiO_2 electrode with high haze," Applied Physics Letters, vol. 88, no. 22, Article ID 223505, 3 pages, 2006.

29. M. Wei, J. Konishi, H. Zhou, M. Janagida, H. Siguhara, and H. Arakawa, "Highly efficient dye-sensitized solar cells composed of mesoporous titanium dioxide," Journal of Materials Chemistry, vol. 16, 1287 pages, 2006.

30. M. Turrion, B. Macht, P. Salvador, and H. Tributsch, "Imaging techniques for the study of photodegradation of dye sensitization cells," Zeitschrift fur Physikalische Chemie, vol. 212, no. 1, 51 pages, 1999.

31. R. Grünwald and H. Tributsch, "Mechanisms of instability in Ru-based dye sensitization solar cells," Journal of Physical Chemistry B, vol. 101, no. 14, 2564 pages, 1997.

32. H. Tributsch, "Function and analytical formula for nanocrystalline dye-sensitization solar cells," Applied Physics A, vol. 73, no. 3, 305 pages, 2001.

33. E. Rijnberg, J. M. Kroon, J. Wienke, et al., "More stability measurements were described," in Proceedings of the 2nd European Photovoltaic Solar Energy Conference and Exhibition (PVSEC '98), p. 47, Vienna, Austria, July 1998.

34. S. A. Haque, Y. Tachibana, R. L. Willis, et al., "Parameters influencing charge recombination kinetics in dye-sensitized nanocrystalline titanium dioxide films," Journal of Physical Chemistry B, vol. 104, no. 3, 538 pages, 2000.

35. A. Hinsch, J. M. Kroon, R. Kern, et al., "Long-term stability of dye-sensitised solar cells," Progress in Photovoltaics: Research and Applications, vol. 9, no. 6, 425 pages, 2001.

36. H. G. Agrell, J. Lindgren, and A. Hagfeldt, "Degradation mechanisms in a dye-sensitized solar cell studied by UV–VIS and IR spectroscopy," Solar Energy, vol. 75, no. 2, 169 pages, 2003.

37. C. Barbé, F. Arendse, P. Comte, et al., "Nanocrystalline titanium oxide electrodes for photovoltaic applications," Journal of the American Ceramic Society, vol. 80, no. 12, 3157 pages, 1997.

38. N. Papageorgiou, W. F. Maier, and M. Grätzel, "An iodine/triiodide reduction electrocatalyst for aqueous and organic media," Journal of the Electrochemical Society, vol. 144, no. 3, 876 pages, 1997.

39. M. A. Anderson, M. J. Gieselmann, and Q. J. Xu, "Titania and alumina ceramic membranes," Journal of Membrane Science, vol. 39, no. 3, 243 pages, 1988.

40. B. O'Regan, J. Moser, J. Anderson, and M. Grätzel, "Vectorial electron injection into transparent semiconductor membranes and electric field effects on the dynamics of light-induced charge separation," Journal of Physical Chemistry, vol. 94, no. 24, 8720 pages, 1990.

41. A. M. Chaparro, P. Salvador, and A. Mir, "Localized photoelectrochemical etching with micrometric lateral resolution on transition metal diselenide photoelectrodes," Journal of Electroanalytical Chemistry, vol. 422, no. 1-2, 35 pages, 1997.

42. R. Grünwald, Ph.D. thesis, Freie Universität, Berlin, Germany, 1997.

43. K. Keis, J. Lindgren, S. E. Lindquist, and A. Hagfeldt, "Studies of the adsorption process of Ru complexes in nanoporous ZnO electrodes," Langmuir, vol. 16, no. 10, 4688 pages, 2000.

44. G. Calzaferri and R. Rytz, "Electronic transition oscillator strength by the extended Hueckel molecular orbital method," Journal of Physical Chemistry, vol. 99, no. 32, 12141 pages, 1995.

45. M. Thomalla and H. Tributsch, "Chromatographic studies of photodegradation of RuL2(SCN)2 in nanostructured dye-sensitization solar cells," Comptes Rendus Chimie, vol. 9, no. 5-6, 659 pages, 2006.

46. R. Kern, N. van der Burg, G. Chmiel, et al., "Long term stability of dye-sensitised solar cells for large area power applications," Opto-Electronics Review, vol. 8, no. 4, 284 pages, 2000.

47. D. Kuang, C. Klein, S. Ito, et al., "High-efficiency and stable mesoscopic dye-sensitized solar cells based on a high molar extinction coefficient ruthenium sensitizer and nonvolatile electrolyte," Advanced Materials, vol. 19, no. 8, 1133 pages, 2007.

48. P. M. Sommerling, M. Späth, H. J. P. Smit, N. J. Bakker, and J. M. Kroon, "Long-term stability testing of dye-sensitized solar cells," Photochemistry and Photobiology A, vol. 164, no. 1–3, 137 pages, 2004.

49. M. Junghaenel and H. Tributsch, "Role of nanochemical environments in porous TiO2 in photocurrent efficiency and degradation in dye sensitized solar cells," Journal of Physical Chemistry B, vol. 109, no. 48, 22876 pages, 2005.

50. T. Dittrich, B. Neumann, and H. Tributsch, "Sensitization via reversibly inducible $Ru(dcbpyH_2)_2(NCS)_2$-TiO_2 charge-transfer complex," Journal of Physical Chemistry C, vol. 111, no. 5, 2265 pages, 2007.

51. H. Tributsch, "Dye sensitization solar cells: a critical assessment of the learning curve," Coordination Chemistry Reviews, vol. 248, no. 13-14, 1511 pages, 2004.

52. J. B. Asbury, Y.-Q. Wang, E. Hao, H. N. Ghosh, and T. Lian, "Evidences of hot excited state electron injection from sensitizer molecules to TiO_2 nanocrystalline thin films," Research on Chemical Intermediates, vol. 27, no. 4-5, 393 pages, 2001.

CITATION

Barkschat A, Moehl T, Macht B, and Tributsch H. The Function of TiO2 with Respect to Sensitizer Stability in Nanocrystalline Dye Solar Cells. International Journal of Photoenergy; Jan 2008 Special Issue, p1. doi:10.1155/2008/814951.

Chromatographic and Spectral Analysis of Two Main Extractable Compounds Present in Aqueous Extracts of Laminated Aluminum Foil Used for Protecting LDPE-Filled Drug Vials

Samuel O. Akapo, Sajid Syed, Anicia Mamangun and Wayne Skinner

ABSTRACT

Laminated aluminum foils are increasingly being used to protect drug products packaged in semipermeable containers (e.g., low-density polyethylene

(LDPE)) from degradation and/or evaporation. The direct contact of such materials with primary packaging containers may potentially lead to adulteration of the drug product by extractable or leachable compounds present in the closure system. In this paper, we described a simple and reliable HPLC method for analysis of an aqueous extract of laminated aluminum foil overwrap used for packaging LDPE vials filled with aqueous pharmaceutical formulations. By means of combined HPLC-UV, GC/MS, LC/MS/MS, and NMR spectroscopy, the two major compounds detected in the aqueous extracts of the representative commercial overwraps were identified as cyclic oligomers with molecular weights of 452 and 472 and are possibly formed from polycondensation of the adhesive components, namely, isophthalic acid, adipic acid, and diethylene glycol. Lower molecular weight compounds that might be associated with the "building blocks" of these compounds were not detected in the aqueous extracts.

Introduction

The potential for adulteration of finished drug products by extractable and leachable compounds from the container or closure systems continues to receive greater attention and scrutiny by regulatory authorities. Consequently, numerous types of guidance including reviews have been published to assist drug manufacturers in providing adequate information regarding the identity, quantity, and control of such compounds to ensure the quality and/or safety of the drug product [1–4]. The focus of these documents is the requirement by the pharmaceutical industry to investigate both analytically and toxicologically compounds that may potentially leach from the packaging materials into the drug products [4, 5].

Several studies have been reported that characterize or identify extractables and leachables from plastic/rubber materials commonly used in packaging and drug delivery devices using various analytical techniques including chromatography, mass spectrometry, and organic synthesis [6–12]. Using both the GC/IR and GC/MS after isolation by Soxhlet extraction, Kim-Kang and Gilbert [13] identified seven unknown compounds that could migrate from plastic laminates into a unit dose injection device. However, little is known about extractables and leachables from preprinted laminated aluminum foil overwrap that is used for protecting drug products (e.g., inhalation solutions and suspensions) packaged in semipermeable containers (e.g., low-density polyethylene (LDPE)) from degradation and/or evaporation.

The components of the protective foil, which include inks, solvents, and unreacted monomers and oligomers derived from the adhesive material, have the potential to permeate through LDPE vials and contaminate the drug product formulations. While the identities of the inks and the associated volatile solvents are often known [14], the identities of compounds that may leach from other components of the packaging material and which may vary in structure depending on the nature of the finished drug product and the condition of use are not even known to the manufacturer. The goal of this study was to structurally identify the compounds obtained from an aqueous extract of preprinted foil laminate overwrap using chromatography, mass spectrometry and nuclear magnetic resonance (NMR) spectroscopic techniques.

Experimental

Materials and Reagents

Analytical grade phthalic acid, acetic acid, and hydrochloric acid were obtained from Mallinckrodt (Phillipsburg, NJ, USA) and used as received. HPLC grade acetonitrile and methanol were from EMD Chemicals (Gibbstown, NJ, USA). Adipic acid, 0.2 M trimethylphenylammonium hydroxide (TMAH) in methanol, and spectroscopy grade trifluoroacetic acid (TFA) were purchased from Sigma-Aldrich (St. Louis, MO, USA). The foil laminate overwrap was obtained from a commercial source, and for proprietary reasons, detailed information pertaining to this material and the drug product evaluated during the course of this study will not be disclosed. BondElut C_{18} solid-phase extraction cartridges (5 gm, 20 cc) were sourced from Varian (Lake Forest, CA, USA).

Sample Preparation

Extractable compounds from the laminated aluminum foil overwrap were extracted into 20 mL of purified water placed in a foil pouch (2 in. × 4 in.) previously rinsed twice with 20 mL of purified water. The pouch was sealed and incubated at 70°C for 24 hours in an oven. The pouched extract was then allowed to cool to room temperature and subsequently analyzed for any extractable compound from the foil.

HPLC

HPLC separation was performed using an Agilent 1100 series liquid chromatographic system (Wilmington, DE, USA) consisting of a quaternary gradient pump, heated column compartment, autosampler, photodiode array, and variable wavelength UV detectors. Data were collected and processed using a Perkin-Elmer Turbochrom Client/Server Data System, Version 6.1.2 Shelton, CT, USA.

The separation employed an Agilent Zorbax, RX-C18 column (4.6 mm i.d. × 15 cm, 5 µm particle size) (Wilmington, DE, USA) and the mobile phase consists of acetonitrile and 0.1% TFA in the ratio 38:62 v/v, filtered through a nylon membrane and degassed under vacuum before use. The column compartment was maintained at 25°C. Using a 50 µL injection volume, the analytes were monitored with UV detection at 210 nm for a total runtime of 20 minutes at a flow rate of 1.0 mL/min.

A typical chromatogram of the laminated foil extract is presented in Figure 1 showing two major extractable compounds, identified as peaks 1 and 2, with the corresponding UV spectra (insert). The blank chromatogram (not shown) showed no peaks in the HPLC beside the solvent front. The retention times for triplicate analysis of the foil extract were 8.4 (0.4% R.S.D) and 10.2 (0.3% R.S.D) minutes, respectively, for peaks 1 and 2, and the resolution (Rs) between the two peaks was 4.6. The limits of detection and quantitation were determined to be 0.02 ppm and 0.06 ppm (1.4% RSD, n=3), respectively, with the corresponding signal-to-noise ratios of approximately 3.8 and 9.9 [15]. Under the described HPLC conditions, several fractions of the extractable peaks 1 and 2 were collected and prepared for structural elucidation.

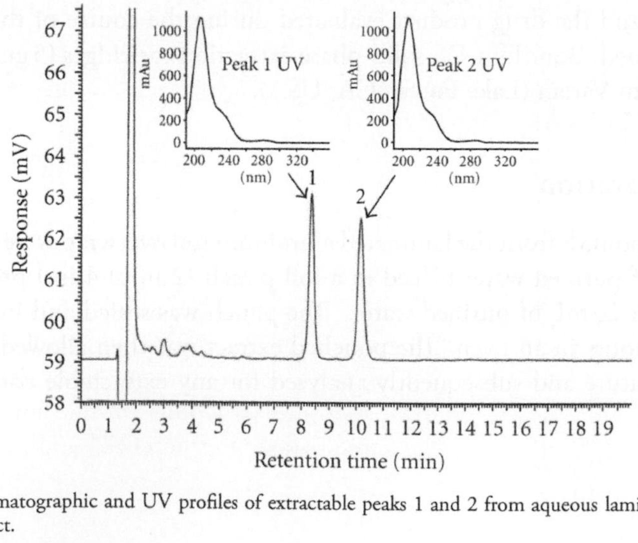

Figure 1. Chromatographic and UV profiles of extractable peaks 1 and 2 from aqueous laminated aluminum foil pouch extract.

GC-MS

Separate HPLC fractions of the extractable peaks were dried using a Büchi Rotavapor R-124 rotary evaporator (Brinkmann Instruments, Inc., Westbury, NY), redissolved in 0.5 mL of 0.5 M methanolic HCl, and the resulting solution was heated for 1 hour at 70°C. The hydrolysates were then dried and dissolved in 40 µL of 0.2 M TMAH in methanol to produce the methyl derivatives, which were subsequently analyzed using Agilent 6890/5973 gas chromatograph-mass spectrometry detector, GC-MSD (Wilmington, DE) equipped with a Gerstel MultiPurposeSampler MPS (Baltimore, MD). The gas chromatograph was fitted with Agilent DB-5MS capillary column (0.25 µm × 30 m, 0.25 µ film) and operated with temperature programming from 50°C (held for 1 min) to 300°C at 10°C/minutes, and held at 300°C for 6 min using helium as the carrier gas at a constant flow rate of 1 mL/min. The GC injector port was set at 280°C in a splitless mode and the MSD was maintained at 280°C. All the GC/MS data were acquired with MSD ChemStation Version D.01.02.16 (Agilent Technologies, Wilmington, DE) in the m/z range of 30–500 at a rate of 1 scan/sec under electron ionization (EI) mode.

LC-MS

A ThermoSeparations HPLC pump (Model P400) and UV detector (Model 600LP) coupled to a Finnigan LCQ Duo ion-trap mass spectrometer with electrospray source (ThermoQuest, San Jose, CA, USA) were used to obtain full-scan MS and MS/MS data of the foil laminate extractables. The column and conditions of HPLC analysis were as described in HPLC section except the mobile phase, which contains 40:60 v/v acetonitrile: 0.1% acetic acid. Foil laminate extract was analyzed by LC/MS in full-scan positive-ion mode using a 50 to 1500 m/z scan range. Following the assignment of MS ions for the identified peaks, the extract was reanalyzed using full-scan MS/MS experiments to obtain product ion spectra.

NMR

Further confirmation of the structure of each extractable compound was performed on a JOEL ECX-400 NMR spectrometer operating at 400 MHz at Acorn NMR Incorporated (Livermore, CA, USA) after isolation using solid-phase extraction (SPE) followed by analytical HPLC purification. For SPE, a BondElut C18 cartridge was washed with 0.1% TFA in methanol and preconditioned with 0.1% TFA in water before loading about 50 mL of the foil extract at approximately 3 mL/min. The retained compounds were eluted with 10 mL of 0.1% TFA in methanol, and the solvent was evaporated to dryness using the Büchi Rotavapor R-124 rotary evaporator. The residue was then redissolved in 0.2 mL methanol and rinsed with 0.8 mL of 0.1% TFA in purified water prior to HPLC purification of the isolates, which were subsequently redissolved in deuterated methanol (CD3OD) containing tetramethylsilane (TMS) for NMR analysis. 1H and COSY spectra were acquired at ambient temperature (25°C), and the resulting FIDs were transferred to a computer and processed using NutsPro NMR software (Acorn NMR Inc., Livermore, CA, USA). 1H chemical shifts were referenced to internal TMS.

Results and Discussion

GC-MS Profiles of Extracted Compounds

The MS total ion chromatograms (TICs) for methylated fractions of hydrolyzed peaks 1 and 2 as illustrated in Figure 2 showed the presence of 3-4 major peaks in addition to several other minor peaks. The signal at m/z 194, 163, and 135 in Figure 3(a) for the main peak at about 13.14 minutes in Figure 2 matched the NIST mass spectral library for 1,3-dimethyl phthalate indicative of a strong preference for isophthalic acid (1,3-benzenedicarboxylic acid) in both the extractable compounds. Other peaks at about 12.37 and 12.98 minutes are identified as phthalic (1,2-benzenedicarboxylic acid) and terephthalic (1,4-benzenedicarboxylic acid) acids and are probably present in trace amounts in the starting materials. Although a fair majority of the peaks observed are very similar for the two compounds, one noticeable difference is the peak at about 9.49 minutes in Figure 2(a) for extractable 1, which was absent in Figure 2(b) for extractable 2. The signal at m/z 143, 114, 101, and 59 in Figure 3(b), which are characteristic of adipic acid, gave a good library match for 1,6-dimethyl hexanoate indicating the presence of adipic acid in extractable 1 alone.

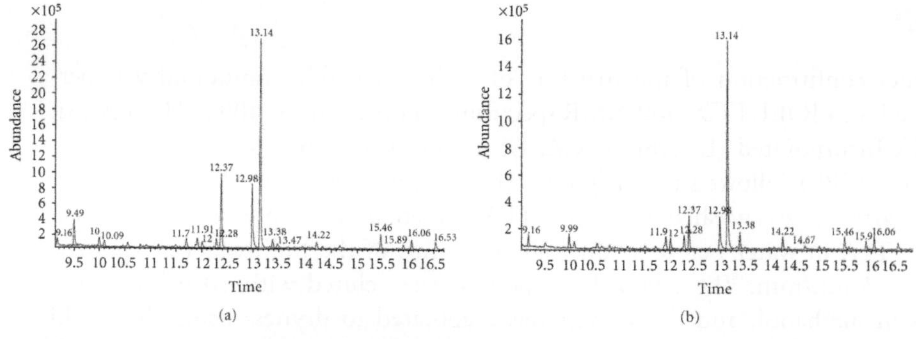

Figure 2. Total ion current GC/MS chromatograms of methylated derivatives of hydrolyzed (a) extractable 1 and (b) extractable 2.

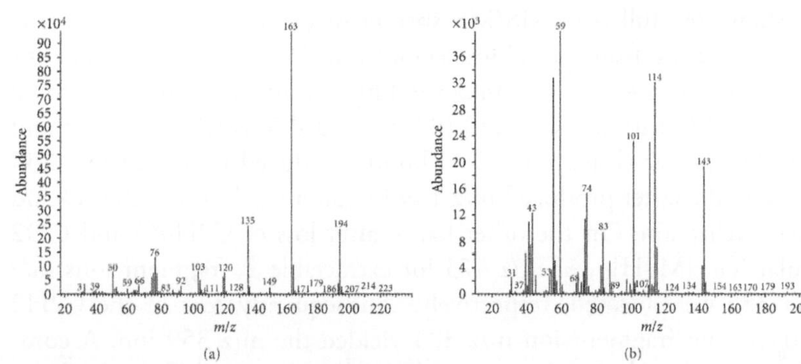

Figure 3. GC/MS spectral for peaks at (a) 13.14 min, and (b) 9.49 min in Figure 2(a).

LC-MS Analysis of Extracted Compounds

The full scan MS spectra of extractables 1 and 2 are shown in Figure 4. The peak at 7.3 minutes produced a base peak at m/z 453 corresponding to the $[M + H]^+$ ion and an ion due to the ammonia adduct $[M + NH_4]^+$ at m/z 470. The peak at 8.3 minutes produced a base peak at m/z 473 corresponding to the $[M+H]^+$ ion, an ion at m/z 490 due to the ammonia adduct $[M + NH_4]^+$, and a sodium adduct of a dimer ion $[2M + Na]^+$ at m/z 967. These full scan spectra enabled molecular weight assignments of 452 and 472 for extractables 1 and 2, respectively.

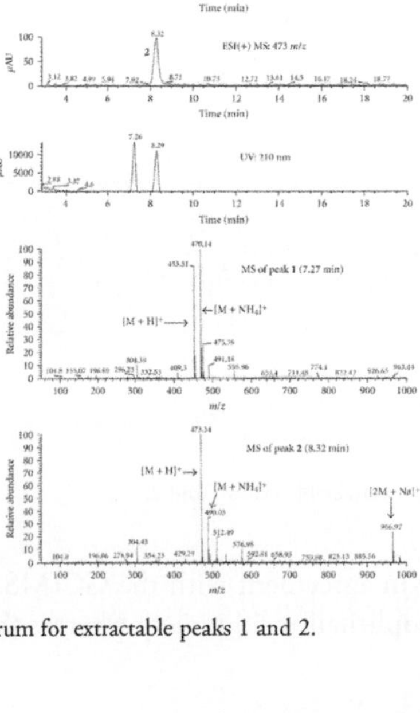

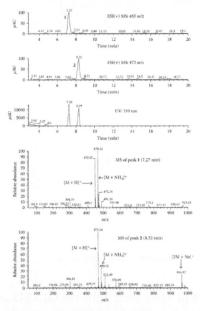

Figure 4. Full scan LC/MS spectrum for extractable peaks 1 and 2.

Figure 5 shows the full scan MS/MS spectra of extractables 1 and 2. The m/z 409 ion originating from pseudomolecular ion, [M+H]+, of m/z 453 for extractable 1 represents a loss of C2H4O yielding the m/z 409 ion. Additional elimination of CO2 (44 u) and C5H8O (84 u) yielded ions with m/z 365 and 281, respectively. Further cleavage of C–O bonds produced ions with m/z 237 and m/z 193, and the latter produced m/z 149 ion after the loss of C2H4O and intramolecular cyclization. On the other hand, after loss of C2H4O and CO2 from molecular ion, [M+H]+, of m/z 473 for extractable 2, fragment ions m/z 429 and m/z 385 were formed, respectively. Subsequently, loss of the C2H2 molecule (26 u) from fragment ion m/z 385 yielded the m/z 359 ion. A comparison of the molecular weights for the two compounds gave a difference of 20 u indicating a possible replacement of an isophthalic acid molecule with adipic acid in extractable 1.

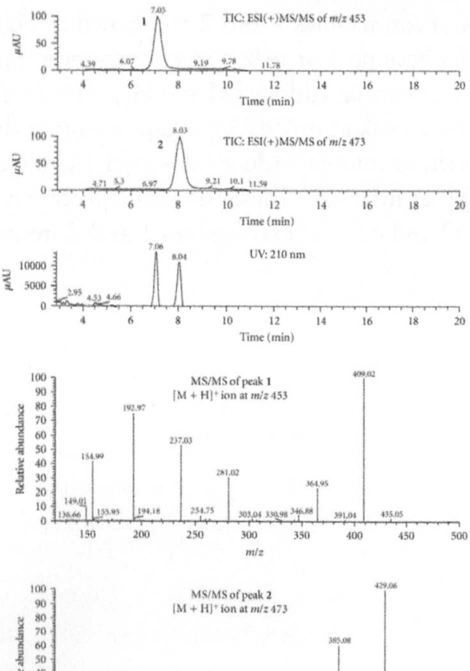

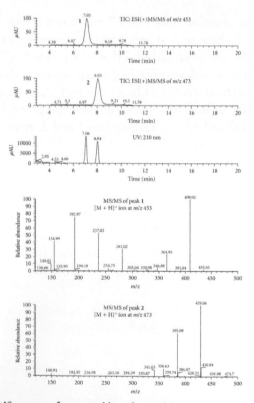

Figure 5. Full scan MS/MS spectrum for extractable peaks 1 and 2.

The LC/MS results were in agreement with the GC/MS data, in which both compounds contained the isophthalic acid moiety whereas the adipic acid moiety was only detected in extractable 1. Thus, extractable 1 appears to be a reaction product of a molecule each of isophthalic and adipic acids with two molecules of diethylene glycol, while extractable 2 is formed by poly-condensation of two molecules of isophthalic acid with two diethylene glycol molecules. The proposed structures and the fragmentation patterns for the two compounds are shown in Figures 6 and 7. Additionally, the MS/MS experiments on the ammonia adducts of extractable peaks 1 and 2 produced only the loss of NH3 to the corresponding pseudomolecular ions, confirming the adduct assignments (data not shown).

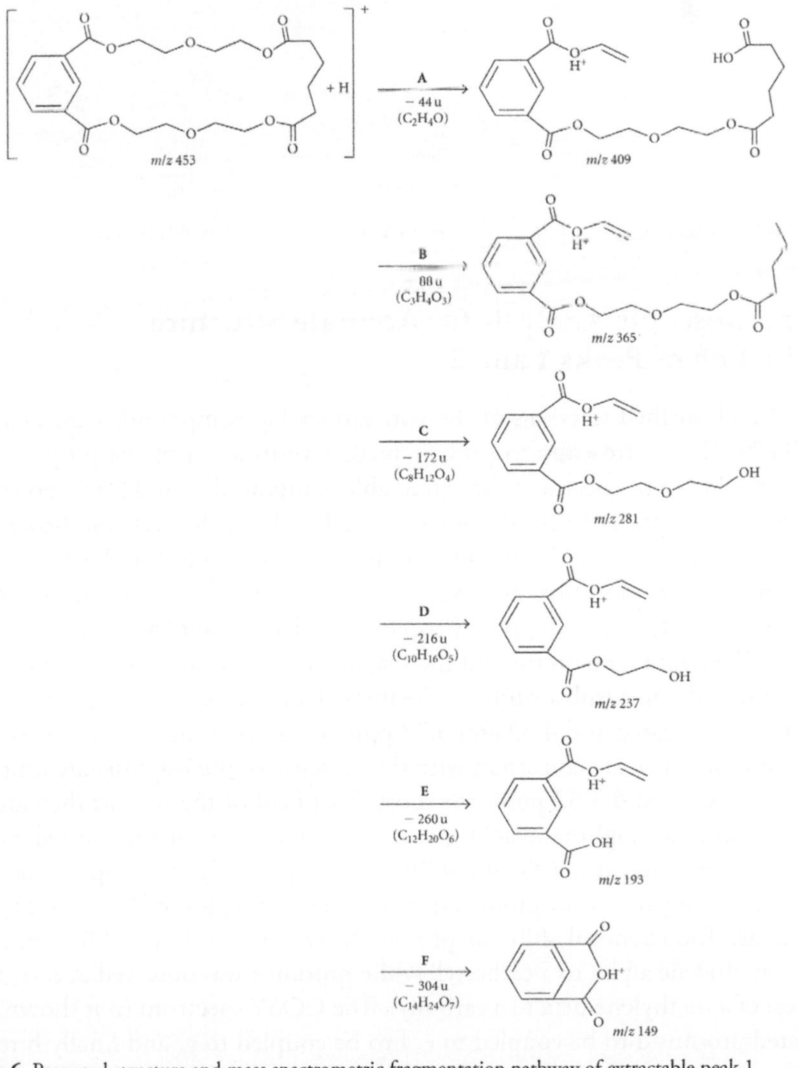

Figure 6. Proposed structure and mass spectrometric fragmentation pathway of extractable peak 1.

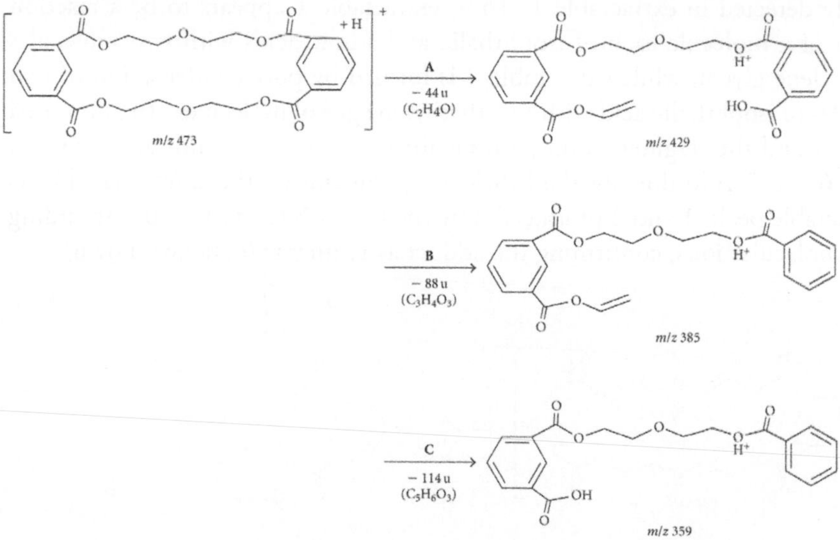

Figure 7. Proposed structure and mass spectrometric fragmentation pathway of extractable peak 2.

NMR Spectroscopic Analysis for Accurate Structure Determination of Peaks 1 and 2

The isolated and purified fractions of the two extractable compounds were analyzed by ^{1}H NMR spectroscopy to provide further verification of the proposed structures. The ^{1}H NMR spectra of the extractable compounds 1 and 2 are shown in Figures 8 and 9, respectively. In addition to several peaks, each spectrum showed the expected solvent peaks at δ 4.9 and 3.3 ppm for HDO and CD_2HOD, respectively. In Figure 8, the three aromatic protons (a, b, and c) could easily be found at δ 8.68, 8.28, and 7.65 ppm as triplet, doublet of doublets, and triplet, respectively. The coupling patterns and the magnitude of the coupling constants are characteristic of a metasubstitution. The methylene protons d and g, next to the ester groups, appeared at δ 4.52 and 3.73 ppm as complex multiplets, respectively, the shifts of which are consistent with the presence of the isophthalate unit. Protons d were assigned δ 4.52 ppm, the most downfield of the pair, as they are esters of the aromatic isophthalic acid group. The resonances of the methylene protons e and f appeared at δ 3.84 and 4.20 ppm, respectively, as complex multiplets. The splitting patterns of protons d, e, f, and g are typical of X-CH_2CH_2-Y spin systems. The chemical shift for protons h was observed at δ 2.00 ppm, typical of a methylene alpha to a carbonyl, while protons i was observed at δ~1.3 ppm, typical of a methylene beta to a carbonyl. The COSY spectrum (not shown) demonstrated protons d to be coupled to e, f to be coupled to g, and finally h to be coupled to i.

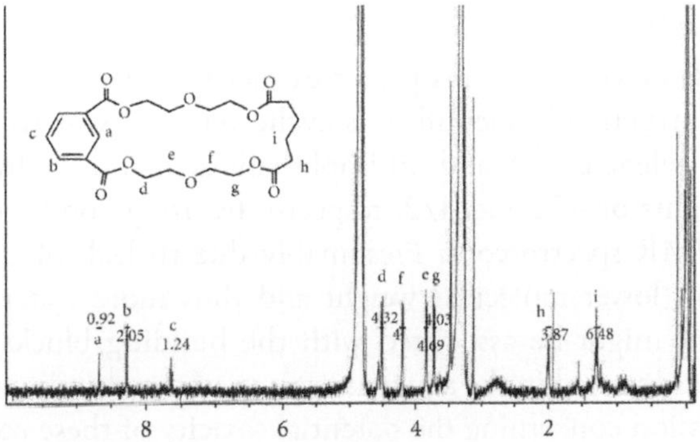

Figure 8. The ^{1}H NMR (400 MHz) spectrum for extractable peak 1 in CD_3OD.

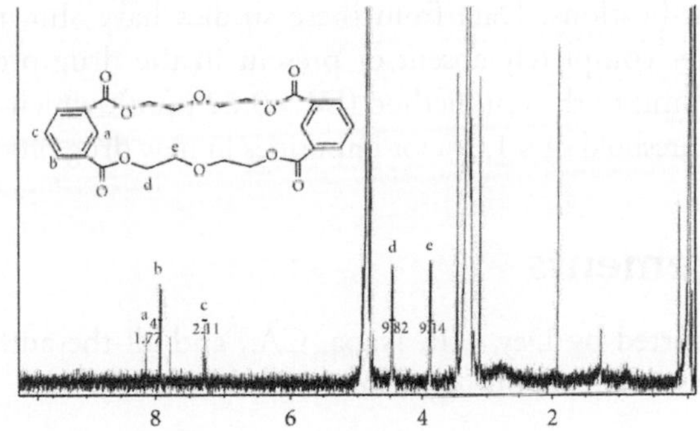

Figure 9. The ^{1}H NMR (400 MHz) spectrum for extractable peak 2 in CD_3OD.

The 1H NMR spectrum of extractable 2 could also be clearly assigned as shown in Figure 9. The resonances of the three aromatic protons (a, b, and c) are similar but appear slightly more upfield than those in extractable 1. The methylene protons d and e appeared at δ 4.48 and 3.89 ppm, respectively, as complex multiplets, as would be expected for ethylene glycols. These shifts and splitting patterns are similar to the analogous protons d and e of extractable 1. Redundant 1H positions in both structures were not labeled due to symmetry. The absence of vinylic and/or carboxylic protons and the fact that the molecules clearly exhibited elements of symmetry, indicate that both extractables are nonlinear contrary to the structures reported by Tiller et al. [12] for two leachable compounds obtained from a custom adhesive used during development of a medical device. Therefore, we propose that extractable 1 is a cyclic oligomer of isophthalic, adipic acid, and diethylene glycol, and extractable 2 is a cyclic oligomer of isophthalic acid and diethylene glycol.

Conclusion

The two compounds detected in the aqueous extract from laminated aluminum foil overwrap were structurally identified as cyclic oligomers of (i) isophthalic, adipic acid and diethylene glycol, and (ii) isophthalic acid and diethylene glycol, with molecular weights of 452 and 472, respectively, using combined GC/MS, LC/MS/MS and NMR spectroscopy. Presumably due to lack of chromophoric functional groups, lower molecular weight and thus more water soluble cyclic compounds that might be associated with the building blocks of the two compounds were not detected in the aqueous extracts of the aluminum foil examined. While a discussion concerning the potential toxicity of these compounds is beyond the scope of this paper, laminated aluminum foils have been certified to be used as packaging materials for LDPE-filled drug vials at our facility through extensive stability studies for several clinical, registration, and commercial batches of aqueous-based medications. Data from these studies have shown that these compounds are either completely absent or present in the drug products at or below the detection limit of the test method (DL <0.02 ppm), which is also lower than the FDA/ICH threshold of ≤ 1.0% for impurities in new drug products [16].

Acknowledgements

This study was supported by Dey, L.P., Napa, CA., and all the authors are full time employees of the company during the course of the study.

References

1. "Guidance for industry: container closure systems for packaging human drugs and biologics," US FDA, Rockville, Md, USA, http://www.fda.gov/cder/guidance/1714fnl.htm.
2. "Guideline on plastic immediate packaging materials," European Medicine Agency (EMEA), London, UK, http://www.emea.europa.eu/htms/human/humanguidelines/quality.htm.
3. C. J. Taborsky, E. B. Sheinin, and D. G. Hunt, "A critical approach to the evaluation of packaging components and the regulatory and scientific considerations in developing a testing strategy," American Pharmaceutical Review, vol. 9, no. 6, pp. 146–150, 2006.
4. J. S. Kauffman, "Identification and risk-assessment of extractables and leachables," http://pharmtech.findpharma.com/pharmtech.
5. V. S. Gaind and K. Jedrzejczak, "HPLC determination of rubber septum contaminants in the iodinated intravenous contrast agent (sodium iothalamate)," Journal of Analytical Toxicology, vol. 17, no. 1, pp. 34–37, 1993.

6. "Guidance for industry: nasal spray and inhalation solution, suspension, and spray drug products—chemistry, manufacturing, and controls documentation," US FDA, Rockville, Md, USA, http://www.fda.gov/cder/guidance/4234fnl.htm.
7. F. Zhang, A. Chang, K. Karaisz, R. Feng, and J. Cai, "Structural identification of extractables from rubber closures used for pre-filled semisolid drug applicator by chromatography, mass spectrometry, and organic synthesis," Journal of Pharmaceutical and Biomedical Analysis, vol. 34, no. 5, pp. 841–849, 2004.
8. J. K. Baker, "Characterization of phthalate plasticizers by HPLC/thermospray mass spectrometry," Journal of Pharmaceutical and Biomedical Analysis, vol. 15, no. 1, pp. 145–148, 1996.
9. D. R. Jenke, J. Story, and R. Lalani, "Extractables/leachables from plastic tubing used in product manufacturing," International Journal of Pharmaceutics, vol. 315, no. 1-2, pp. 75–92, 2006.
10. D. L. Norwood, L. Nagao, S. Lyapustina, and M. Munos, "Application of modern analytical technologies to the identification of extractables and leachables," American Pharmaceutical Review, vol. 8, no. 1, pp. 78–87, 2005.
11. D. R. Jenke, J. M. Jene, M. Poss, et al., "Accumulation of extractables in buffer solutions from a polyolefin plastic container," International Journal of Pharmaceutics, vol. 297, no. 1-2, pp. 120–133, 2005.
12. P. R. Tiller, Z. El Fallah, V. Wilson, et al., "Qualitative assessment of leachables using data-dependent liquid chromatography/mass spectrometry and liquid chromatography/tandem mass spectrometry," Rapid Communications in Mass Spectrometry, vol. 11, no. 14, pp. 1570–1573, 1997.
13. H. Kim-Kang and S. G. Gilbert, "Isolation and identification of potential migrants in gamma-irradiated plastic laminates by using GC/MS and GC/IR," Applied Spectroscopy, vol. 45, no. 4, pp. 521–714, 1991.
14. S. O. Akapo and C. M. McCrea, "SPME-GC determination of potential volatile organic leachables in aqueous-based pharmaceutical formulations packaged in overwrapped LDPE vials," Journal of Pharmaceutical and Biomedical Analysis, vol. 47, no. 3, pp. 526–534, 2008.
15. "Method validation report of HPLC limit test for potential leachables from foil laminate overwrap," Analytical Test Report, Dey L.P., Napa CA., USA.
16. "Guidance for industry: Q3B(R2): impurities in new drug products," US FDA, Rockville, Md, USA, http://www.fda.gov/cder/guidance/7385fnl.pdf.

CITATION

Akapo SO, Syed S, Mamangun A, and Skinner W. Chromatographic and Spectral Analysis of Two Main Extractable Compounds Present in Aqueous Extracts of Laminated Aluminum Foil Used for Protecting LDPE-Filled Drug Vials. International Journal of Analytical Chemistry, Volume 2009 (2009). http://dx.doi.org/10.1155/2009/693210.

The Cold Contact Method as a Simple Drug Interaction Detection System

Ilma Nugrahani, Sukmadjaja Asyarie, Sundani Nurono Soewandhi and Slamet Ibrahim

ABSTRACT

The physical interaction between 2 substances frequently occurs along the mixing and manufacturing of solid drug dosage forms. The physical interaction is generally based on coarrangement of crystal lattice of drug combination. The cold contact method has been developed as a simple technique to detect physical interaction between 2 drugs. This method is performed by observing new habits of cocrystal that appear on contact area of crystallization by polarization microscope and characterize this cocrystal behavior by melting point determination. Has been evaluated by DSC, this method is proved suitable to identify eutecticum interaction of pseudoephedrine HCl-acetaminophen, peritecticum interaction of methampyrone-phenylbutazon, and solid solution interaction of amoxicillin-clavulanate, respectively.

Introduction

In the pharmaceutical area, investigation of physical interaction which is very possible to occur along the mixing manufacturing process of dosage forms is an important issue because a lot of physical properties of drugs especially in solid state dosage forms could be influenced. Such changes as stability, drug performance, dissolution profile, pharmacokinetics profile, and, moreover, the pharmacological effect should be much impacted by the interactions [1–3]. Physical interactions in the solid state dosage form frequently occur even in storage and distribution time of dosage forms [4, 5]. Therefore, the simple technique to detect the interaction might be very useful. Along last decades, Kofler's hot stage contact method, which observed the co-recrystallization from 2 compounds from their hot molten state, has been used as a simple method to detect the cocrystal formation as an indicator for physical interaction occurrence [6]. Unfortunately, in pharmaceutical area, there are a lot of thermolabile compounds which cannot be crystallized after melting, because we have arranged a simple method used to detect physical interaction of the thermolabile compounds. This method was conducted by observing the process of cocrystallization and its behavior on crystallization contact area of 2 drugs from their solution under microscope polarization and melting point determination. Briefly, this method is developed from Kofler's contact methods by changing the comelting technique to the cosolvating crystallization. Recently, we have investigated the cold contact suitability to detect and identify the physical interaction of 3 drug combinations which are usually found in the dosage forms. Acetaminophen-pseudoephedrine HCl is found in antiinfluenza dosage forms, methampyrone-phenylbutazon in analgesic dosage forms, while amoxicillin-clavulanate in antibiotic combination dosage forms. All of these combinations have brought some troubles in mixing and compounding which caused high variability on their dosage forms quality. We used the cold contact method to detect the possibility of physical interaction of these drug combinations and evaluated the method by differential scanning calorimeter (DSC) as a primary thermal analysis method.

Method and Results

The cocrystallization from solution in ambient temperature was suggested to be mentioned as the cold contact method [7–9]. In this report, differential scanning calorimeter (DSC) was used to evaluate the results by observing the thermal profile and arranging phase diagrams [10–12]. From DSC data of varies molar or weight ratios (0 : 1, 1 : 9, 2 : 8, 3 : 7, 4 : 6, 5 : 5, 6 : 4, 7 : 3, 9 : 1, and 10 : 0), phase diagrams were arranged. In purpose to abbreviate the paper, phase diagrams are not presented.

The first drug combination which was examined is acetaminophen-pseudoephedrine HCl. Under polarization microscope, a black area was observed which melted at 113°C, while pseudoephedrine HCl alone melted at 184°C and acetaminophen alone at 170°C (Figure 1(a)). It could be predicted that the binary system composed an eutectic mixture. By DSC, the prediction was evaluated and proved coherency. To clarify, the thermograms of 3 : 7, 5 : 5, and 7 : 3 weight ratios are described in Figure 1(b) which indicate the eutectic point at 112.3°C, appropriate with cold contact data. Secondly, phenylbutazon-methampyrone combination showed a peritectic interaction which was proved by cold contact method and has been proven to be coherent with DSC analysis data. Last, amoxicillin-clavulanate mixture showed strong interaction with single exothermic transition curve at 202°C, equal to its melting point which was observed by cold contact method.

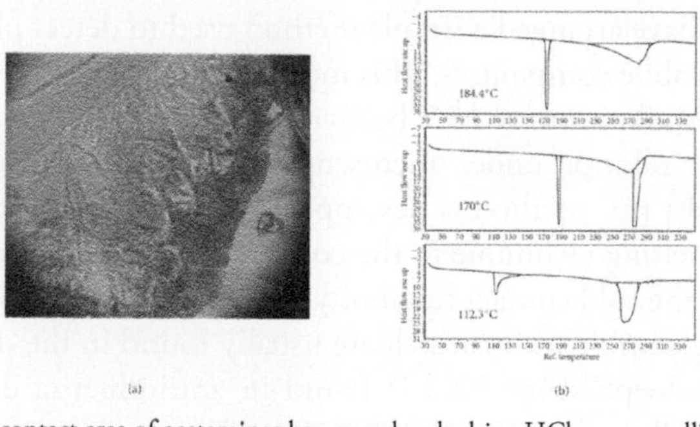

Figure 1. (a) The cold contact area of acetaminophen-pseudoephedrine HCl corecrystallization showed an area which melted at (1) 113°C, compared to the starting materials at (2) 170°C and (3) 184°C. (b) Thermograms of pseudoephedrine HCl showed melting point at 184.4°C (top), acetaminophen at 169.5°C (middle), and the eutectic mixture in weight ratios 5 : 5 at 112.3°C (bottom).

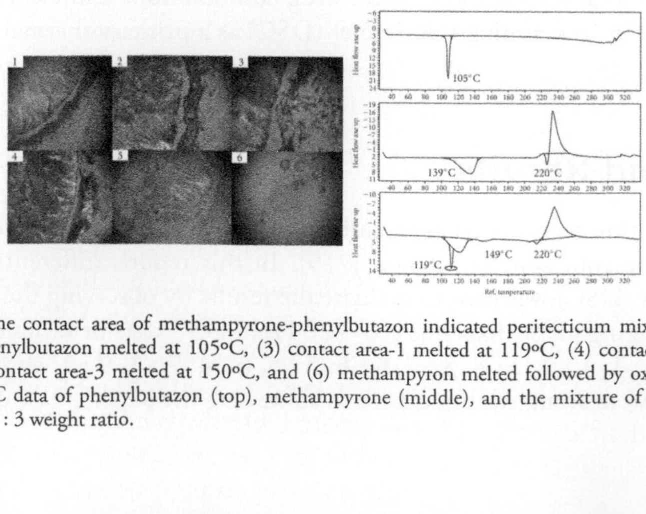

Figure 2. (a) The contact area of methampyrone-phenylbutazon indicated peritecticum mixture: (1) before heating, (2) phenylbutazon melted at 105°C, (3) contact area-1 melted at 119°C, (4) contact area-2 melted at 140°C, (5) contact area-3 melted at 150°C, and (6) methampyron melted followed by oxidation at 220–230°C. (b) DSC data of phenylbutazon (top), methampyrone (middle), and the mixture of methampyrone-phenybutazon 7 : 3 weight ratio.

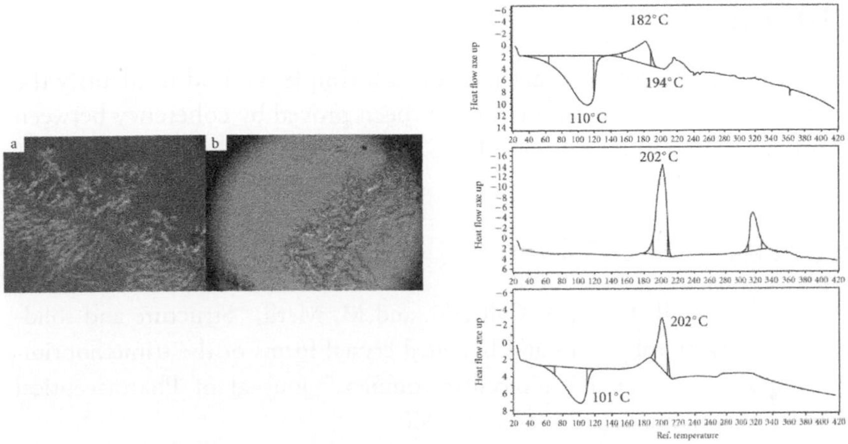

Figure 3. (a) Cold contact observation results: (i) the cocrystal grew from clavulanate crystal to amoxicillin solution in NaOH, (ii) amoxicillin melted at 194°C, clavulanate oxidized at 203°C, while the cocrystal oxidized least. (b) DSC data of amoxicillin trihydrate (top), potassium clavulanate (middle), and the physical mixture 1 : 1 (bottom). The exothermic peak of 1 : 1 mixture shows that amoxicillin and clavulanate overlay and become 1 peak at 202°C which indicates a solid solution interaction.

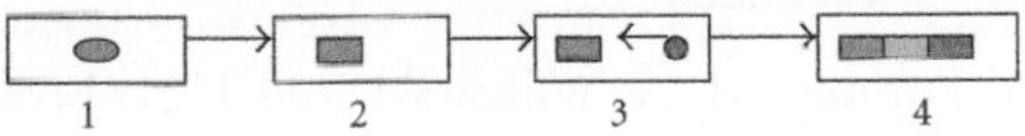

Figure 4. Cold contact preparation: (1) solution A dropped on clean object glass, (2) let it be crystallized, (3) then solution B dropped near the crystal A, (4) if it interacted, the new habit with different melting point will be formed on the contact area.

The results prove strong relation between cold contact method data and DSC. The simple eutectic interaction of pseudoephedrine hcl-acetaminophen, the peritectic interaction between methampyrone-phenylbutazon, and solid solution formation between amoxicillin trihydrate-clavulanate have been early detected by this simple method. Therefore, this method has high possibility to be used as a simple method to evaluate the other drug interactions [10–13].

Briefly, the experiment could be performed as follows:

(i) Each of the components is dissolved in the same solvent. Drop the solution 1 on cleaned object glass, evaporate the solvent and let it crystallize. The second solution is dropped near the formed crystal, let it diffuse slowly toward the crystal then quickly evaporate the solvent. Let second crystallization be performed and observe the contact area.

(ii) Heat the cold contact preparation on hot plate and observe the melting point. Differences of melting points indicate the cocrystallization or physical interaction.

(iii) The observation result could be confirmed with DSC.

Conclusion

The acceptability of the cold contact method as a simple method to identify the physical interaction of drug combination has been proved by coherency between the cold contact method data with the DSC evaluation.

References

1. G. Bettinetti, M. R. Caira, A. Callegari, and M. Merli, "Structure and solid-state chemistry of anhydrous and hydrated crystal forms of the trimethoprim-sulfamethoxypyridazine 1:1 molecular complex," Journal of Pharmaceutical Sciences, vol. 89, no. 4, pp. 478–489, 2000.
2. M. R. Caira, "Sulfa drugs as model co-crystals former," Molecular Pharmaceutics, vol. 4, no. 3, pp. 310–316, 2007.
3. G. P. Stahly, "Diversity in single- and multiple-component crystals. The search for and prevalence of polymorphs and cocrystals," Crystal Growth & Design, vol. 7, no. 6, pp. 1007–1026, 2007.
4. R. E. Davis, K. A. Lorimer, M. A. Wilkowski, and J. H. Rivers, "Studies of phase relationships in cocrystal systems," Transaction of the American Crystallographic Association, vol. 39, pp. 41–61, 2004.
5. N. Rodriguez-Hornedo, "Crystallization and the properties of crystals," in Encyclopedia of Pharmaceutical Technology, J. Swarbrick and J. C. Boylan, Eds., vol. 3, pp. 399–434, Marcel Dekker, New York, NY, USA, 1990.
6. Wikipedia, "Recrystallization," 2007, http://en.wikipedia.org/.
7. I. Nugrahani, S. N. Soewandhi, S. Asyarie, and S. Ibrahim, "The cold contact methods to detect physical interaction of paracetamol-pseudoephedrine HCl," Artocarpus, vol. 6, no. 1, pp. 18–29, 2007.
8. I. Nugrahani, S. N. Soewandhi, S. Asyarie, and S. Ibrahim, "The cold contact method to detect physical interaction of amoxicillin-clavulanate," in Proceeding of International Chemical Conference and Seminar, Yogyakarta, Indonesia, 2007.
9. I. Nugrahani, S. N. Soewandhi, S. Asyarie, and S. Ibrahim, "Study of levodopa-benserazide interaction by cold contact method," Indonesian Journal of Pharmaceutical Science, vol. 18, no. 2, 2007.
10. S. R. Byrn, R. R. Pfeiffer, and J. G. Stowell, Solid State Chemistry of Drugs, SSCI, West Lafayette, Ind, USA, 2nd edition, 2000.

11. J. T. Cartensen, Advanced Pharmaceutical Solids, Taylor & Francis, New York, NY, USA, 2001.
12. F. Giordano and A. Rossi, "Phase diagrams in the binary system," Bollettino Chimico Farmaceutico, vol. 139, no. 4, pp. 345–349, 2000.
13. Drugbank, "Acetaminophen, Pseudoephedrine, Antalgine, Phenylbutazone, Amoxicillin, and Clavulanate," March 2006, http://en.wikipedia.org/wiki/DrugBank/.

CITATION

Nugrahani I, Asyarie S, Soewandhi SN, and Ibrahim S. The Cold Contact Method as a Simple Drug Interaction Detection System. Research Letters in Physical Chemistry, Volume 2008 (2008). http://dx.doi.org/10.1155/2008/169247.

NMR and Molecular Modelling Studies on the Interaction of Fluconazole with β-Cyclodextrin

Santosh Kumar Upadhyay and Gyanendra Kumar

ABSTRACT

Background

Fluconazole (FLZ) is a synthetic, bistriazole antifungal agent, effective in treating superficial and systemic infections caused by Candida species. Major challenges in formulating this drug for clinical applications include solubility enhancement and improving stability in biological systems. Cyclodextrins (CDs) are chiral, truncated cone shaped macrocyles, and can easily encapsulate fluconazole inside their hydrophobic cavity. NMR spectroscopy has been recognized as an important tool for the interaction study of cyclodextrin and pharmaceutical compounds in solution state.

Results

Inclusion complex of fluconazole with β-cyclodextrins (β-CD) were investigated by applying NMR and molecular modelling methods. The 1:1 stoichiometry of FLZ:β-CD complex was determined by continuous variation (Job's plot) method and the overall association constant was determined by using Scott's method. The association constant was determined to be 68.7 M-1 which is consistent with efficient FLZ:β-CD complexation. The shielding of cavity protons of β-CD and deshielding of aromatic protons of FLZ in various ^{1}H-NMR experiments show complexation between β-CD and FLZ. Based on spectral data obtained from 2D ROESY, a reasonable geometry for the complex could be proposed implicating the insertion of the m-difluorophenyl ring of FLZ into the wide end of the torus cavity of β-CD. Molecular modelling studies were conducted to further interpret the NMR data. Indeed the best docked complex in terms of binding free energy supports the model proposed from NMR experiments and the m-difluorophenyl ring of FLZ is observed to enter into the torus cavity of β-CD from the wider end.

Conclusion

Various NMR spectroscopic studies of FLZ in the presence of β-CD in D_2O at room temperature confirmed the formation of a 1:1 (FLZ:β-CD) inclusion complex in which m-difluorophenyl ring acts as guest. The induced shift changes as well as splitting of most of the signals of FLZ in the presence of β-CD suggest some chiral differentiation of guest by β-CD.

Background

Fluconazole (FLZ), diflucan, chemically known as α-(2,4)-difluorophenyl)-α-(1H-1,2, 4-triazol-1-ylmethyl)-1H-1,2,4-triazole-1-ethanol (figure 1), used to treat variety of fungal pathogens that cause systemic mycoses. It is an orally active bistriazole antifungal agent, well absorbed and has been found to be safe and effective in treating superficial and systemic infections with Candida species and maintenance therapy for cryptococcal meningitis, particularly in patients with AIDS [1]. It is also effective in preventing invasive Candida infections in patients with acute leukaemia without increasing non-albicans infections [2]. FLZ is commonly used to prevent yeast infections in patients undergoing bone marrow transplantation [3].

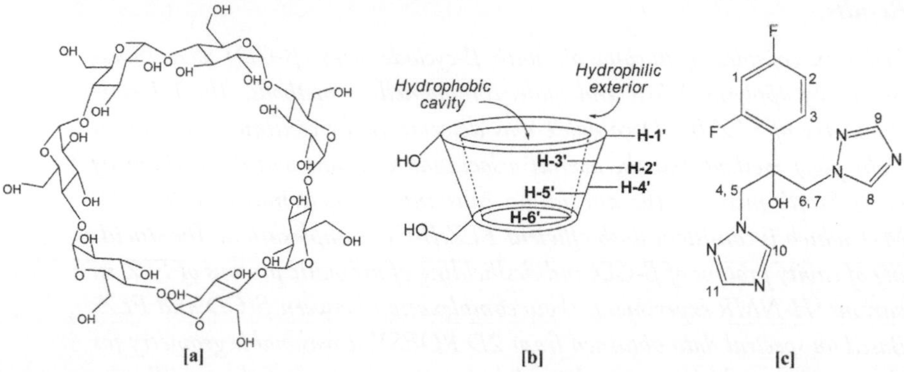

Figure 1. (a) Chemical structure of the host β-cyclodextrin (β-CD); (b) truncated cone shape of β-CD; and (c) guest fluconazole (FLZ).

The principle objective of the work embodied in this article was to study the complexation of fluconazole with β-cyclodextrin in aqueous solution. Complexation of pharmaceutical compounds with cyclodextrins (CDs), to form a host-guest complex in solution as well as solid states, results in altered physicochemical properties of the guest, like solubility, stability, dispersibility, volatility, masking of undesirable properties and so on, which are desirable for their use as pharmaceuticals [4]. Moreover, these host-guest complexes are considered as new entities and are required to be characterized for their approval as drugs. CDs are chiral saccharides that exhibit chiral recognition i.e. they form diastereomeric complexes, usually of different stability, with enantiomeric species [5]. The separation of enantiomers of a racemic drug is of great importance to the pharmaceutical industry because in a racemic drug, only one enantiomer is usually desired. CDs are all-purpose molecular containers for organic, inorganic, organometallic, and metal organic compounds that may be neutral, cationic, anionic or even radical.

The CDs belongs to the family of cyclic oligosaccharides, and have been studied extensively as a host in supramolecular chemistry. The three major types of CDs are crystalline, homogeneous, nonhygroscopic, consisting of six (α-), seven (β-), and eight (γ-) D-glucose units, respectively, attached by α-(1→4) glycosidic linkages (figure 1) [4,5]. Each of the chiral glucose units is in the rigid 4C1-chair conformation, giving the macrocycle shape of a hollow truncated cone with all the secondary hydroxyl groups located on the wider rim, while all the primary hydroxyl groups on the narrower rim (figure 1). The primary and secondary hydroxyls on the outside of the CDs make it water-soluble. The non-bonding electron pairs of the glycosidic oxygen bridges are directed toward the inside of the cavity, producing a high electron density and lending it a Lewis base character.

These features suggest that the CD cavity is relatively hydrophobic compared to the exterior faces which are hydrophilic.

Structure determination is of particular importance for supramolecular host-guest complexes, which are the basis of most cyclodextrin applications in medicine, catalysis, separation and sensor technology and also food chemistry. NMR spectroscopy has been recognized as an important tool for the structural elucidation of organic compounds, particularly in solution state in view of its application in drug discovery. This technique also gives information on the topology of the interaction between the guest and β-CD; furnishing information not only on the structure of inclusion complexes but also deriving the stoichiometry and association or binding constant of guest:β-CD complexes [6]. 2D COSY and ROESY experiments are important in cyclodextrin related studies, as they complement each other, COSY provides information on coupling of protons while 2D ROESY gives same information through space, i.e., the two nuclei are at 3–4 Å from each other (intermolecular distance) [5,7].

Here we present a study highlighting the interaction between antifungal drug fluconazole and β-cyclodextrin, using NMR as a spectroscopic tool. In order to better understand the structure of the inclusion complex between the two chemical species, we have complemented these studies with molecular modelling simulations.

Experimental

Materials

Fluconazole (FLZ) was obtained from Dr. Reddy's Ltd., India while β-cyclodextrin was a generous gift from DKSH India Pvt. Ltd. and these were used as received. All other reagents were of analytical reagent grade.

NMR Studies

Samples were prepared in 99.96% D_2O for NMR analysis. All the ^1H NMR spectra were recorded with a Bruker Avance 400 MHz spectrometer operating at 300 K and were acquired with a spectral width of 5995.204 Hz, 128 scans and 65536 data-points. Both 2D COSY and ROESY experiments were acquired on a Bruker DRX 500 MHz using 5 mm BBI ^1H-BB probe or a Varian Inova 500 MHz, equipped with a triple resonance, Z pulsed field gradient probe. 2D COSY spectra displaying ^1H-^1H cross correlation for free FLZ, free β-CD and FLZ:β-CD mixture were acquired using 2048 data-points with 128 increments and 18 scans for each increment. 2D ROESY spectrum of FLZ:β-CD was acquired using 2048 data-points, 256 increments, 8 scans for each increment and 500 ms mixing time. ^{1}H NMR spectra of five samples containing mixtures of β-CD and FLZ with FLZ/β-CD molar ratios ranging from 0.2 to 1.2 was recorded. As there was no separate peak for free as well as complexed form of FLZ, we presume that it undergoes rapid exchange between free and bound state on the NMR time scale. The resonance at 4.7 ppm due to residual solvents (H_2O and HDO) was used as internal reference. Chemical shift (δ) reported in ppm and chemical shift changes (Δδ) was calculated by using the formula: $\Delta\delta = \delta_{(complex)} - \delta_{(free)}$.

Molecular Modelling Studies

Molecular modelling studies were done using the software package MOE 2005.06 (Molecular Operating Environment, Version 2005.06). 3D coordinates for FLZ were generated using builder module of MOE and the molecule was energy-minimized using MMFF94x force field. The 3d coordinates of β-CD were taken from the Protein Data Bank file with PDB id: 1DMB. Since the positions of hydrogen atoms are not included in the PDB files, these were added using MOE, and energy-minimized with MMFF94x force field. The dock application built in MOE was used to dock FLZ with β-CD. This application is divided into three stages: 1. Conformational analysis during which ligands are treated in a flexible manner by rotating rotatable bonds; ring conformations are not searched. 2. Placement during which a collection of orientation is generated from the pool of ligand conformations. In this case, the alpha triangle placement method was used, it generates orientations by superposition of ligand atom triplets and triplets points in the receptor site. The receptor site points are alpha sphere centers which represent locations of tight packing. At each iteration a random conformation is selected, a random triplet of ligand atoms and a random triplet of alpha sphere centers are used to determine the orientation. 3. Scoring, during which each orientation generated by the placement methodology is subjected to scoring in an effort to identify the most favorable orientations. The dock application provides a framework for the integration of multiple scoring methodologies; each such scoring methodology will have different properties. Typically, scoring functions emphasize favorable hydrophobic, ionic and hydrogen bond contacts. The default method uses "affinity dG" scoring function to assess candidate orientations. This function estimates the enthalpic contribution to the free energy of binding using a linear function:

$$G = C_{hb} f_{hb} + C_{ion} f_{ion} C_{mlig} f_{mlig} + C_{hh} f_{hh} + C_{hp} f_{hp} + C_{aa} f_{aa}$$

where the f terms fractionally count atomic contacts of specific types and the C's are coefficients that weight the term contributions to affinity estimate. The individual terms are: hb, interactions between hydrogen bond donor-acceptor pairs; ion, ionic interactions; mlig, metal ligation; hh, hydrophobic interactions; hp, interactions between hydrophobic and polar atoms; aa, an interaction between any two atoms. The docked complex was finally energy-minimized keeping both the ligand and the receptor flexible.

Results and Discussion

Determination of the FLZ:β-CD Inclusion Complex Stoichiometry by Continuous Variation Method (Job's Plot)

The stoichiometry of host-guest complex was determined by the continuous variation (Job's plot) method [8]. It involves preparing a series of solutions containing both the host and the guest in varying proportions so that a complete range of mole ratios is sampled (0 > [H]/[H] + [G] < 1) and where the total concentration [H] + [G] is kept constant for each solution. The experimentally observed parameter is a host or guest chemical shift that is sensitive to complex formation. The plot of Δδ × [β-CD] against the mole fraction of β-CD, (r = m/[m/n]), where m and n represent the stoichiometric ratios of β-CD and FLZ, respectively, for the H-6' protons of β-CD is presented in figure 2. As presented in figure 2, the plot shows a maximum value at r = 0.5 and a highly symmetrical shape, which demonstrates the existence of a FLZ:β-CD complex with a 1:1 stoichiometry.

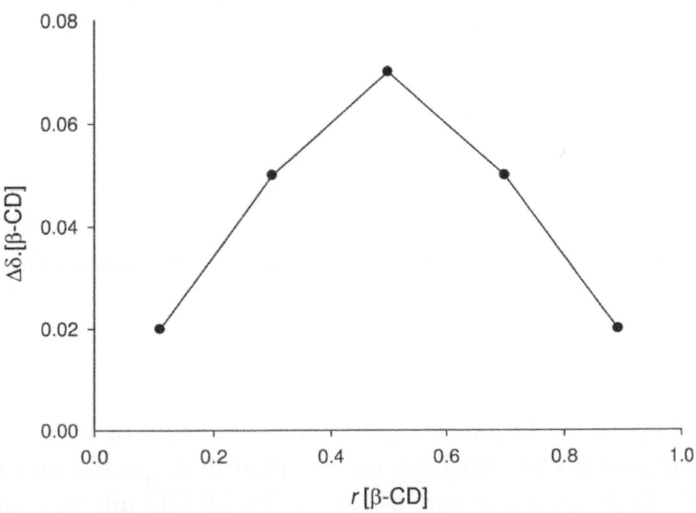

Figure 2. Job's plot (continuous variation method) of FLZ:β-CD inclusion complex showing 1:1 stoichiometry. $\Delta\delta = \delta_{(complex)} - \delta_{(free)}$.

Determination of Association Constant (K_a)

The association constant (K_a) of the FLZ:β-CD complex were determined by using well known Scott's method [9] which is a modification of Benesi-Hildebrand equation [10]. Equation (1) refers the Scott's equation:

$$[\text{FLZ}]/\Delta\delta_{obs} = [\text{FLZ}]/\Delta\delta_{\max} + \Delta\delta_{\max}/K_a \quad (1)$$

where [FLZ] is the molar concentration of the guest, $\Delta\delta_{obs}$ the observed chemical shift change for a given [FLZ] concentration, $\Delta\delta_{max}$ the chemical shift change between a pure sample of complex and the free component at saturation.

In this procedure, the plot of chemical shift changes (Δδ) for the β-CD protons against [FLZ] in the form of [FLZ]/Δδobs versus [FLZ] (referred to as y-reciprocal plot) should be linear for 1:1 inclusion complex. The slope of the plot thus equals to 1/Δδmax and the intercept with the vertical axis to Δδmax/Ka, allowing the estimation of association constant (Ka). A typical Scott's plot for FLZ:β-CD inclusion complex is shown in figure 3 and the association constant (Ka) was calculated to be 68.7 M-1.

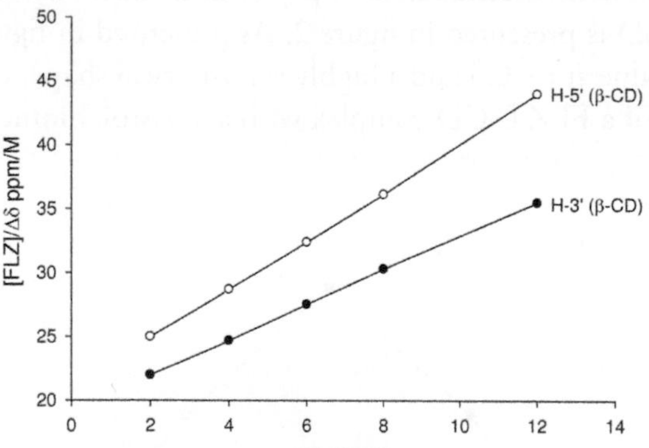

Figure 3. A typical Scott's plot for FLZ:β-CD inclusion complex showing overall association constant (K_a) = 68.7 M-1.

NMR Studies

The assignment of the β-CD protons was made on the basis of their splitting pattern, ^{1}H NMR and 2D ^{1}H-^{1}H COSY and ^{1}H-^{1}H ROESY spectral data. Comparison the 1H NMR spectra of pure β-CD and FLZ:β-CD mixture, suggests that the chemical shift of β-CD proton is changes upon bonding to FLZ. β-CD cavity protons H-3' and H-5' show maximum shielding compared to H-1', H-2' and H-4' in presence of varying amounts of FLZ. However, H-6' proton of β-CD shows a deshielding of a similar magnitude, confirming that guest FLZ only interacts with the cavity of truncated cone of β-CD and forms an inclusion complex. The shielding β-CD cavity proton arises due to magnetic anisotropic effects in the β-CD cavity, as a result of the inclusion of a π-electron rich group. Such a group in FLZ molecule being a phenyl or triazole ring suggests its inclusion in the β-CD cavity. Furthermore, magnitude of the change in the chemical shift of these β-CD protons increased with increase in concentration of FLZ. H-3' and H-5' protons of β-CD show a greater shift than H-6' (positioned at narrow mouth), indicating the insertion of FLZ molecule most plausibly from wider rim side of β-CD cavity. The chemical shift (δ) data of β-CD protons in the presence as well as in the absence of varying amounts of FLZ is shown in table 1. Expanded region of part of 1H NMR spectra displaying β-CD protons in the presence, as well as absence, of FLZ is shown in figure 4.

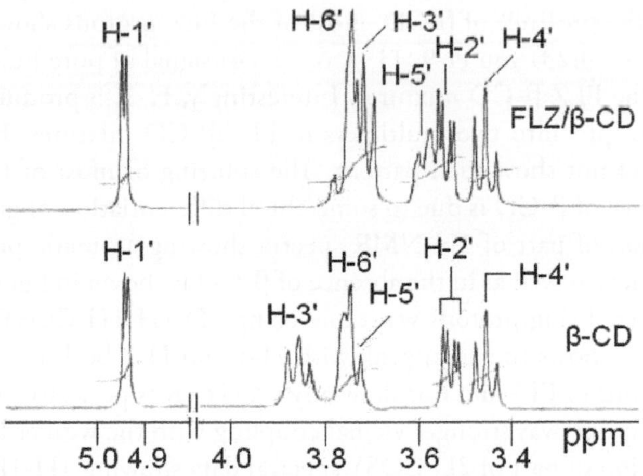

Figure 4. Expansion of a part of 1H NMR data (500 MHz), showing β-CD protons in the presence as well as in the absence of FLZ.

Table 1. 1H NMR (500 MHz) chemical shift (δ) data of pure β-CD and β-CD:FLZ mixtures

Sample	H-1'	H-2'	H-3'	H-4'	H-5'	H-6'
Pure β-CD	4.948	3.556	3.882	3.486	3.754	3.782
β-CD/FLZ = 0.2	4.947	3.554	3.791	3.482	3.676	3.784
β-CD/FLZ = 0.4	4.948	3.554	3.718	3.479	3.615	3.787
β-CD/FLZ = 0.6	4.946	3.555	3.659	3.485	3.566	3.789
β-CD/FLZ = 0.8	4.948	3.553	3.609	3.480	3.530	3.787
β-CD/FLZ = 1.2	4.947	3.551	3.529	3.485	3.471	3.784

Unambiguous resonance assignments of FLZ protons were made on the basis of 1H NMR data, 2D COSY and 2D ROESY spectra. Some of the peaks that were completely not distinct in the spectrum of the pure FLZ separated well in the presence of β-CD aiding the assignment. FLZ has one difluoro substituted phenyl ring and two triazole rings and its 1H NMR chemical shift appeared separately. Here, we are giving the 1H NMR assignment of pure FLZ. The two singlets at δ = 8.23 and 7.74 were assigned as H-8, 10 and H-9, 11 resonance, consistent with the two protons of the two equivalent triazole rings of FLZ. Two highly coupled resonance patterns (probably multiplet) appeared near δ = 6.9 (H-1, 3) and 6.7 (H-2), integrating with the ratio of 2:1 protons of three consistent aromatic protons of the m-difluorophenyl ring of FLZ. An AB pattern appeared at δ = 4.9 (H-4, 5, 6, 7), totally integrating for the four protons consistent for the two sets of β-methylene protons of FLZ.

The 1H NMR spectrum of FLZ:β-CD mixture is different compared to that of pure FLZ. In the presence of β-CD, most of the FLZ protons showed splitting. H-8, 10 (s, δ = 8.23) and H-9, 11 (s, δ = 7.74) signal of pure FLZ display a doublet in all the FLZ:β-CD mixtures. Interestingly, H-1, 3 proton (m, δ = 6.9) of pure FLZ split into two multiplets in FLZ:β-CD mixtures. However, the H-2 signal did not show such pattern. The splitting of most of FLZ protons in the presence of β-CD is due to some chiral differentiation of guest FLZ protons. Expansion of part of 1H NMR spectra showing aromatic protons of FLZ in the presence, as well as in the absence of β-CD is shown in figure 5. The assignment of phenyl ring protons was done using 2D 1H-1H COSY spectral data in which H-2 shows the cross peak with H-1 and H-3 both but H-8, 10 and H-9, 11 protons of FLZ did not show any COSY cross peaks to any signal. The H-2/H-3 coupling was stronger vicinal coupling with the weaker H-1/H-2 coupling. Expansion of part of 2D COSY spectral data showing 1H-1H COSY cross peak of aromatic protons is illustrated in figure 6. The assignment of triazole ring was challenging through it was resolved using NOEs correlation in 2D 1H-1H ROESY spectrum. Expanded region of a 2D ROESY spectra showing 1H-1H NOEs is shown in figure 7. The H-8, 10 proton of triazole ring is very close to β-methylene proton (H-4, 5 and 6, 7) and therefore we expect see the NOEs between them. On inspecting the ROESY data a cross peak between H-8, 10 and H-4, 5 and 6, 7 was observed confirming that the former is close in space to the latter. No NOEs to triazole ring (H-9, 11) of FLZ was observed to any other protons.

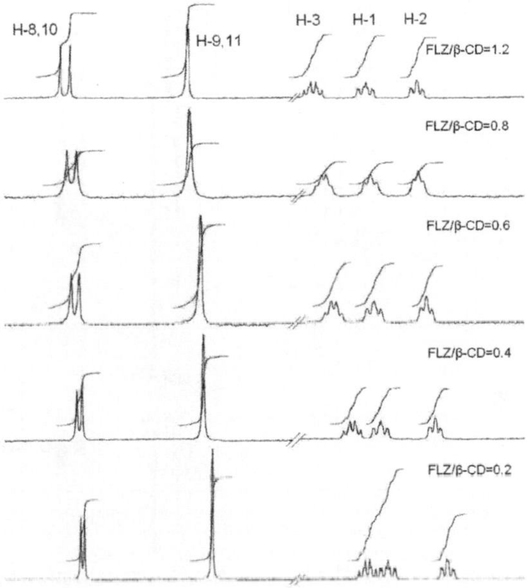

Figure 5. Expansion of a part of 1H NMR data (500 MHz), of FLZ protons showing deshielding in the presence of varying amount of β-CD.

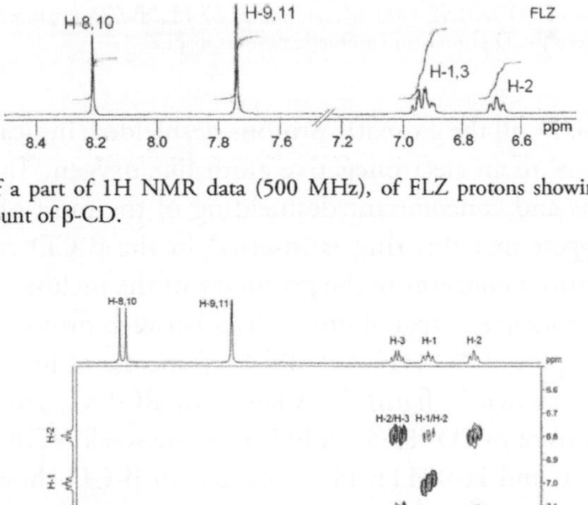

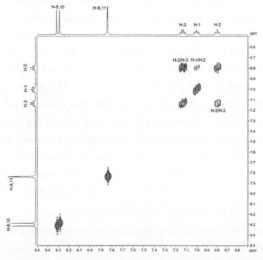

Figure 6. 2D COSY (500 MHz) spectra of FLZ:β-CD mixture (1:1) showing the cross peak between aromatic protons of FLZ.

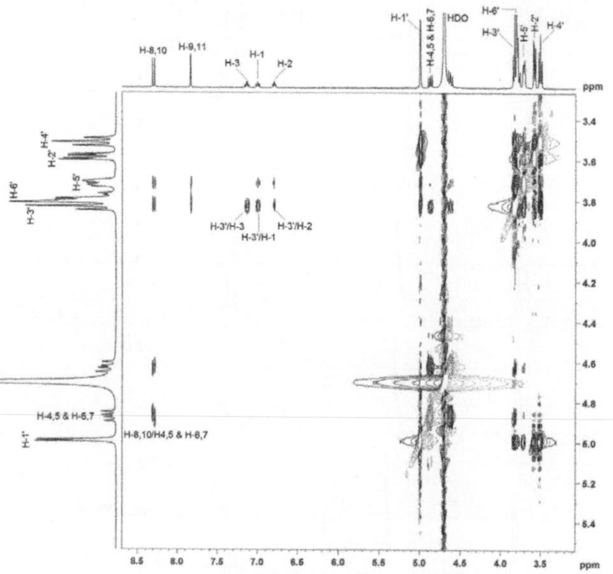

Figure 7. Expanded region of 2D ROESY (500 MHz) spectra of FLZ:β-CD mixture (1:1) showing the intermolecular NOE's between β-CD protons and aromatic protons of FLZ.

In presence of β-CD, all the aromatic protons deshielded, indicating the proximity of these protons' to an electronegative atom like oxygen. The shielding of β-CD cavity protons and concomitant deshielding of the most of the aromatic protons of FLZ suggest that this ring is inserted in the β-CD cavity [11-13]. More detailed indications concerning the geometry of the inclusion complex can be derived from the evidence of spatial proximities between protons of β-CD and FLZ. Partial contour plot of 2D (1H-1H) ROESY spectra of inclusion complex of β-CD and FLZ is shown in figure 7. Analysis of ROESY data revealed that both the cavity protons of β-CD (H-5' and H-3') share weak NOE's with triazole ring protons (H-8, 10 and H-9, 11). H-5' protons of β-CD show NOE's with H-1 (weak), H-2 (weak) and H-3 (very weak). On the other hand H-3' protons of β-CD show NOE's with H-1 (strong), H-3 (strong) and H-2 (weak-to-strong). Collectively, these data suggest that the m-difluorophenyl moiety penetrates deep into the β-CD cavity from the wider rim side. The inclusion of m-diflurophenyl ring from narrower rim side is not possible based on the weak NOE's with H-5'. The penetration of triazole ring was ruled out on the basis of weak NOE's cross peaks between triazole ring protons and β-CD protons.

We performed molecular modelling studies, to rationalize the NMR experiment results described above. The Dock module of MOE was used to dock flexible FLZ into a rigid β-CD (25 runs). From the resulting complex, the one with the best binding energy was then energy-minimized keeping both FLZ and β-CD flexible. The energy-minimized complex of FLZ and β-CD is shown in figure 8. It is interesting to note that starting in a random location and orientation, the docked minimum energy conformation of FLZ with β-CD is in very good agreement with the distances obtained from 2D ROESY spectra. The complex has the m-difluorophenyl ring of FLZ nestled in the truncated cone of β-CD entering from the secondary rim (the wider opening), and the triazole rings are oriented towards the primary not secondary rim. A Hydrogen bond is also observed between the OH of FLZ and an OH of β-CD at position 3 that would contribute to the stability of the host-guest complex.

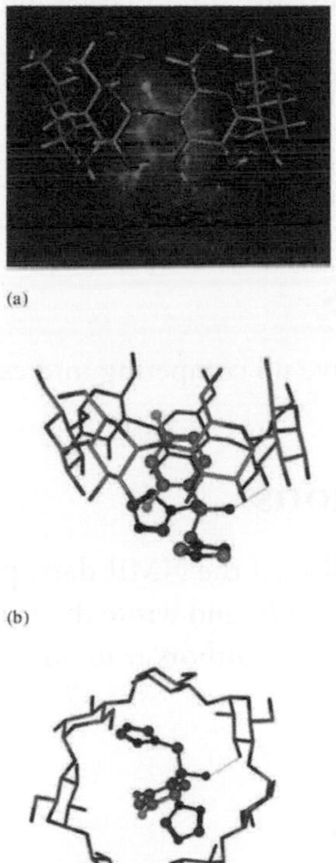

Figure 8. FLZ docked with β-CD, FLZ is shown in Ball and Stick, β-CD in sticks (a) an all atom model with Connolly surface representation of FLZ, (b) FLZ and β-CD (side-view) and (c) FLZ and β-CD (top-view). Hydrogens are not shown in (b) and (c) for sake of clarity. All other atoms are shown in their elemental colour.

Conclusion

In summary, our NMR experiments confirm that FLZ forms 1:1 complex with β-CD in aqueous medium. Employing experimental and theoretical methods, the present work unambiguously determined the geometrical inclusion parameters of FLZ with β-CD. The ROESY experiments showed that the complex is formed with the m-difluorophenyl ring of FLZ inside the β-CD torus cavity. These results were also supported by molecular modelling which also highlight a hydrogen bond formation between the host and the guest providing stability to the complex.

List of Abbreviations

FLZ: fluconazole; CD: cyclodextrin; β-CD: β-cyclodextrin; 1H NMR: proton Nucleic Magnetic Resonance spectroscopy; 2D: two dimensional; COSY: COrrelation SpectroscopY; ROESY: Rotating frame Overhauser Effect SpectroscopY; NOE: Nuclear Overhauser Effect; [H]: host; [G]: guest; MOE: Molecular Operating Environment.

Competing Interests

The authors declare that they have no competing interests.

Authors' Contributions

SKU conceived of the study, collected the NMR data, processed and analysed all NMR spectra, discussion of the results and wrote the manuscript. GK carried out the molecular docking studies. Both authors read and approved the final manuscript.

Acknowledgements

SKU and GK are recipients of Research Associateship from the centre of excellence, Department of Biotechnology, Government of India, awarded Prof. A. Surolia. SKU is grateful to Prof. A. Surolia and Dr. Monica Sundd for teaching and training me in NMR.

References

1. Meunier F., Aoun M., Gerard M.: Therapy for oropharyngeal candidiasis in the immunocompromised host: a randomized double-blind study of fluconazole vs. ketoconazole. Rev. Infect. Dis. 1990, 12:S364–S368.

2. Nihtinen A., Anttila V. J., Elonen E., Juvonen E., Volin L., Ruutu T.: Effect of fluconazole prophylaxis on the incidence of invasive candida infections and bacteraemias in patients with acute leukaemia. Eur. J. Haematol. 2008, 80:391–396.
3. MacMillan M. L., Goodman J. L., DeFor T. E., Weisdorf D. J.: Fluconazole to prevent yeast infections in bone marrow transplantation patients: a randomized trial of high versus reduced dose, and determination of the value of maintenance therapy. Am. J. Med. 2002, 112:369–379.
4. Szejtli J.: Introduction and general overview of cyclodextrin chemistry. Chem. Rev. 1998, 98:1743–1754.
5. Dodziuk H.: Cyclodextrins and Their Complexes. Chemistry, Analytical Methods, Applications. Wiley-VCH: London; 2006.
6. Fielding L.: Determination of association constant (Ka) from solution NMR data. Tetrahedron 2000, 56:6151–6170.
7. Neuhaus D. L., Williamson M. P.: The Nuclear Overhauser Effect in Structural and Conformational. Analysis. 2nd edition. Wiley-VCH, New York; 2000.
8. Job P.: Formation and stability of inorganic complexes in solution. Ann. Chim. 1925, 9:113–125.
9. Scott R. L.: Some comments on the Benesi-Hildebrand equation. Recl. Trav. Chim. Pays-Bas 1956, 75:787–789.
10. Benesi H. A., Hildebrand J. H.: A spectrophotometric investigation of the interaction of iodine with aromatic hydrocarbons. J. Am. Chem. Soc. 1949, 71:2703–2707.
11. Schneider H. J., Hacket F., Rudiger V., Ikeda H.: NMR studies of cyclodextrins and cyclodextrin complexes. Chem. Rev. 1998, 98:1755–1785.
12. Ali S. M., Upadhyay S. K.: Complexation study of midazolam hydrochloride with β-cyclodextrin: NMR spectroscopic study in solution. Magn. Reson. Chem. 2008, 46:676–679.
13. Upadhyay S. K., Ali S. M.: Solution structure of loperamide and β-cyclodextrin inclusion complexes using NMR spectroscopy. J. Chem. Sci. 2009, 121:521–527.

CITATION

Upadhyay SK and Kumar G. NMR and Molecular Modelling Studies on the Interaction of Fluconazole with β-Cyclodextrin. Chemical Central Journal, 2009 Aug 10;3:9. doi: 10.1186/1752-153X-3-9.

Spectrophotometric Determination of Etodolac in Pure Form and Pharmaceutical Formulations

Ayman A. Gouda and Wafaa S. Hassan

ABSTRACT

Background

Etodolac (ETD) is a non-steroidal anti-inflamatory antirheumatic drug. A survey of the literature reveals that there is no method available for the determination of ETD in pure form and pharmaceutical formulations by oxidation-reduction reactions.

Results

We describe three simple, sensitive and reproducible spectrophotometric assays (A-C) for the determination of etodolac in pure form and in pharmaceutical formulations. Methods A and B are based on the oxidation of *etodolac by Fe^{3+} in the presence of o-phenanthroline (o-phen) or bipyridyl (bipy). The formation of the tris-complex on reaction with Fe^{3+}-o-phen and/or Fe^{3+}-bipy mixtures in acetate buffer solution at optimum pH was demonstrated at 510 and 520 nm with o-phen and bipy. Method C is based on the oxidation of etodolac by Fe^{3+} in acidic medium, and the subsequent interaction of iron(II) with ferricyanide to form Prussian blue, with the product exhibiting an absorption maximum at 726 nm. The concentration ranges are 0.5–8, 1.0–10 and 2–18 $\mu g\ mL^{-1}$ respectively for methods A, B and C. For more accurate analysis, Ringbom optimum concentration ranges were calculated, in addition to molar absorptivity, Sandell sensitivity, detection and quantification limits.*

Conclusion

Our methods were successfully applied to the determination of etodolac in bulk and pharmaceutical formulations without any interference from common excipients. The relative standard deviations were ≤ 0.76 %, with recoveries of 99.87 % – 100.21 %.

Background

Etodolac (ETD), 1,8-diethyl-1,3,4,9-tetrahydropyrano- [3,4-b]indole-1-acetic acid [1], is a non-steroidal anti-inflamatory antirheumatic drug (Scheme 1). A survey of the literature reveals that there are very few reported methods for the determination of ETD in biological fluids, pharmaceutical formulations and in presence of its enantiomer. Of those studies reported, the techniques used include chromatography, HPLC [2-5], GC [6-8], in addition to spectrofluorimetric [9] and spectrophotometric methods [9-11]. However, an extensive survey of the literature revealed that there is no method available for the simultaneous determination of ETD in pure form and pharmaceutical formulations by oxidation-reduction reactions.

Scheme 1: The chemical structure of Etodolac (ETD).

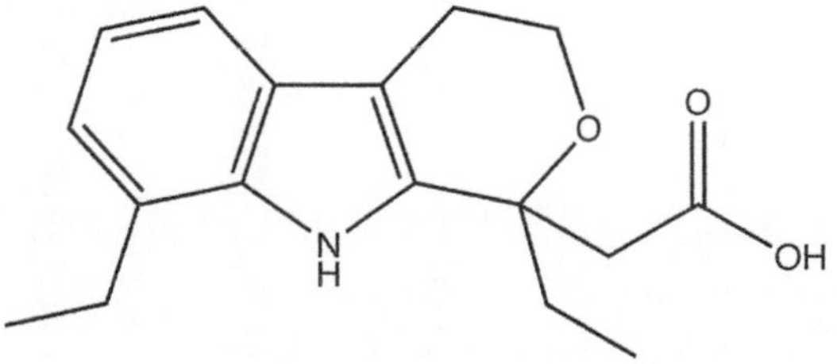

The aim of this study was to apply redox reactions in developing simple, accurate, sensitive and reproducible assays to analyse ETD in pure form and pharmaceutical formulations, by employing iron(III) with o-phenanthroline (o-phen), bipyridyl (bipy) and ferricyanide. This study describes spectrophotometric methods that can be used in laboratories where modern and expensive equipment, such as that required for GC or HPLC, is not available.

Results and Discussion

Methods A and B

Absorption Spectra

1, 10-Phenantholine and 2,2'-bipyridyl are organic bases whose chemical structures are similar, and contain the iron(II) specific group (18). Methods A and B are based on the formation of tris(o-phenanthroline)-iron(II) or tris(2,2'-bipyridyl)-iron(II) following the reaction of ETD with Fe^{3+}-o-phen or Fe^{3+}-bipy respectively. The reaction proceeds through the reduction of Fe^{3+} to Fe^{2+}, with the subsequent formation of an intense orange-red colouration attributable to the complex (12–20). The absorption spectra of the coloured complexes under optimum conditions were scanned in double beam mode against a reagent blank over the range 400–900 nm, and recorded according to general procedures. Characteristic λ_{max} values were obtained at 510 and 521 nm for methods A and B, respectively (Figure 1). The experimental conditions were established by varying each parameter individually (21) and observing the effect on the absorbance of the coloured species. All the spectral characteristics, as well as measured or calculated factors and parameters, are summarised in Table 1.

Table 1. Quantitative parameters for methods A-C.

Parameter	A	B	C
λ_{max}, nm	510	521	726
Beer's conc. Range (μg mL^{-1})	0.5–8.0	1.0–10	2.0–18
Ringbom conc. Range (μg mL^{-1})	0.85–7.5	2.0–8.5	3.0–16.5
Detection limits (μg mL^{-1})	0.065	0.104	0.228
Quantification limit (μg mL^{-1})	0.217	0.347	0.76
Molar absorpitivity × 10^{4}(L mol^{-1} cm^{-1})	1.812	1.876	1.039
Sandell sensitivity (ng cm^{-2})	15.86	15.32	27.66
Regression equation [a]			
$S_{y/x}$	0.202	0.285	0.246
Intercept	0.0078	- 0.0018	0.0011
S_a	0.142	0.201	0.2007
$\pm tS_a$	0.365	0.517	0.516
Slope	0.0582	0.066	0.036
S_b	0.0285	0.0365	0.018
$\pm tS_b$	0.0733	0.094	0.0462
Correlation coefficient (*r*)	0.9999	0.9999	0.9997
Mean ± SD%	99.87 ± 0.659	100.21 ± 0.727	99.875 ± 0.759
RSD%	0.66	0.73	0.76
Variance	0.434	0.528	0.576
SE	0.269	0.297	0.31
Student, s *t*-value[b] (2.571)	1.16	0.323	1.07
Variance ratio *F*-test [b] (6.256)	1.06	1.29	1.406

[a] A = a + b c, where c is the concentration in μg mL^{-1}.
[b] The theoretical t- and F-values, 2.571 and 6.256, respectively for five degree of freedom at 95% confidence level.

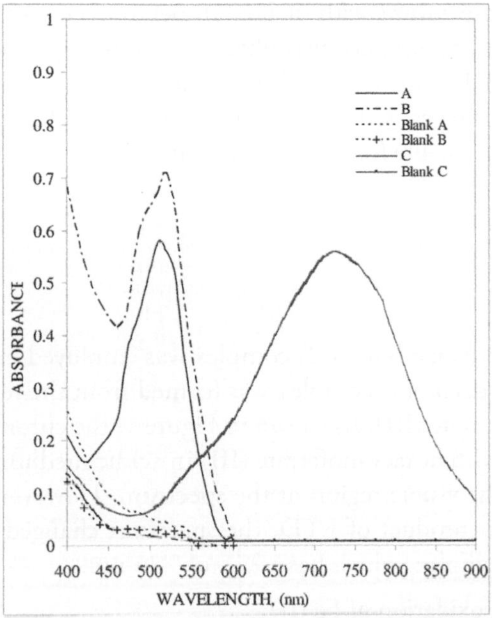

Figure 1. Absorption spectra of: (A) Fe(III)-1,10-phenanthroline with ETD (6.0 μg mL^{-1}); (B) Fe(III)-2,2'-bipyridyl with ETD (6.0 μg mL^{-1}) and (C) Fe(III)- ferricyanide with ETD (6.0 μg mL^{-1}) versus reagent blanks for each method.

Effect of pH

Of the buffers investigated, acetate solution proved to be the optimal (universal, phosphate, thiel, borate and acetate). A pH adjustment was necessary especially given the acidic medium, because the reaction was affected by change in pH over the range 2.5–5.6. The optimum pH was 4.5 for both methods A and B; moreover, 4.0 mL of buffer solution was sufficient for complete colour development.

Effect of Reagent Concentration

The addition of 1.0 mL of Fe^{3+}-o-phen (method A) or 1.5 mL of Fe^{3+}-bipy (method B) solutions was sufficient to obtain the maximum and reproducible absorbance for 6.0 μg mL^{-1} of ETD. Smaller amounts resulted in incomplete complex formation. Increased concentration had no effect on complex formation, although absorbance increased slightly owing to the reagent background used.

Effects of Temperature and Heating Time

The effects of temperature and heating time on the formation of the coloured complex were also studied. The reaction of ETD with both reagents proceeded very slowly at room temperature, with higher temperature employed to accelerate the reaction. Maximum absorbance was obtained following heating on a water bath at 80°C for about 10 min with both Fe^{2+}-phen and Fe^{2+}-bipyridyl coloured complexes. Further heating caused no appreciable change in the colour. The complex obtained was highly stable for more than 12 h.

Method C

Absorption Spectra

The formation of the Prussian blue (PB) complex was employed in the qualitative detection of Fe (II). A deep blue complex was formed from the reaction between Fe (II) and hexacyanoferrate (III). As shown in Figure 1, the chromogenic reagent blank (Fe (III) mixed with hexacyanoferrate (III) in acidic medium) did not show strong absorption in the visible region of the spectrum. However, after addition of the acidic hydrolysis product of ETD, the spectrum changed because of the formation of the PB complex, which has a λ_{max} of 726 nm.

The first step is the oxidation of Fe (II):

$$Fe^{2+} + [Fe(CN)_6]^{3-} \rightarrow Fe^{3+} + [Fe(CN)_6]^{4-}.$$

The second step is the formation of hexacyanoferrate (II) complex (PB):

$$4Fe^{3+} + 3\,[Fe(CN)_6]^{4-} + \text{acidic hydrolysis product of (ETD)}$$
$$\rightarrow Fe4\,[Fe(CN)6]3$$

The complex formed is highly insoluble (Ksp = 3 × 10-41) (22). Neither iron (III) nor ferricyanide solution absorb at 726 nm. Hence, the use of measured volumes of reagent, and measurement against the corresponding reagent blank gave a linear calibration for the drug. We herefore report the formation and application of the Prussian blue complex in the development of a sensitive spectrophotometric method for the determination of ETD.

The optimum conditions were established by varying parameters, such as iron (III), ferricyanide and acid concentrations, reaction time and the order in which the reagents were added:

Optimum Iron (III) and Ferricyanide Concentrations

Having studied the effect of iron (III) chloride concentration on colour development, it was observed that the absorbance increased as the volume of 0.2% iron (III) solution increased, reaching a maximum upon the addition of 2.0 mL of the 0.2% iron (III) solution to 6.0 μg mL^{-1} of ETD and 2.0 mL of 0.2%

ferricyanide solution in a total volume of 10 mL. These results indicate that maximum absorbance was obtained when the final iron (III) chloride concentration was 0.04%. Larger volumes of iron (III) chloride (up to 4.0 mL) had no effect on the sensitivity of the reaction. Similar observations were made when varying volumes of 0.2% ferricyanide solution were added to fixed amounts of ETD (6.0 μg mL^{-1}) and iron (III) chloride (2 mL; 0.2%), and diluted to 10 mL after full colour development. The results of this study reveal that the concentrations of iron (III) and ferricyanide reagents are not critical. However, 2.0 mL of 0.2% reagent solutions in a total volume of 10 mL were used to ensure adequate reagent concentrations for higher drug concentrations.

Effect of Nature of Acid, and Its Concentration

The reaction product, Prussian blue, was found to flocculate within 20–30 min of colour development. To delay the flocculation, acid was added after full colour development and before dilution. Sulphuric acid was found to give more stable colour and reproducible results compared to hydrochloric acid. A 1.0 mL volume of 10 M sulphuric acid in a total volume of 10 mL was found to be adequate.

Effect of Reaction Time and Stability of Coloured Species

The reaction is slow at 30 ± °2°C, but absorbance increased with time and reached a maximum in 10 min, with colouration remaining stable for at least 6 h.

Effect of Order in Which Reagents Were Added

After fixing all other parameters, a few other experiments were performed to ascertain the influence of the order in which reagents were added. The following order: drug, ferricyanide and iron (III), followed by sulphuric acid after full colour development, gave maximum absorbance and stability, and for this reason the same order of addition was followed throughout the investigation.

Method Validation

Linearity

Under the optimum conditions described, Beer's law holds over the concentration ranges 0.5–8, 1–10 and 2–18 μg mL^{-1} respectively for methods A, B and C. The optimum concentration ranges of ETD that can be measured accurately, as evaluated from the Ringbom plot, are 0.85–7.5, 2.0–9.0 and 3.0–16.5 μg mL^{-1} using methods A, B and C, respectively. We calculated the apparent molar absorptivity,

Sandell sensitivity, relative standard deviations for six replicate determinations of 6.0 μg mL^{-1} and the regression equations (Table 1).

Sensitivity

The detection limit (LOD) for the proposed methods was calculated using the following equation according to IUPAC definition (23):

$$LOD = 3s/k;$$

where s is the standard deviation of replicate determination values under the same conditions as the sample analysis in the absence of the analyte, and k is the sensitivity, namely the slope of the calibration graph. In accordance with the formula, the detection limits obtained for the absorbance were found to be 0.065, 0.104 and 0.228 μg mL^{-1} for the ETD-Fe^{2+}-phen, Fe^{2+}-bipyridyl and Fe^{2+}-ferricyanide methods, respectively.

The limits of quantitation, LOQ, is defined as;

$$LOQ = 10\ s/k.$$

According to this equation, the LOQs were found to be 0.217, 0.347 and 0.76 μg mL-1 for the Fe2+-phen, Fe2+-bipyridyl and Fe2+-ferricyanide methods, respectively.

Accuracy and Precision

In order to determine the accuracy and precision of the proposed methods, solutions containing four different concentrations of ETD were prepared and analysed in six replicates. The relative standard deviation as precision and percentage relative error (Er %) as accuracy of the suggested methods were calculated at 95% confidence levels, and can be considered satisfactory. Precision was carried out by six determinations at four different concentrations in these spectrophotometric methods. The percentage relative error was calculated according the following equation:

$$Er\ \% = [(found - added)/added] \times 100.$$

The inter- and intra-day precision and accuracy results are shown in (Table 2). The analytical results for accuracy and precision show that the methods proposed have good repeatability and reproducibility.

Table 2. The intra-day and inter-day precision and accuracy data for ETD obtained by the proposed methods (A – C).

		Intra-day			Inter-day		
Method	Added (μg mL⁻¹)	Found ± SE [a, b] (μg mL⁻¹)	Precision RSD %	Accuracy R.M.E %	Found[a, b] (μg mL⁻¹)	Precision RSD %	Accuracy R.M.E %
A	1	0.99 ± 0.45	1.11	-1.00	1.01 ± 0.20	1.49	1.00
	3	3.02 ± 0.49	1.20	0.67	2.97 ± 0.37	0.91	-1.00
	5	4.99 ± 0.26	0.64	-0.20	4.94 ± 0.31	0.77	-1.20
	7	6.97 ± 0.24	0.59	-0.43	6.94 ± 0.34	0.85	-0.86
B	2	2.02 ± 0.55	1.34	1.00	2.01 ± 0.63	1.54	0.50
	4	3.97 ± 0.31	0.76	-0.75	3.99 ± 0.46	1.13	-0.25
	6	5.93 ± 0.29	0.71	-1.17	5.96 ± 0.43	1.06	-0.67
	8	8.05 ± 0.36	0.88	0.625	7.97 ± 0.38	0.93	-0.38
C	4	3.98 ± 0.18	0.45	-0.50	4.01 ± 0.46	1.12	-0.50
	8	7.95 ± 0.20	0.50	-0.625	8.03 ± 0.40	0.98	0.38
	12	11.94 ± 0.33	0.82	-0.50	12.04 ± 0.48	1.16	0.33
	16	16.02 ± 0.49	1.19	0.125	15.91 ± 0.54	1.32	-0.56

[a] Average of six determinations.
[b] Mean ± standard error.
RSD%, percentage relative standard deviation;
R.M.E %, percentage relative mean error.

Effects of Interference

The interference criterion was an error of not more than ± 3.0 % in the absorbance. To test the efficiency and selectivity of the proposed analytical methods (A-C) to pharmaceutical formulations, we carried out a systematic study of additives and excipients (e.g. lactose, glucose, dextrose, talc, calcium hydrogen phosphate, magnesium stearate and starch) that usually present in dosage forms. Experiments showed that there was no interference from additives or excipients for methods A-C (Table 3).

Table 3. Determination of ETD in presence of additives or excipients.

Material	Amount (mg)	Method A	Method B	Method C
		Recovery [a] %± SD [b]	Recovery [a] %± SD [b]	Recovery [a] %± SD [b]
Lactose	50	99.63 ± 0.82	99.20 ± 0.78	100.2 ± 0.51
Glucose	50	98.84 ± 0.67	100.35 ± 1.22	99.55 ± 0.88
Dextrose	50	99.30 ± 0.78	98.70 ± 0.56	98.65 ± 0.46
Magnesium stearate	30	99.25 ± 0.75	99.55 ± 0.89	100.15 ± 1.30
Calcium hydrogen phosphate	50	99.50 ± 0.96	98.95 ± 0.73	99.40 ± 0.84
Talc	40	99.80 ± 0.61	100.40 ± 1.05	100.10 ± 0.95
Starch	50	100.05 ± 1.14	98.80 ± 0.69	99.75 ± 0.62

[a] 6.0 μg mL⁻¹ of ETD was taken.
[b] Average of five determinations.
SD: Standard deviation.

The main degradation products of etodolac are identified as 7-ethyl-2(1-methylene propyl)-1H-indole-3-ethanol, 1,8 diethyl-1-methyl-1,3,4,9-tetrahydropyrano- [3,4-b]indole and 7-ethyl tryptophol [24]. Consequently, there was no interference of the degradation products in the determination of the ETD in proposed methods.

Analytical Applications

Because our three methods were successfully applied to the determination of ETD in its pharmaceutical formulations, they could therefore be used easily for the routine analysis of pure ETD and its dosage forms. Moreover, to check the methods' validity, dosage forms [Napilac capsules (200 mg ETD per capsule) and Etodine capsules (300 mg ETD per capsule)] were tested for possible interference with the standard addition method (Table 4). The methods' performance was assessed using the t-test (for accuracy) and a variance ratio F-value (for precision) compared with the reference method [9] (for 95% confidence level with five degrees of freedom (25)). The results showed that the t- and F-values were less than the critical value, indicating that there was no significant difference between the proposed and reference method for ETD (Table 5). Because the proposed methods were more reproducible and had higher recoveries than the reference method, they can be recommended for adoption in routine analysis in the majority of drug quality control laboratories.

Table 4. Determination of ETD in its pharmaceutical dosage form applying the standard addition technique.

Dosage forms	Taken (μg mL⁻¹)	Added (μg mL⁻¹)	Proposed methods Recovery [a] % ± RSD	
			Napilac capsules	Etodine capsules
A	2.0	-	99.98 ± 0.19	100.01 ± 0.16
		1.0	100.05 ± 0.20	99.99 ± 0.18
		3.0	99.93 ± 0.32	99.60 ± 0.26
		5.0	99.84 ± 0.47	100.35 ± 0.32
B	2.0	-	100.09 ± 0.21	99.95 ± 0.27
		2.0	99.74 ± 0.26	99.60 ± 0.18
		4.0	99.30 ± 0.38	100.10 ± 0.22
		6.0	99.55 ± 0.55	99.70 ± 0.56
C	4.0	-	100.25 ± 0.19	100.08 ± 0.23
		4.0	99.80 ± 0.34	100.40 ± 0.51
		8.0	100.15 ± 0.45	98.90 ± 0.49
		12	99.79 ± 0.62	100.05 ± 0.54

[a] Average of six determinations.

Table 5. Determination of ETD in pharmaceutical preparations using the proposed methods.

Sample		Recovery[a] ± SD (%)			
		Official method	Proposed methods		
			A	B	C
Napilac capsules (200 mg ETD/Capsule)	X ± SD	99.70 ± 1.16	98.52 ± 0.79	100.65 ± 1.19	100.27 ± 0.92
	t[b]		1.87	0.74	0.86
	F[b]		2.17	1.05	1.60
Etodine capsules (300 mg ETD/Capsule)	X ± SD	100.50 ± 1.36	99.95 ± 0.81	100.30 ± 0.96	100.04 ± 1.02
	t[b]		0.78	0.27	0.61
	F[b]		2.82	2.01	1.78

[a] Average of six determinations.
[b] The theoretical t- and F-values, 2.571 and 6.256, respectively for five degree of freedom at 95% confidence level.

Conclusion

The methods proposed are simpler, less time consuming and more sensitive than those previously published. All the proposed methods were more advantageous than other reported visible spectrophotometric [9-11] methods with respect to sensitivity, simplicity, reproducibility, precision, accuracy and stability of the coloured species for ≥ 12 h. The proposed methods are suitable for the determination of ETD in pure form and pharmaceutical formulations without interference from excipients such as starch and glucose, suggesting potential applications in bulk drug analysis.

Experimental

Apparatus

All absorption spectra were recorded using a Kontron 930 (UV-Visible) spectro photometer (German) with a scanning speed of 200 nm/min and a band width of 2.0 nm, equipped with 10 mm matched quartz cells. A Hanna pH meter (USA) was used for checking the pH of buffer solutions.

Materials and Reagents

All chemicals and materials were of analytical grade, and all solutions were freshly prepared in bidistilled water.

Pure Samples

Etodolac (ETD) pure grade was supplied by Pharco, Egypt. The purity was found to be 100.35 ± 0.64 % according to the Pharco method [26] in which the absorbance of 0.002% w/v etodolac solution in 0.1 N sodium hydroxide was measured at 276 nm.

Standard Stock Solutions

Stock solutions of ETD were prepared by dissolving 100 mg of pure drug in methanol, followed by dilution to 100 mL with the same solvent to obtain 1 mg mL^{-1}standard solutions. Working solutions were prepared by an appropriate dilution of the stock standard solution.

Market Samples

Napilac capsules (200 mg ETD/Capsule) were provided by Global Napi Co (Egypt) and Etodine capsules (300 mg ETD/Capsule) were provided by Pharco (Egypt).

Reagents

1. Iron (III)-o-phenanthroline (27) was prepared by mixing 0.198 g of 1,10 phenanthroline monohydrate (Fluka, Swiss), 2 mL of 1 M HCl and 0.16 g of ferric ammonium sulphate dodecahydrate (Aldrich, Germany) before dilution with bidistilled water to 100 mL in a calibrated flask.
2. Iron (III)-bipyridyl (27) was prepared by dissolving 0.16 g of 2, 2'-bipyridyl (Fluka, Switzerland) in 2 mL of 1 M HCl and 0.16 g of ferric ammonium sulphate dodecahydrate, before dilution with bidistilled water to 100 mL in a calibrated flask.
3. Anhydrous FeCl3 (Merck) and K3 [Fe(CN)6] (BDH Lab. Chemicals, Poole, England) 0.2% (w/v) were prepared in bidistilled water. Sulphuric acid (10 M) was prepared by adding 555 mL of concentrated acid, (Sp. Gr. 1.83) to 445 mL of bidistilled water with cooling (28).
4. The acetate buffer solutions, with pH ranges from 2.56 – 5.6, were prepared by mixing appropriate quantities of 0.2 M sodium acetate with 0.2 M acetic acid to obtain the desired pH as previously recommended (29).

Recommended Analytical Procedure

Methods A and B

100 μg mL^{-1} aliquots of the standard solutions (A: 0.05–0.8 and B: 0.1–1.0 mL) were transferred to a series of 10 mL calibrated flasks. To these were added 1.0 mL of Fe^{3+}-o-phen (method A) or 1.5 mL of Fe^{3+}-bipy (method B) reagent solutions and 4.0 mL of acetate buffer (pH 4.5), before heating on a water bath at 80°C for 10 min. The mixture was cooled to room temperature (25 ± 1°C) and the volume made up to the mark with bidistilled water. The coloured complexes formed were measured at 510 and 521 nm against a reagent blank treated similarly according to methods A and B.

Method C

Into a series of 10 mL calibrated flasks, aliquots (0.2–1.8 mL) of 100 μg mL^{-1} standard solutions were transferred using a micro burette and the total volume adjusted to 3 mL by adding bidistilled water. Then, 2 mL each of $FeCl_3$ (0.2 %) and $K_3Fe(CN)_6$ (0.2 %) were added to each flask, mixed well and left to stand for 10 min. Finally, 1 mL of 10 M H_2SO_4 was added to each flask, diluted to the mark with bidistilled water and mixed well. The absorbance of the resulting solution was measured at 726 nm for ETD against a reagent blank prepared similarly. A calibration graph was constructed by plotting the absorbance against the drug concentration or calculated regression equation.

Analysis of Pharmaceutical Formulations

Ten tablets were accurately weighed and powdered. An accurately weighed quantity equivalent to 20 mg ETD was dissolved in 20 mL of methanol and transferred to a 100 mL calibrated flask. The contents of the flask were shaken for 10 min, and then made up to the mark with methanol. The general procedure was then followed for concentration ranges already mentioned for methods A, B and C.

References

1. British Pharmacopiea 2004.
2. Becker S. U., Blaschke G.: Evaluation of the stereoselective metabolism of the chiral analgesic drug etodolac by high-performance liquid chromatography. J. Chromatogr 1993, 621:199–207.
3. Wright M. R., Jamali F.: Limited extent of stereochemical conversion of chiral non-steroidal anti-inflammatory drugs induced by derivatization methods employing ethyl chloroformate. J. Chromatogr 1993, 616:59–65.
4. Jamali F., Mehavan R., Lemko C., Eradiri O.: Application of a stereospecific high-performance liquid chromatography assay to a pharmacokinetic study of etodolac enantiomers in humans. J. Pharm. Sci. 1988, 77:963–966.
5. Ficarra R., Ficarra P., Calauro M. L., Costantino D.: Quantitative high-performance liquid chromatographic determination of etodolac in pharmaceutical formulations. Farmaco 1991, 46:403–407.
6. Singh N. N., Jamali F., Pasutto F. M., Coutts R. T., Russell A. S.: Stereoselective gas chromatographic analysis of etodolac enantiomers in human plasma and urine. J. Chromatogr 1986, 382:331–337.

7. Giachetti C., Assandri A., Zanolo G., Brembilla E.: Gas chromatography-mass spectrometry determination of etodolac in human plasma following single epicutaneous administration. Biomed Chromatogr 1994, 8:180–183.
8. Srinivas N. R., Shyu R. W. C., Barbhaiya H.: Gas chromatographic determination of enantiomers as diastereomers following pre-column derivatization and applications to pharmacokinetic studies: A review. Biomed Chromatogr 1995, 9:1–9.
9. El Kousy N. M.: Spectrophotometric and spectrofluorimetric determination of etodolac and aceclofenac. J. Pharm. Biomed. Anal. 1999, 20:185–194.
10. Duymus H., Arslan M., Kucukislamoglu M., Zengin M.: Charge transfer complex studies between some non-steroidal anti-inflammatory drugs and π-electron acceptors. Spectrochimica Acta Part A 2006, 65:1120–1124.
11. Amer S. M., El-Saharty Y. S., Metwally F. H., Younes K. M.: Spectrophotometric study of etodolac complexes with copper (II) and iron (III). J AOAC Int 2005, 88:1637–1643.
12. Rahman N., Singh M., Hoda M. N.: Application of oxidants to the spectrophotometric determination of amlodipine besylate in pharmaceutical formulations. IL Farmaco 2004, 59:913–919.
13. Farhadi K., Ghadamgahi S., Maleki R., Asgari F. S.: Spectrophotometric determination of selected antibiotics using Prussian Blue reaction. J Chin Chem Soc 2002, 49:993–997.
14. El-Didamony A. M., Amin A. S.: Adaptation of a color reaction for spectrophotometric determination of diclofenac sodium and piroxicam in pure form and in pharmaceutical formulations. Anal Lett 2004, 37:1151–1162.
15. Saleh H. M., Amin A. S., El-Mammli M.: New colorimetric methods for the determination of indapamide and its formulations. Mikrochim Acta 2001, 137:185–189.
16. Syeda A., Mahesh H. R. K., Syed A. A.: 2,2'-Bipyridine as a new and sensitive spectrophotometric reagent for the determination of nanoamounts of certain dibenzazepine class of tricyclic antidepressant drugs. IL Farmaco 2005, 60:47–51.
17. Nagarali B. S., Seetharamppa J., Melwanki M. B.: Sensitive spectrophotometric methods for the determination of amoxycillin, ciprofloxacin and piroxicam in pure and pharmaceutical formulations. J Pharm Biomed Anal 2002, 29:859–864.
18. Basavaiah K., Chandrashekar U., Prameela H. C.: Sensitive spectrophotometric determination of amlodipine and felodipine using iron (III) and ferricyanide. IL Farmaco 2003, 58:141–148.

19. Nagaraja P., Dinesh N. D., Gowada N. M. M., Rangappa KS: A simple spectrophotometric determination of some phenothiazine drugs in pharmaceutical samples. Anal. Sci. 2000, 16:1127–1131.

20. EL-Sheikh R., Amin A. S., Gouda A. A.: Spectrophotometric determination of pipazethate hydrochloride in pure form and in pharmaceutical formulations. J. AOAC Int 2007, 90:686–692.

21. Massart D. L., Vandeginste B. G. M., Deming S. N., Michtte Y., Kaufmann L.: "Chemometries, A Text Book". Elsevier, Amsterdam; 1988:390–393.

22. Pesez M., Bartos: J. Ann. Pharm. Fr. 1965, 23:218–221.

23. Lurie J.: Hand book of Analytical Chemistry. Mir Publishers: Moscow; 1975.

24. IUPAC: Spectrochim Acta part B. 1978, 33:242–248.

25. Miller J. C., Miller J. N.: In Statistics for Analytical Chemistry. Volume 3. 3rd edition. Ellis Horwood: Chichester, New York, NY; 1993:53–62.

26. Through personal communications with Pharco, Egypt 2000.

27. Amin A. S., Zaky M., Khater H. M., El-Beshbeshy A. M.: New colorimetric methods for microdetermination of melatonin in pure and in dosage forms. Anal. Lett. 1999, 32:1421–1433.

28. Vogel A. I.: Text book of Macro and Semimicro Qualitative Inorganic Analysis. 5th edition. Longman: London; 1979.

29. Britton H. T. S.: "Hydrogen Ions". 4th edition. Chapman and Hall, London; 1952.

CITATION

Gouda AA and Hassan WS. Spectrophotometric Determination of Etodolac in Pure Form and Pharmaceutical Formulations. Chemical Central Journal. 2008; 2: 7. doi: 10.1186/1752-153X-2-7.

Lead Optimization in Discovery Drug Metabolism and Pharmacokinetics/Case Study: The Hepatitis C Virus (HCV) Protease Inhibitor SCH 503034

K.-C. Cheng, Walter A. Korfmacher,
Ronald E. White and F. George Njoroge

ABSTRACT

Lead optimization using drug metabolism and pharmacokinetics (DMPK) parameters has become one of the primary focuses of research organizations involved in drug discovery in the last decade. Using a combination of rapid in vivo and in vitro DMPK screening procedures on a large array of compounds during the lead optimization process has resulted in development of compounds that have acceptable DMPK properties. In this review, we present a general screening paradigm that is currently being used as part of drug discovery at Schering-Plough and we describe a case study using the Hepatitis C Virus (HCV) protease inhibitor program as an example. By using the DMPK optimization tools, a potent HCV protease inhibitor, SCH 503034, was selected for development as a candidate drug.

Introduction

Lead optimization in a drug metabolism environment is a multifaceted operation. It typically involves the use of various in vitro and in vivo screens to assess the drug metabolism and pharmacokinetic (DMPK) properties of multiple compounds, as well as to provide an early check on the safety issues that can be assessed in a higher throughput manner [1–4]. This process involves interaction between DMPK scientists, biologists/pharmacologists and medicinal/physical chemists. The goal of the interaction is to find a molecule that has the desired biological activity as well as DMPK properties and a safety profile appropriate for the targeted therapeutic indication. In this paper, we provide an overview of the DMPK lead optimization process that is used to support drug discovery projects at Schering-Plough. In addition, we will demonstrate how the process was used in a particular program (HCV protease inhibitor) as a case study.

Lead Generation as a Part of New Drug Discovery

Contemporary parallel and combinatorial chemical synthesis produces large arrays of compounds that are available for evaluation in new drug discovery. Furthermore, other improvements by structural chemists using a variety of tools, such as X-ray crystallography, structural modeling and ligand/substrate docking algorithms, and by molecular biologists developing high-throughput binding targets and cell-based activity assays provide drug discovery scientists with an unprecedented level of structural-based rational designs to guide the synthesis of new chemotypes as potential drug leads. Along with the advancement of chemistry and biology, new automated in vitro activity screening tools have become commercially available which can carry out complex, programmable and adaptable robotic operations to test hundred of thousands of compounds in a speedy and precise manner. As a result, these new forces have worked together to increase our ability to create new chemical entities (NCEs) that exhibit the targeted pharmacological activity.

Drug Metabolism as a Part of New Drug Discovery

Several review articles in recent years have described the role that a drug metabolism and pharmacokinetics (DMPK) department can play in the process of new drug discovery [1, 2, 4–11]. As shown in Figure 1, DMPK provides the tools and the assays to assess various new chemical entities (NCEs) in terms of their absorption, distribution, metabolism and excretion (ADME) properties as well as their pharmacokinetic (PK) parameters. In addition, DMPK scientists may also use various screens to understand the potential of NCEs for preclinical or clinical toxicity [12]. The goal of these efforts is to find a compound that is suitable for development.

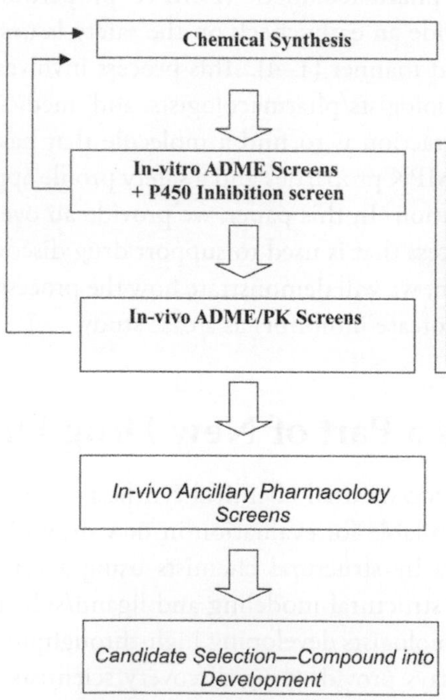

Figure 1. Scheme showing the iterative nature of lead optimization leading to candidate.

Lead Optimization in a DMPK Environment

In order to understand the needs of lead optimization, it is important to define the basic characteristics of drug-like leads [7]. As shown in Table 1, there are at least five essential properties that need to be considered in order for a compound to be drug-like: potency, bioavailability, duration, safety and reasonable pharmaceutical properties. In addition, there are some other important properties, such as selectivity, efficacy and dose-proportionality, to be considered. A successful clinical drug candidate must at least meet the minimal acceptance criteria for each of these five properties for the type of drug program that is being developed. A major deficit in any one of the properties may prevent the compound from progressing from the drug development stage to the clinical phase or to the market.

Table 1. General properties of drug-like lead compounds.

Property	Definition/Requirement
Potency	The intrinsic ability of a compound to produce a desirable pharmacological response (usually measured via high throughput in vitro screens)
Oral Bioavailability	The ability of a compound to pass through multiples barriers, such as the GI tract and the liver in order to reach the target
Duration (Half-life)	The ability of the compound to remain in circulation (or at the target site) for sufficient time to provide a meaningful pharmacological response
Safety	The compound has sufficient selectivity for the targeted response relative to non-targeted responses so that an adequate therapeutic index exists
Pharmaceutical Acceptability	The compound has suitable pharmaceutical properties, such as a reasonable synthetic pathway, adequate aqueous solubility, reasonable rate of dissolution, good chemical stability, etc.

During the discovery phase of lead optimization, the goal of the process is to find NCEs that fall into the acceptable range for each of these five properties. Among the five essential properties, three belong to the domain of DMPK: oral bioavailability, duration and safety issues. Hence, the lead optimization in discovery DMPK could be divided into three categories. First, for a drug to be given by oral administration (as is most often the case), the primary goal would be improving the oral bioavailability. This could be achieved by improving either the oral absorption or reducing the first-pass effect, or a combination of these. Secondly, improving the duration of the drug in the body could reduce the dose and the frequency of the dosing regimen. The duration of the drug in the body as measured by the half-life is inversely related to the systemic clearance of the compound. Therefore, improving (reducing) the systemic clearance of a series of compounds should extend their in vivo half-lives. Lastly, reducing any DMPK-related toxicity involves the use of multiple tests. For example, various tests are used in order to evaluate the potential for drug-drug interactions due to inhibition or induction of major CYPs, such as 3A4, 2D6, 1A2, 2C8, and 2C9. Another goal is to minimize the generation of reactive metabolites that may cause covalent binding. However, it is not totally clear whether the covalent binding may elicit any significant toxicity.

Many well-established assays/screens are now available for lead optimization in the DMPK environment (Table 2). These screens include both in vitro and in vivo assays. In the sections below, a more detailed discussion will focus on how many of these tools can be used in the lead optimization stage of new drug discovery.

Improving Oral Bioavailability

Oral bioavailability (F) is governed by the absorption in the gastrointestinal (GI) tract and the fraction of the dose that is not metabolized by the GI tract or the liver (the first-pass effect) before it enters the systemic circulation. In a report by Chatuverdi et al. [1], oral bioavailability was defined as:

$$F = Fa \cdot Fg \cdot Fh \cdot Fl.$$

Fa is defined as the fraction of the drug that is absorbed across the intestinal wall, while Fg, Fh and Fl represent the fraction of the dosed drug that gets through the GI tract, the liver and the lung, respectively. A combination of in vitro and in vivo screens may be employed to assess preclinical absorption and used to predict human absorption.

The in vitro approach typically relies on using the Caco-2 system for screening the permeability of the NCEs. In addition to the caco-2 system, other types of membrane preparations or artificial membranes, such as isolated intestinal membrane and PAMPA, may be also suitable for the screening of permeability. Numerous reports have documented the utility of Caco-2 screening as well as the correlation between the Caco-2 permeability and the absorption in humans [13–16]. The Caco-2 system appears to be most predictable for compounds that are absorbed by the transcellular mechanism. Due to the small pore size of the tight junction, the Caco-2 system is less permeable to compounds that are absorbed by the paracellular mechanism. However, treatment of the Caco-2 cells with calcium-chelating reagents, such as EDTA, can increase the pore size of the tight junctions. This approach has been used to understand the potential paracellular permeability of lead compounds [17]. One of the drawbacks of using the caco-2 system is that the passive permeability may be underestimated for p-glycoprotein (p-GP) substrates due to effl ux. The alternative is to use the PAMPA system which utilizes an artificial membrane, for the evaluation of passive permeability across membrane.

Table 2. In vitro and In vivo DMPK screening tools.

Assay Type	Assay	Species Relevance	References
In-vitro	Caco-2	Human	[13–16]
In-vitro	Plasma Protein binding	Multiple	[39]
In-vitro	Intrinsic Clearance (microsomes or hepatocytes)	Multiple	[20–23]
In-vitro	CYP P450 Inhibition	Human	[25–29]
In-vitro	CYP P450 Induction	Human	[30–31]
In-vitro	CYP P450 Profiling	Human	[40]
In-vitro	Metabolite Profiling (microsomes or hepatocytes)	Multiple	
In-vitro	Transporter profiling	Human	
In-vivo	Rapid rat PK (CARRS)	Rat	
In-vivo	Single dose PK	Rat, Dog, Monkey	
In-vivo	Single rising dose PK	Rat and Dog or Monkey	
In-vivo	Metabolite Identification	Rat and Dog or Monkey	
In-vivo	Rat Mass Balance	Rat	
In-vivo	Multiple Rising Dose	Rat	

The in vivo approach to measure absorption in the discovery phase relies on animal pharmacokinetics. For instance, if the absorption of the lead compound is within the acceptable range in rodent and non-rodent species, such as dogs and monkeys, it is likely that human absorption may be within the acceptable range as well. To support this hypothesis, several publications [18, 19] have suggested that there is a correlation between the animal and human absorption, despite the fact that some distinct differences, such as the transit time, exist in the GI physiology between species.

The second element involved in the oral bioavailability is the first-pass effect. A compound entering the systematic circulation from the GI tract needs to first pass through two barriers—intestinal wall and the liver (this is often called the "first-pass effect"). Both the intestinal mucosa and the liver are enriched in drug metabolism enzymes. It has been well accepted that, due to species differences, animal metabolism may not be suitable for predicting the first-pass effect in humans. One common screen for estimating the human first-pass effect is to use microsomal preparations from human livers. The extraction ratio calculated from how quickly the NCE disappears in the microsomal incubation, may be used to estimate the extent of the liver first pass. With improving cryopreservation technologies, human hepatocytes have become a very useful tool in evaluating the metabolic clearance of test articles. A major advantage of using hepatocytes is that they contain both phase I and phase II metabolic enzymes. A fairly interesting approach to estimate the oral bioavailability was recently presented by using a Caco-2/hepatocyte hybrid system [20]. This novel system combines the Caco-2 permeability assay and the liver first pass assay into one system. As shown before [20], the Caco-2/hepatocyte system could provide a reasonable prediction of the oral bioavailability in humans. This approach could be used in conjunction with animal pharmacokinetic evaluation in lead optimization.

Optimizing the Half-life of a Compound Series

The half-life of a compound is determined by both the clearance and the volume of distribution. As the clearance increases, the half-life decreases. Conversely, for a given clearance, a higher volume of distribution results in a longer half-life. For orally administered compounds, the apparent half-life is a combination of the elimination half-life and/or the absorption half-life. Hence, it is sometimes possible to develop a slow release formulation to extend the apparent half-life of a compound. However, the primary goal in the drug discovery phase is to optimize the half-life of the series of compounds. Frequently, the goal is to increase the half-life, while in certain cases shortening the half-life is the goal; half-life increase is generally done by trying to reduce the clearance of a compound. Since there are several ways to predict human clearance, the screening assays are designed according to these approaches. For example, allometry using animal clearance data has been often used to estimate the human clearance. This approach requires one to measure a compound's clearance in at least three animal species and to use the animal clearances for allometric scaling. Therefore, using allometry to predict human clearance is hardly a high-throughput process.

In the lead optimization process, some of the most frequently used high-throughput in-vitro screens aim to predict hepatic clearance. These assays use either hepatic microsomes or primary hepatocytes derived from humans or animals. The general screening procedure monitors the disappearance of the NCE in an incubation mixture containing the compound and a fixed amount of microsomes or hepatocytes [1, 10, 21–23]. If the disappearance of the test compound follows first order kinetics, the rate of the process can be used to calculate the intrinsic clearance. A number of recent publications have suggested that using pooled hepatocytes from human donors results in a reasonable correlation between the measured intrinsic clearance and the in vivo clearance for a number of marketed compounds [24]. The predictive value of the hepatocyte clearance was demonstrated by a very good correlation (R2 = 0.86) when comparing the hepatocyte intrinsic clearance with in vivo clearance. Hence, by reducing the intrinsic clearance in an in vitro hepatocyte assay may predict improvement of the half-life.

Safety/Toxicity Screening

Several potential safety issues can be related to DMPK properties. For example, toxicity due to drug-drug interactions that result from CYP isozyme inhibition or induction may cause a candidate drug to fail in development. In order to avoid these problems, NCEs are usually screened for their ability to inhibit major human CYP isozymes using either pooled human microsomes or supersomes which contain individual isozymes [17, 25–29]. It is also important to differentiate whether the observed inhibition is direct, metabolism or mechanism-based. The difference between the direct and the metabolism-based inhibition is that the direct inhibition by the parent compound is reversible whereas the metabolism-based inhibition is that a metabolite is a reversible inhibitor. A mechanism-based inhibition occurs when a reactive intermediate covalently modifies the CYP enzyme. The mechanism-based inhibition is usually irreversible. Practically, NCEs in drug discovery may encounter frequent CYP inhibition issues due to the wide substrate specificities of major human CYP isozmes, such as 3A4, 2D6, and 2C families.

CYP induction may cause opposite effects of CYP inhibition in that the exposure of the drug may be reduced. In rodents, the major induction issue appears to be the induction of the CYP 1A, CYP 2B, and CYP 3A family. There are considerably literatures suggesting that nuclear receptors, such as AhR, CAR and PXR, are involved in the induction of the respective CYPs [30]. In humans, the major induction pathway appears to be controlled by PXR. Administration of PXR agonists, such as rifampicin, causes elevated levels of intestinal and hepatic CYP 3A4, resulting in the reduction of oral bioavailability. CYP induction potential can be measured by certain in vitro assays, such as PXR-reporter gene assay and hepatocyte induction assay [4, 31].

Human PK Prediction

The human PK and dose regimen prediction is also performed in order to facilitate the design and implementation of a clinical program. An evolving paradigm of human PK prediction combines allometry and scaling of the in vitro hepatic clearance. It is generally accepted that allometry using animal PK data may more accurately predict the volume of distribution and renal clearance, while it may be less accurate in predicting the hepatic clearance, since the metabolic enzymes, especially CYP enzymes, show significant differences between animals and humans. Hence, the use of intrinsic clearance obtained by microsomal or hepatocyte clearance assay for the scaling of in vivo clearance serves as an alternative way to predict the human hepatic clearance. When projected clearance values from allometry and in vitro-in vivo scaling are in reasonable agreement, one can feel confident that preclinical predictions of dosing regimen will not be highly inaccurate.

Case study: HCV ProteaseScreening Paradigm

Hepatitis C Virus

Hepatis C virus (HCV), the etiologic agent of non-A, non-B hepatitis, represents a world wide-health problem, with approximately 170 million people infected with the virus [32]. Infection with HCV often leads to a chronic form of hepatitis. Without therapeutic intervention it could lead to cirrhosis, hepatic failure or hepatocellular carcinoma [33]. The current therapy for chronic HCV infection is subcutaneous injection of pegylated-interferon αalone or in combination with oral ribavirin [34]. HCV belongs to the family of flaviviridae, which includes other human pathogens, such as Yellow Fever and West Nile Virus. It is an enveloped positive stranded RNA virus. Upon entering a suitable host cell, the HCV genome serves as a template for cap-independent translation through its 5' internal ribosome entry sites. The resulting 3000 amino acid polypeptide undergoes both co- and post-translational proteolytic maturation by host and virus-encoded proteases [35]. The virus-encoded protease responsible for processing the non-structural (NS) portion of the polypeptide is located in the N-terminal region of the NS3 protein. The NS3 protease structure provided necessary details to permit rational structure-assisted inhibitor design. This endeavor targeting the enzyme-substrate binding site resulted in the discovery of SCH 503034, a structurally novel ketoamide protease inhibitor. Recent proof-of-concept clinical studies with SCH 503034 and other HCV protease inhibitors BILN2061 [36] and VX-950 [37] demonstrated the feasibility of targeting the protease.

DMPK Screening Paradigm

The following is a brief summary of the drug metabolism/pharmacokinetics process involved in the discovery of SCH 503034. About 10,000 compounds were synthesized and went through the cell-based assay (Replicon assay) for HCV protease inhibition activity. More than 1,000 compounds met the cut-off criterion of an IC90 of 1 μM or lower. These compounds were further optimized by DMPK (Fig. 2). Within the DMPK screening, several tiers were employed. In the first level, several higher throughput screenings were deployed: cassette-accelerated rapid rat screen (CARRS) screening [38], human hepatocyte clearance, Caco-2 permeability screening, and CYP enzyme inhibition (including mechanism-based inhibition) screening. In addition, some special screenings are also employed: plasma esterase/amidase screening and liver uptake screening. In the second level screening, more labor intensive assays were employed, such as full pharmacokinetic (PK) studies in rodent species and non-rodent species (monkey and dog). Of the 1000 compounds tested, three emerged as advance leads including SCH 503034. They appeared to meet the following acceptable DMPK criteria: moderate oral bioavailability in rats and dogs, absence of reactive metabolite, IC50 > 5 uM for CYPs 3A4, 2D6, 2C8, and 2C9, moderate huan hepatocyte clearance, and no CYP induction liability.

Following the DMPK screening process, a few advanced leads were identified that had acceptable DMPK characteristics. These advanced leads went through a DMPK profiling process for the final selection of the best compound for drug development (Level 3). These processes involved, for example, single rising dose studies in both rodent and non-rodent species to determine if desirable exposure multiples could be reached in the preclinical toxicology program. Then, multiple dosing in rats was performed in order to determine whether the circulating compound accumulates or produces auto-induction. Accumulation of the compound may indicate that the elimination of the compound, whether is due to hepatic clearance or renal clearance, or both, has been hampered.

Figure 3 shows the three chemotypes that were submitted for lead optimization in DMPK. A multitude of factors were involved in selection or elimination of certain chemotypes. A brief description of the DMPK properties of each of the chemotypes is discussed below:

Macrocyclic Compounds

An example compound from this series is SCH 416538 (Fig. 3). This type of compound showed resistance to peptidases and amidases. Since a more rigid conformation is maintained, this chemotype appears to have better potency. Some compounds in this class had good oral bioavailability in rats. However, the PK in dogs and monkeys was poor for most of the compounds.

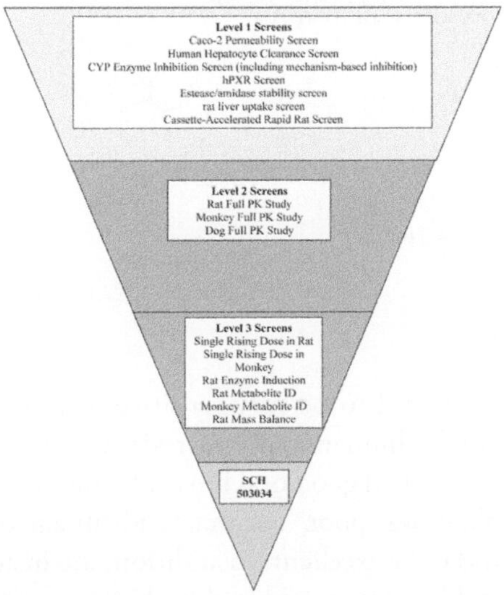

Figure 2. DMPK screening paradigm as part of the lead optimization and candidate selection process—application to HCV compound selection.

Secondary Amides

An example compound from this series is SCH 446211 (Fig. 3). This chemotype showed reasonable half-life and excellent bioavailability after sub-cutaneous (SC) dosing, suggesting a potential dosing regimen through the SC route. Another advantage of this chemotype was the resistance to peptidases and amidases. Again, however, the oral bioavailability for this chemotype was very poor in rats and monkeys.

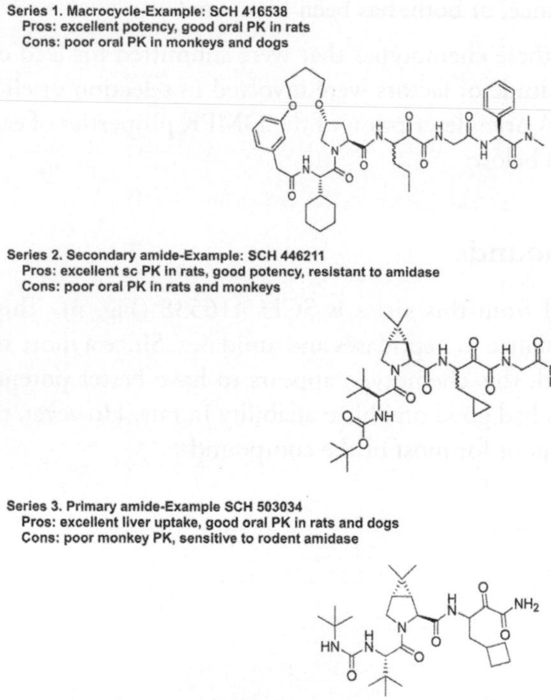

Series 1. Macrocycle-Example: SCH 416538
Pros: excellent potency, good oral PK in rats
Cons: poor oral PK in monkeys and dogs
Series 2. Secondary amide-Example: SCH 446211
Pros: excellent sc PK in rats, good potency, resistant to amidase
Cons: poor oral PK in rats and monkeys
Series 3. Primary amide-Example SCH 503034
Pros: excellent liver uptake, good oral PK in rats and dogs
Cons: poor monkey PK, sensitive to rodent amidase

Figure 3. HCV compounds leading to SCH 503034.

Primary Ketoamides

This class of compounds appeared to be very sensitive to rodent amidase. They were generally more resistant to human and non-rodent amidases in the plasma. Certain compounds in this class had good oral bioavailability in rats and dogs, but oral bioavailability in monkeys was poor. The major advantage of this chemotype is that liver uptake was found to be excellent. In addition, intrinsic clearance in human hepatocyte was acceptable. One compound in this series, SCH 503034, met the acceptance criteria for this program and was advanced into development.

Conclusion

Higher-throughput DMPK screens using multiple in vitro and in vivo techniques are now in place and have become an essential part of the lead optimization process in new drug discovery. Future improvements in this lead optimization arena may be achieved by using automated systems to enhance the speed of these in-vitro screens. There is a continuing need in the area to improve the ability to predict in vivo pharmacokinetics by using in vitro evaluations of oral absorption, intestinal and hepatic first pass, hepatic intrinsic clearance, organ uptake and efflux mediated by transporters as well as plasma and cellular protein binding. In the near term, we will continue to use the combination of in vitro systems and fast in vivo screening for the selection of early discovery leads. In addition, in vivo studies for evaluation of the exposure multiple and metabolism and disposition may be accelerated in order to reduce the time to final candidate selection. Ultimately, it may become possible to use pharmacologically based in silico DMPK computer model parameters to support the rapid screening of drug candidates in order to shorten the time-frame of the lead optimization process while still discovering candidate drugs with acceptable DMPK properties.

References

1. Chaturvedi, P.R., Decker, C.J. and Odinecs, A. 2001. Curr. Opin. Chem. Biol., 5:452–63.
2. Korfmacher, W.A. 2003. Curr. Opin. Drug. Discov. Devel., 6:481–5.
3. Korfmacher, W. 2005. In Integrated Strategies for Drug Discovery using Mass Spectrometry (Ed, Lee, M. S.) John Wiley and Sons, Inc., Hoboken, NJ, pp. 359–378.
4. Roberts, S.A. 2001. Xenobiotica, 31:557–89.
5. Eddershaw, P.J., Beresford, A.P. and Bayliss, M.K. 2000. Drug. Discov. Today, 5:409–414.
6. Eddershaw, P.J. and Dickins, M. 1999. Pharm. Sci. Technol. Today, 2:13–19.
7. Egan, W.J., Walters, W.P. and Murcko, M.A. 2002. Curr. Opin. Drug. Discov. Devel., 5:540–9.
8. Riley, R.J., Martin, I.J. and Cooper, A.E. 2002. Curr. Drug. Metab., 3:527–50.
9. Kumar, G.N. and Surapaneni, S. 2001. Med. Res. Rev., 21:397–411.
10. Bertrand, M., Jackson, P. and Walther, B. 2000. Eur. J. Pharm. Sci., 11 Suppl 2:S61–72.

11. Kariv, I., Rourick, R.A., Kassel, D.B. and Chung, T.D. 2002. Comb. Chem. High, Throughput, Screen, 5:459–72.
12. Meador, V., Jordan, W. and Zimmermann, J. 2002. Curr. Opin. Drug. Discov. Devel., 5:72–8.
13. Mandagere, A.K., Thompson, T.N. and Hwang, K.K. 2002. J. Med. Chem., 45:304–11.
14. Hidalgo, I.J. 2001. Curr. Top. Med. Chem., 1:385–401.
15. van De Waterbeemd, H., Smith, D.A., Beaumont, K. and Walker, D.K. 2001. J. Med. Chem., 44:1313–33.
16. Wang, Z., Hop, C.E., Leung, K.H. and Pang, J. 2000. J. Mass Spectrom, 35:71–6.
17. Cogburn, J.N., Donovan, M.G. and Schasteen, C.S. 1991. Pharmaceut Res., 8:210–216.
18. Chiou, W.L. and Barve, A. 1998. Pharmaceut Res., 15:1792–1794.
19. Chiou, W.L. and Buehler, P.W. 2002. Pharmaceut Res., 19: 868–874.
20. Lau, Y.Y., Chen, Y.-H., Liu, T., Li, C., Cui, X., White, R.E. and Cheng, K.-C. 2004. Drug Metab. Disp., 32:937–942.
21. Ansede, J.H. and Thakker, D.R. 2004 J. Pharm. Sci., 93:239–55.
22. Thompson, T.N. 2000. Curr. Drug. Metab., 1:215–41.
23. Thompson, T.N. 2001. Med. Res. Rev., 21:412–49.
24. Lau, J., Sapidou, E., Cui, X., White, R.E. and Cheng, K.-C. 2002. Drug Metab. Disp. 30:1446–1454.
25. Bjornsson, T.D., Callaghan, J.T., Einolf, H.J., Fischer, V., Gan, L., Grimm, S., Kao, J., King, S.P., Miwa, G., Ni, L., Kumar, G., McLeod, J., Obach, S.R., Roberts, S., Roe, A., Shah, A., Snikeris, F., Sullivan, J.T., Tweedie, D., Vega, J.M., Walsh, J. and Wrighton, S.A. 2003. J. Clin. Pharmacol., 43:443–69.
26. Jenkins, K.M., Angeles, R., Quintos, M.T., Xu, R., Kassel, D.B. and Rourick, R.A. 2004. J. Pharm. Biomed. Anal., 34:989–1004.
27. Kim, M.J., Kim, H., Cha, I.J., Park, J.S., Shon, J.H., Liu, K.H. and Shin, J.G. 2005. Rapid Commun. Mass Spectrom, 19:2651–8.
28. Walsky, R.L. and Obach, R.S. 2004. Drug. Metab. Dispos., 32:647–60.
29. Obach, R.S., Walsky, R.L., Venkatakrishnan, K., Gaman, E.A., Houston, J.B. and Tremaine, L.M. 2005. J. Pharmacol. Exp. Ther. 316:336–348.
30. Moore, D.D., Kat, S., Xie, W., Mangelsdorf, D.J., Schmidt, D.R., Xiao, R. and Kliewer, S.A. 2006. Pharmacol. Rev. 58:742–759.

31. Luo, G., Cunningham, M., Kim, S., Burn, T., Lin, J., Sinz, M., Hamilton, G., Rizzo, C., Jolley, S., Gilbert, D., Downey, A., Mudra, D., Graham, R., Carroll, K., Xie, J., Madan, A., Parkinson, A., Christ, D., Selling, B., LeCluyse, E. and Gan, L.S. 2002. Drug Metab Dispos, 30:795–804.

32. Wasley, A. and Alter, M.J. 2000. Semin Liver Dis., 20:1–16.

33. Alter, M.J. and Seeff, L.B. 2000. Semin Liver Dis., 20:17–35.

34. Manns, M.P., McHutchison, J.G., Gordon, S.C., Rustgi, V.K., Shiffman, M., Reindollar, R., Goodman, Z.D., Koury, K., Ling, M. and Albrecht, J.K. 2001. Lancet, 358:958–965.

35. Reed, K.E. and Rice, C.M. 2000. Curr. Top. Microbiol. Immunol., 242:55–84.

36. Lamarre, D., Anderson, P.C., Bailey, M., Beaulieu, P., Bolger, G., Bonneau, P., Bos, M., Cameron, D.R., Cartier, M., Cordingley, M. G., Faucher, A.-M., Goudreau, N.; Kawai, S. H.; Kukolj, G.; lagace, L., LaPlante, S. R., Narjes, H., Poupart, M.-A., Rancourt, J., Sentjens, R.E., St George, R., Simoneau, B., Steinmann, G., Thibeault, D., Tsantrizos, Y.S., Weldon, S.M., Yong, C.-L. and Llinas-Brunet, M. 2003. Science, 426:186–189.

37. Resnick, H. W., Zeuzem, S., van Vliet, A., McNair, L., Purdy, S., Chu, H.-M. and Jansen, P. L. 2005. Digestive Dis Week, May 14–19, Chicago, IL, Abs #527.

38. Cox, K.A., White, R.E. and Korfmacher, W.A. 2002. Comb. Chem. High, Throughput Screen, 5:29–37.

39. Artmann, T., Schmitt, J., Rohring, C., Nimptsch, D., Noller, J. and Mohr, C. 2006. Current. Drug. Deliv., 3:181–192.

40. Venkatakrishnan, K., Von Moltke, L.L. and Greenblatt, D.J. 2001. J. Clin. Pharmacol., 41:1149–1179.

CITATION

Cheng K-C, Korfmacher WA, White RE, and Njoroge GF. Lead Optimization in Discovery Drug Metabolism and Pharmacokinetics/Case Study: The Hepatitis C Virus (HCV) Protease Inhibitor SCH 503034. Perspectives in Medicinal Chemistry. 2007; 1: 1–9.

Pressure-Tuning Raman Spectra of Diiodine Thioamide Compounds: Models for Antithyroid Drug Activity

Ghada J. Corban, Constantinos Antoniadis, Sotiris K. Hadjikakou, Nick Hadjiliadis, Jin-Fang Meng and Ian S. Butler

ABSTRACT

The pressure-tuning Raman spectra of five solid, diiodine heterocyclic thioamide compounds (mbztS)I_2 (mbztS = N-methyl-2-mercaptobenzothiazole) (1); $[(mbztS)_2I]^+[I_7]^-$ (2); (pySH)I_2 (pySH = 2-mercaptopyridine) (3); $[(pySH)(pyS]^+[I_3]^-$ (4); (thpm)$(I_2)_2$ or possibly $[(thpm)I_2]+[I_3]-$ (thpm = 2-mercapto-3,4,5,6-tertahydropyrimidine (5) have been measured for pressures up to ~ 50 kbar using a diamond-anvil cell. Compounds 1, 4, and 5 undergo pressure-induced phase transitions at ~ 35, ~ 25, and ~ 32 kbar, *respectively. Following the phase transition in 1, the pressure dependences of the vibrational modes, which were originally located at 84, 111, and 161 cm^{-1} and are associated with the S···I–I linkage, are 2.08, 1.78, and 0.57 cm^{-1}/kbar, respectively. These pressure dependences are typical of low-energy vibrations. The pressure-tuning FT-Raman results for the pairs of compounds 1, 2, 3, and 4 are remarkably similar to each other suggesting that the compounds are most probably perturbed diiodide compounds rather than ionic ones. The Raman data for 5 show that it is best formulated as (thpm)$(I_2)_2$ rather than $[(thpm)_2I]^+[I_3]^-$.*

Introduction

Iodine chemistry is proving to be of considerable interest lately, in part, because of the discovery of low-temperature, semi- and super-conducting polyiodides, which quickly led to the deliberate doping of conjugated polymers with elemental iodine, and ultimately resulted in the award of the 2000 Nobel Prize in Chemistry to Professor Heeger et al. for their research on these materials [1]. A second important field in which iodine chemistry plays a pivotal role is in the activity of antithyroid drugs and there is now a major research effort focused on determining structure-activity relationships of thioamides with iodine [2, 3]. The antithyroid drugs that are most commonly used today are the thioamide derivatives, 6-n-propylthiouracil, N-methyl-imidazoline-2-thione (methimazole), and 3-methyl-2-thioxo-4-imidazoline-1-carboxylate (carbimazole). For the past few years, we have been exploring the iodine chemistry of thioamides in an effort to bring about a clearer understanding of the interactions involved in antithyroid drug treatment. As part of this research, we have reported the results of a fundamental study on the effect of high external pressures (up to ~ 50 kbar) on the ambient-temperature FT-Raman spectra of four solid, diiodine-heterocyclic thioamide compounds [4]. In the case of one of these compounds, [(bztzdtH)I_2} • I_2 (bztzdtH = benzothiazole-2-thione), we discovered that I_2 disproportionation occurs with increasing pressure and I_3 – ions are produced. In addition, empirical correlations were established between the wavenumber of the I–I stretching vibration, the I–I bond length, and the applied external pressure. In this present paper, we have extended these pressure-tuning Raman spectroscopic studies to an examination of a second series of five, solid diiodineheterocyclic thioamide compounds: (mbztS)I_2 (mbztS = Nmethyl-2-mercaptobenzothiazole) (1); $[(mbztS)_2I]^+[I_7]^-$ (2); (pySH)I_2 (pySH = 2-mercaptopyridine) (3); $[(pySH)(pyS]^+[I_3]^-$ (4); (thpm)$(I_2)_2$ or $[(thpm)I_2]^+[I_3]^-$(thpm = 2-mercapto-3,4,5,6-tertahydropyrimidine) (5). These systems are also of potential importance as model compounds for the development of structure-activity relationships associated with the interaction of antithyroid drugs with iodine. They contain iodine atoms in several different structural arrangements and three of them have already been characterized by single-crystal X-ray diffraction, namely, compounds 1 [5, 6], 2 [5–7], and 4 [8].

Experimental

The diiodine-heterocyclic thioamides 1–5 were prepared according to the literature procedures [2, 5–8]. Complete details of the pressure-tuning FT-Raman measurements, including the ruby R_1 fluorescence-pressure calibration procedure, have been described elsewhere [4, 9]. A diamondanvil cell (High Pressure Diamond Optics, Inc, Tucson, Ariz, USA) was used to generate the applied pressures. FT-Raman spectra were recorded on a Bruker IFS-88 FT-IR spectrometer equipped with an FRA-105 Raman module connected via two 1m photo-optic cables to a Nikon Optiphot-II optical microscope using a Nikon 20X super-long-range objective. A near-IR (Nd^{3+}: YAG) laser, emitting at 1064.1nm with a power ~ 25mW, was used to excite the Raman spectra and typically a 2.6 cm^{-1} spectral resolution was employed while collecting 1000 scans. The band positions are considered to be accurate to at least ± 1 cm^{-1}.

Results and Discussion

In our previous pressure-tuning FT-Raman work on diiodine thioamide compounds [4], we pointed out that electron charge-transfer between the S atom of a thioamide and diiodine will result in stabilization of the lone pair of electrons on the S atom by overlapping of the S donor orbital with the σ*-orbital of diiodine. This situation will lead to a lengthening of the thione double bond in the thioamide and subsequent bond formation between the S atom and diiodine, together with a concomitant lengthening of the I–I bond. It was also shown that there are direct empirical correlations between the S–I, C–S, and I–I bond distances, namely, if the S–I distances are shorter than the C–S ones are, then the I–I distances will be longer than normal and there will be strong electron donation and vice versa. The application of high external pressures to the diiodine thioamide compounds would, therefore, be expected to produce some significant changes in bond distances and particularly on the positions of the low-frequency modes associated with the iodine atoms. There have been only a few high-pressure vibrational studies reported on diiodine systems and these have involved chiefly oxides [10, 11] and sulfides [12]. The interaction of polyiodides with polyvinyl alcohols has also been investigated under high pressure [13]. In addition, the effect of high pressures on the Raman spectrum of solid diiodine itself has been the subject of several investigations [14, 15]. In the present work, the FT-Raman spectra of compounds 1–5 were first recorded under ambient conditions and then pressure-tuning FT-Raman studies were undertaken on each of the species. An excellent review of the Raman spectra expected for different diiodine compounds has been published recently by Deplano et al. [16].

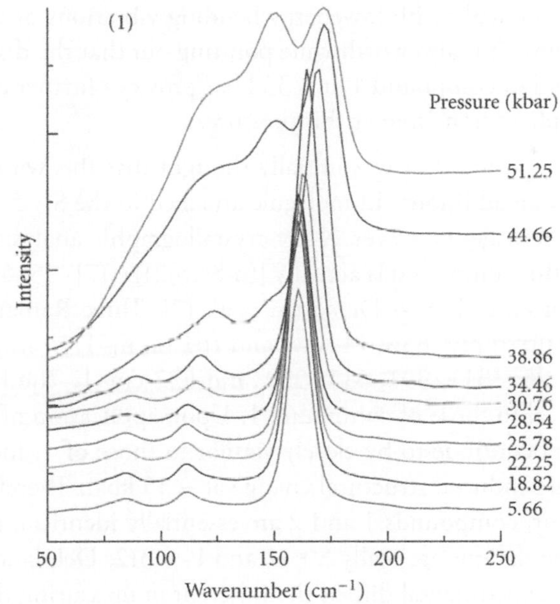

Figure 1. Pressure-tuning FT-Raman spectra of 1 in the low-energy region.

FT-Raman spectra of compound 1 in the 50–250 cm−1 region are shown in Figure 1 for selected pressures up to ~ 51.3 kbar. Initially, there are three bands observed at 84 w, 111mw, and 161 vs cm−1. The Raman spectrum is closely similar to that reported for the (dtt)I2 (dtt = 1,3-ditholane-2-thione) for which the presence of a S…I–I linkage has been established [17]. Upon increasing the pressure, the three bands for compound 1 gradually shift to higher wavenumbers, as is typically the case [18]. At ~ 35 kbar, the band originally at 84 cm−1 vanishes and a new band develops at ~ 150 cm−1 which, eventually at ~ 51 kbar, has a comparable intensity to that of the originally very strong peak, which has now shifted to 171 cm−1. These spectral changes can also be seen quite in the ν (cm−1) versus P (pressure, kbar) plot (Figure 2), which indicates that compound 1 undergoes a pressure-induced structural change at~35 kbar. From X-ray crystallographic data for 1 [5, 6], the I–I distance is 2.7912(9) Å which, from the linear plot of I–I bond length versus position of the ν(I–I) mode given in [4], would lead to an I–I stretching vibration being expected at ~ 160 cm−1 at ambient pressure, just as it is observed for this adduct. Before the phase transition for compound 1 occurs at ~ 35 kbar, the three vibrations originally located at 84, 111, and 161 cm−1 are completely pressure insensitive. Following the phase transition, however, the pressure dependences are 2.08, 1.78, and 0.57cm−1/kbar, respectively, which are typical of low-energy vibrations. There are other weak Raman features detected at ambient pressure at 255, 311, 392, 511, 539, 636, and 708 cm−1, which are presumably associated chiefly with lowenergy bending vibrations of the thioamide group. Furthermore, it is also worthwhile pointing out that the disappearance of the 311 cm−1 band of compound 1 at ~ 35 kbar provides further evidence for the existence of the phase transition at this pressure.

In the case of compound 2, it was initially thought that this would prove to be compound 1 with an additional I2 molecule attached to the S…I–I unit leading to a S…I–I–I–I linkage. However, X-ray crystallographic analysis has subsequently shown that this compound is actually [(mbztS)2I]+[I7]− [5, 6]. The same compound was reported earlier by Demartin et al. [7]. Three Raman bands are observed at ambient pressure at 85w, 112mw, and 161 vs cm−1. In addition, there are weak features at 257, 311, 391, 511, 539, and 637 cm−1. The Raman data are essentially identical to those of compound 1. Upon application of pressure to 2, the Raman spectra continue to be closely similar to those of 1, including the existence of a pressure-induced structural change at ~ 35 kbar. Therefore, from a vibrational standpoint, compounds 1 and 2 are essentially identical, not surprisingly since their formulations are really S • I2 and I− • 3I2. Delplano et al. [16] have emphasized the experimental difficulties inherent in measuring the laser Raman spectra of polyiodides, that is, some I2 may be lost during the measurements, for example, I7− → I5− + I2; → I5− →→ I3− + I2. There was no spectroscopic evidence, however, for the formation of any free I2 in this particular case—a strong Raman band would have been observed at ~ 180 cm−1 [1, 14, 15].

The low-energy FT-Raman spectra of (pySH)I2 (pySH = 2-mercaptopyridine) (3) [(pySH)(pyS]+[I3]− (4) are also quite similar to one another. It is possible that compound 3 may have disproportionated to 4 upon initial pressurization in the DAC, as has been reported previously for [(bztzdtH)I2} • I2 (bztzdtH = benzothiazole-2-thione) [4]. On the basis of X-ray crystallographic data for 4, the I3 moiety is slightly bent. For a linear I3 − entity, only one very strong Raman band would be expected at ~ 110 cm−1 from the symmetric inphase ν(I–I) stretching mode [17]. In the case of 4, however, there is an intense doublet at 155 and 164 cm−1, and several other weaker features are observed at 77 sh, 110 vw, 176 sh, and 234 vvw cm−1. The appearance of this strong band at 164 cm−1, together with the observation of the other weaker bands, suggests that the I3 moiety in 4 is better formulated as an I− • I2 entity, containing a slightly perturbed diiodine molecule, rather than as an I3 – species. A comparable situation exists for [(EtNH2)2dt]I3, which exhibits a very strong Raman band at 167 cm−1 and is considered to have a perturbed diiodine molecule [17]. The I– • I2 formulation for 4 could also be another reason for the similarity of the spectra to those of 3. The second intense Raman feature observed for 4 at 155 cm−1 may be the result of partial decomposition to another polyiodide species or possibly even site or factor-group splitting of the symmetric ν(I–I) vibrational mode. Under pressure, the doublet at 155 and 164 cm−1 shifts gradually to higher wavenumbers, with the higher energy component eventually merging with the lower energy one (Figure 3). The versus P plot in Figure 4 suggests that there is a pressure-induced structural change occurring at ~ 25 kbar. The signals in the C–H stretching region are much too weak to be analytically useful at low pressures, but there is a significant increase in intensity occurring in this region, beginning just after the phase transition near 30 kbar, that continues until the highest pressure (~ 46 kbar) is reached. This dramatic increase in intensity may be associated with pressure-induced fluorescence involving the I– • I2 entity. Such pressure-induced fluores cence effects have been reported previously for diamond anvils [19], ZnS : Mn2+ nanoparticles [20], and acetone [21]. In addition, the acridine diiodine (2 : 3) adduct has been shown to a yellow-green fluorescence upon exposure to 632.8 He–Ne laser excitation [22].

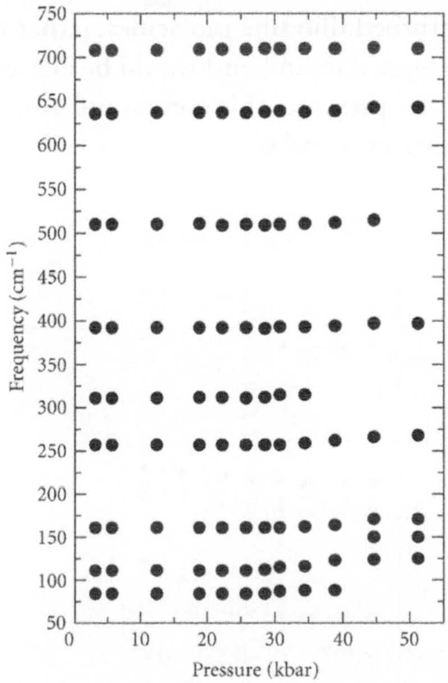

Figure 2. Wavenumber versus pressure plot for the FT-Raman spectra of 1 in the low-energy region.

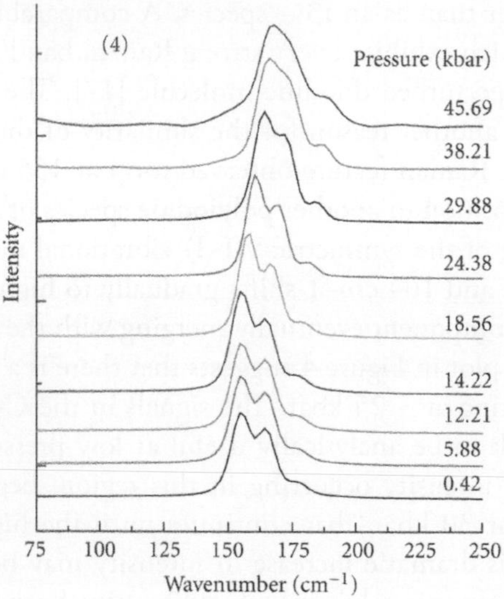

Figure 3. Pressure-tuning FT-Raman spectra of 4 in the low-energy region.

Finally, a FT-Raman study was also conducted for compound 5, which it was thought could be formulated as either (thpm)(I2)2 or [(thpm)I2]+[I3]-(thpm = 2-mercapto-3,4,5,6-tertahydropyrimidine). The principal, low-energy features appear at 91 w, 161 vvs and 197w cm-1 suggesting the presence of neutral (thpm)(I2)2, containing perturbed diiodine molecules, rather than ionic [(thpm)I2]+[I3]- for which the strongest Raman band would be expected at ~ 110 cm-1 and not at ~ 160 cm-1. Under pressure, this compound also undergoes a structural change at ~ 32 kbar (Figures 5 and 6).

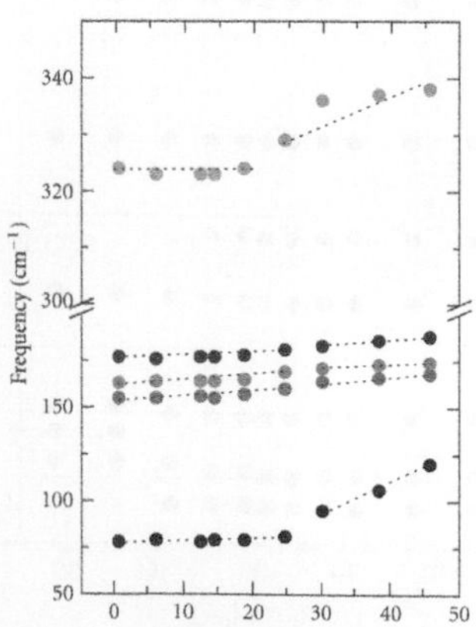

Figure 4. Wavenumber versus pressure plot for the FT-Raman spectra of 4 in the low-energy region.

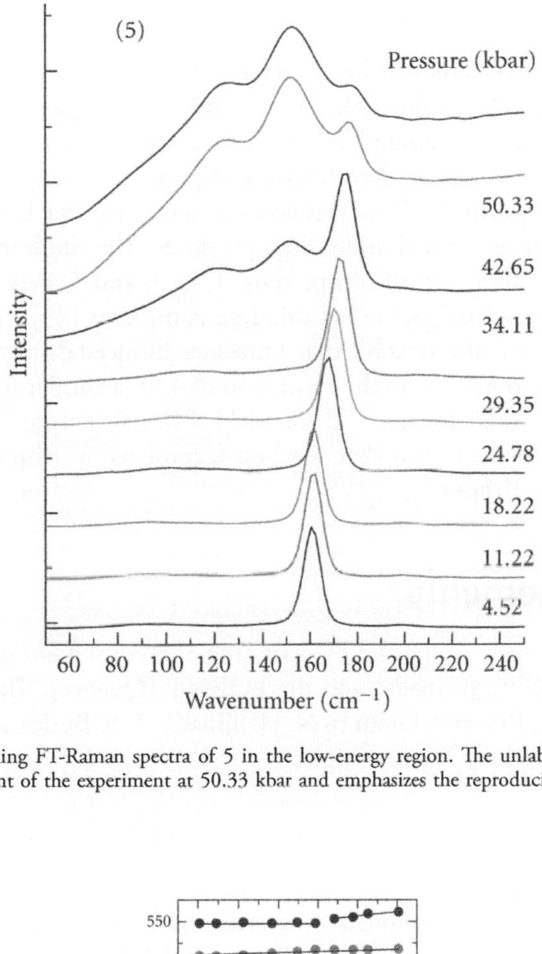

Figure 5. Pressure-tuning FT-Raman spectra of 5 in the low-energy region. The unlabeled pressure value is a repeated measurement of the experiment at 50.33 kbar and emphasizes the reproducibility of the pressure-tuning work.

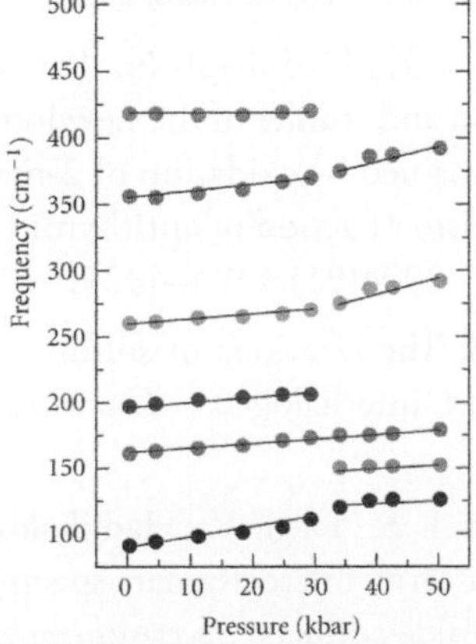

Figure 6. Wavenumber versus pressure plot for the FT-Raman spectra of 5 in the low-energy region.

Conclusions

Raman spectroscopy continues to be an extremely useful probe for examining the structural features of diiodine [14, 15, 23] and interhalogen [24] thioamide compounds. In the pressure-tuning FT-Raman spectroscopic work reported here, structural phase transitions were detected for compounds 1, 4, and 5 at ~ 35, ~ 25, and ~ 32 kbar, respectively. There was no evidence of any free I_2 formation for any of the compounds examined under high pressures. The similarity of the FT-Raman spectra of the two pairs of compounds, 1, 2, 3, and 4, may be the result of them being closely related perturbed diiodine complexes [25]. In the case of compounds 3 and 4, it is also possible that a pressure-induced disproportionation of 3 may have occurred leading to the formation of 4, in a similar manner to the effect of pressure on $[(bztzdtH)I_2\}$ • I2 (bztzdtH = benzothiazole-2-thione) [4]. Finally, the FT-Raman data show that 5 is best formulated as $(thpm)(I_2)_2$ rather than the ionic species $[(thpm)I_2]^+[I_3]^-$.

Acknowledgements

This work was generously supported by operating and equipment grants to I. S. Butler from the NSERC (Canada) and the FQRNT (Quebec). The award of a NATO Collaborative Research Grant to N. Hadjiliadis, I. S. Butler, and I. Sovago (Hungary) is also gratefully acknowledged.

References

1. Svensson P. H., Kloo L. Synthesis, structure, and bonding in polyiodide and metal iodide-iodine systems. *Chemical Reviews.* 2003; 103(5):1649–1684.
2. Antoniadis C., Hadjikakou S. K., Hadjiliadis N., Kubicki M., Butler I. S. Synthesis, X-ray characterisation and studies of the new ionic complex [bis(pyridin-2-yl) disulfide] triiodide, obtained by oxidation of 2-mercaptopyridine with I_2 - implications in the mechanism of action of antithyroid drugs. *European Journal of Inorganic Chemistry.* 2004;2004(21):4324–4329.
3. Boyle P. D., Godfrey S. M. The reactions of sulfur and selenium donor molecules with dihalogens and interhalogens. *Coordination Chemistry Reviews.* 2001;223(1):265–299.
4. dos Santos J. H. Z., Butler I. S., Daga V., Hadjikakou S. K., Hadjiliadis N. High-pressure Fourier transform micro-Raman spectroscopic investigation of diiodine-heterocyclic thioamide adducts. *Spectrochimica Acta - Part A:Molecular and Biomolecular Spectroscopy.* 2002;58(12):2725–2735.

5. Corban G. D. *Synthesis, characterization and study of new compounds of diiodine and/or iron with ligands which act or may act as antithyroid drugs. Possible implications of the new compounds in the mechanism of thyroid hormones synthesis* [PhD thesis]. Ioannina, Greece: University of Ioannina; 2006.
6. Corban G. J., Hadjikakou S. K., Hadjiliadis N., et al. Synthesis, structural characterization, and computational studies of novel diiodine adducts with the heterocyclic thioamides Nmethylbenzothiazole-2-thione and benzimidazole-2-thione: Implications with the mechanism of action of antithyroid drugs. *Inorganic Chemistry.* 2005; 44(23): 8617–8627.
7. Demartin F., Deplano P., Devillanova F. A., Isaia F., Lippolis V., Verani G. Conductivity, FT-Raman spectra, and X-ray crystal structures of two novel [D2I]In (n = 3 and D = N-methylbenzothiazole-2(3H)-selone; n = 7 and D = N-methylbenzothiazole-2(3H)-thione) iodonium salts. First example of I-.3I2 heptaiodide. *Inorganic Chemistry.* 1993; 32(17):3694–3699.
8. Antoniadis C. *Synthesis, characterization and study of new compounds of diiodine with ligands with possible antithyroid activity (e.g., thioamides). Contribution in the study of the mechanism of action of antithyroid drugs* [PhD thesis]. Ioannina, Greece: University of Ioannina; 2005.
9. Butler I. S. High-pressure Raman techniques. In: McCleverty J. A., Meyer T. J., eds. *Comprehensive Coordination Chemistry II.* Chapter 2.9. Amsterdam, The Netherlands: Elsevier; 2004:113–120.
10. Tezuka T., Nunoue S.-Y., Yoshida H., Noda T. Spontaneous emission enhancement in Pillar-type microcavities. *Japanese Journal of Applied Physics. Part 2.* 1993; 32(1):L54–L57.
11. Qiu C. H., Ahrenkiel S. P, Wada N., Ciszek T. F. X-ray diffraction and high-pressure Raman scattering study of iodineintercalated $Bi_2Sr_2CaCu_2O_{8+x}$. *Physica C: Superconductivity.* 1991; 185–189:825–826.
12. Zhao X.-S., Schroeder J., Bilodeau T. G., Hwa L.-G. Spectroscopic investigations of CdS at high pressure. *Physical Review B (Condensed Matter and Materials Physics).* 1989; 40(2):1257–1264.
13. Sengupta A., Quitevis E. L., Holtz M. W. Effect of high pressure on vibrational modes of polyiodides in poly(vinyl alcohol) films. *Journal of Physical Chemistry B.* 1997; 101(51):11092–11098.
14. Congeduti A., Postorino P., Nardone M., Buontempo U. Raman spectra of a high-pressure iodine single crystal. *Physical Review B (Condensed Matter and Materials Physics).* 2002;65(1):014302-1–014302-6.
15. Olijnyk H., Li W., Wokaun A. High-pressure studies of solid iodine by Raman spectroscopy. *Physical Review B (Condensed Matter and Materials Physics).* 1994;50(2):712–716.

16. Deplano P., Ferraro J. R., Mercuri M. L., Trogu EF. Structural and Raman spectroscopic studies as complementary tools in elucidating the nature of the bonding in polyiodides and in donor-I_2 adducts. *Coordination Chemistry Reviews.* 1999; 188(1):71–95.
17. Deplano P., Devillanova F. A., Ferraro J. R., Mercuri M. L., Lippolis V., Trogu E. F. FT-Raman study on charge-transfer polyiodide complexes and comparison with resonance Raman results. *Applied Spectroscopy.* 1994; 48(10):1236–1241.
18. Ferraro J. R. *Vibrational Spectroscopy at High External Pressures: The Diamond Anvil Cell.* New York, NY: Academic Press; 1984.
19. Eggert J. H., Goettel K. A., Silvera I. F. Elimination of pressureinduced fluorescence in diamond anvils. *Applied Physics Letters.* 1988; 53(25):2489–2491.
20. Chen W., Li G., Malm J.-O., et al. Pressure dependence of MN^{2+} fluorescence in ZnS:MN^{2+} nanoparticles. *Journal of Luminescence.* 2000; 91(3-4):139–145.
21. Thurber M. C., Hanson R. K. Pressure and composition dependences of acetone laser-induced fluorescence with excitation at 248, 266, and 308 nm. *Applied Physics B: Lasers and Optics.* 1999; 69(3):229–240.
22. Bowmaker G., Knappstein R. J. The low-frequency vibrational spectra of the 1: 1 acridine-bromine and 2 : 3 acridine-iodine complexes. *Australian Journal of Chemistry.* 1978; 31(10):2131–2136.
23. Aragoni M. C., Arca M., Demartin F., et al. C.T. complexes and related compounds between S and Se containing donors and I_2, Br2, IBr, ICL. *Trends in Inorganic Chemistry.* 1999; 6:1.
24. Boyle P. D., Christie J., Dyer T., et al. Further structural motifs from the reactions of thioamides with diiodine and the interhalogens iodine monobromide and iodine monochloride: an FT-Raman and crystallographic study. *Journal of the Chemical Society, Dalton Transactions.* 2000:3106–3112.
25. Deplano P., Devillanova F. A., Ferraro J. R., Isaia F., Lippolis V., Mercuri M. L. On the use of Raman spectroscopy in the characterization of iodine in charge-transfer complexes. *Applied Spectroscopy.* 1992; 46(11):1625–1629.

CITATION

Corban GJ, Antioniadis C, Hadjikakou SK, Hadjiliadis N, Meng F-F, and Butler IS. Pressure-Tuning Raman Spectra of Diiodine Thioamide Compounds: Models for Antithyroid Drug Activity. Bioinorganic Chemistry and Applications. 2006; 2006: 68542. doi: 10.1155/BCA/2006/68542.

Uncertainty Analysis of Drug Concentration in Pharmaceutical Mixtures

Michalakis Savva

ABSTRACT

Using a Taylor expansion to first order, a novel method was developed to calculate the uncertainty of drug concentration in pharmaceutical dosage forms. The method allows, in principle, calculation of the maximum potential error in drug concentration in a mixture composed of an infinite number of ingredients that are measured on multiple balances of variable sensitivity requirements.

Keywords: uncertainty propagation analysis; powder mixtures; dosage forms; drug compounding.

Introduction

Unlike mathematics which is based on numbers-and numbers are exact- science is based on measurements. Measurements are composed of a number, a unit and an uncertainty that denotes the deviation of the measurement from the true or ideal value. Advances made in basic and applied sciences are intimately related to the level of accuracy and precision with which experiments are carried out and to the accuracy with which the levels of confidence associated with the measured quantities are estimated.

Quite frequently experimentally measured quantities have to be combined in some way in order to determine some other derived quantity. For example, in order to find the uncertainty in drug concentration of a drug that was measured on a balance of a particular sensitivity and mixed with excipients measured on a balance of same or different sensitivity, the uncertainties in the measured quantities have to be combined appropriately. Although the rules that govern uncertainty propagation calculations are well developed [1], they were never applied to carry out uncertainty analysis of ingredient concentration in mixtures. In this paper, the author uses differential calculus and develops a method that combines uncertainties in the measurements of drug and excipients to effectively calculate the uncertainty of drug concentration in pharmaceutical mixtures.

The Method

The key idea in determining the uncertainty associated with drug concentration in drug-excipient mixtures is to produce a function that describes the variation of the drug quantity in the whole mixture, as shown below:

$$f = \frac{x}{x + y + z + ...} \tag{1}$$

x = drug quantity of interest

y, z .... = other ingredients

f = fraction or weight concentration of the drug in the total mixture.

Equation 1, is continuous in the domain of our interest since x, y, z... are real positive numbers. The potential error associated with f can be found from the total differential of the function above, as shown below:

$$f = f\,(x, y, z, ...)$$

$$\Rightarrow df = \left|\frac{\partial f}{\partial x}\right|_{y,z,...} dx + \left|\frac{\partial f}{\partial y}\right|_{x,z,...} dy + \left|\frac{\partial f}{\partial z}\right|_{x,y,...} dz + ... \tag{2}$$

df, dx, dy, dz are the uncertainties associated with the measurement of the corresponding quantities, f, x, y and z. More specifically, dx, dy, dz are equal to the sensitivity or readability of the balances utilized to measure the corresponding quantities, x, y and z. Performing the calculus of partial derivatives in Equation 1, yields:

$$\frac{df}{f} = \frac{y+z+...}{x} \bullet \frac{dx}{(x+y+z+...)} + \frac{dy+dz+...}{(x+y+z+...)}, \tag{3}$$

$\frac{df}{f}$ is the relative uncertainty in f, i.e. drug concentration by weight in the mixture.

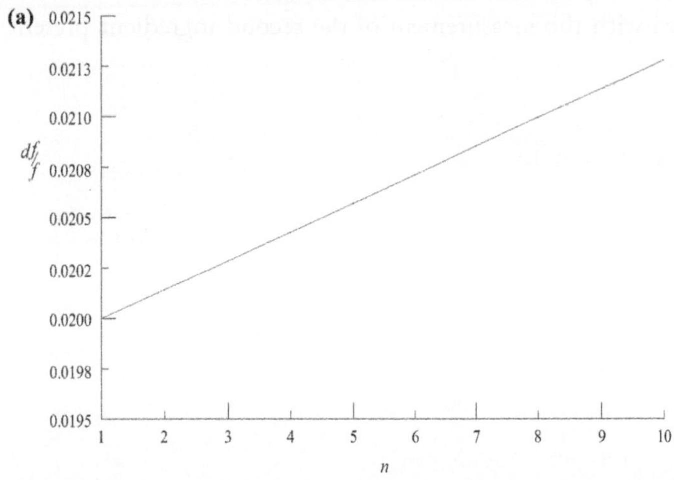

Figure 1a. Relative uncertainty in drug concentration at constant drug quantity in the mixture utilizing a single balance for measurement of all ingredients, as a function of the total # of ingredients n, using equation 5. The slope of the line is equal to dx/x.

Equation 3, denotes the fractional or relative uncertainty of drug concentration in the mixture, that is,

$$\frac{df}{f} = \left|\frac{\text{uncertainty in } f}{\text{ideal value } f}\right|. \tag{4}$$

To better investigate the effect of the # of ingredients composing the mixture n, on the uncertainty of drug concentration, Equation 4 was expanded for fixed balance sensitivity.

Starting from Equation 3 and assuming use of same balance to measure all ingredients, i.e. dx = dy = dz = …

$$\Rightarrow \frac{df}{f} = \frac{dx}{x} \bullet [1+(n-1)f] \tag{5}$$

n = total # of ingredients composing the mixture; drug x is excluded

A number of conclusions can be summarized from Equation 5:

1. $\frac{df}{f} = \frac{dx}{x}$ if the drug is mixed with a single ingredient (n = 1) and measured on the same balance. This result suggests that for a two-component mixture we only need to calculate the uncertainty associated with the drug measurement and we need not be concerned with the uncertainty associated with the measurement of the second ingredient present in the mixture.

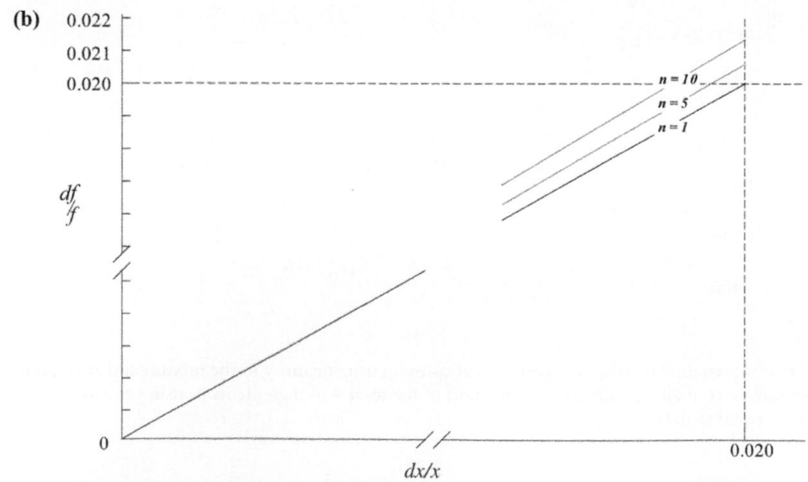

Figure 1b. Contour lines of constant n using equation 5. The value of n for each contour is also shown. Equation 5 indicates that the slope of the contour lines is given by [1+(n–1)f].

2. The uncertainty in drug concentration in the mixture increases with the number of other ingredients assuming that f, x and dx are constant as

shown in Figure 1a and b. Recognize, however, that both Figures indicate that the uncertainty in drug concentration is predominantly affected by the sensitivity of the balance.

The values for x, dx, and f were 0.5 g, 0.01 g and 0.0071 g, respectively for both figures. Both plots clearly indicate that as n → 1, $\frac{df}{f} \to \frac{dx}{x} = ..0\ 02$. These conclusions are valid for virtually all possible drug fractional masses, f, i.e. within the domain [1,0].

Conclusions

Using a rational approach and differential calculus, a function was created and a 1st degree Taylor polynomial was derived, respectively, to calculate the uncertainty of drug quantity in a mixture. This novel method can, in principle, calculate the "maximum interval" uncertainty in drug concentration in a mixture that is composed of an infinite number of ingredients and the ingredients are measured using balances of same or different sensitivity. This "core" equation can be used to: (1) study the maximum interval uncertainty variation as a function of the number and corresponding masses of ingredients and the sensitivities of the balances used (2) determine a balance of appropriate sensitivity needed to measure drug concentrations within the range of a maximum allowable uncertainty and (3) to determine the least allowable weight of an ingredient within a maximum interval uncertainty.

Reference

1. Bevington, P. R. Data Reduction and Error Analysis for the Physical Sciences. McGraw-Hill, New York, U.S.A., 1969. Analytical Chemistry Insights 2006:1

CITATION

Savva M. Uncertainty Analysis of Drug Concentration in Pharmaceutical Mixtures Analyical Chemistry Insights. 2006; 1: 1–3.

Characterization of Thermally Stable Dye-Doped Polyimide Based Electrooptic Materials

**Michael B. Meinhardt, Paul A. Cahill, Carl H. Seager,
Allyson J. Beuhler and David A. Wargowski**

ABSTRACT

Preparation and characterization of novel dye-doped polyimide films for electrooptics is described. Thermal stabilities of donor-acceptor 2, 5-diaryl oxazoles were evaluated by differential scanning calorimetry. Absorptive losses in thin films of Ultradel 9000D doped with donor-acceptor oxazoles were measured by photothermal deflection spectroscopy. Absorptive losses at high doping levels may be explainable by dye-dye aggregation or dye degradation during the curing process. Lower doping levels, however, show losses of ≤ 3.0 dB/cm at 830 nm and ≤ 2.4dB/cm at 1320 nm.*

Introduction

Polymeric electrooptic materials have the potential to replace electronic switches in applications which require minimization of heat dissipation while maintaining high switching speeds. Polyimide matrices incorporating electrooptic dyes are promising materials for such applications due to their low cost and compatibility with existing processing environments. Requirements for practical systems include large electrooptic coefficients, thermal stability, high Tg, refractive index differences (waveguide formation), conductivity (poling and data impression) and low optical loss (transmission) (Figure 1).

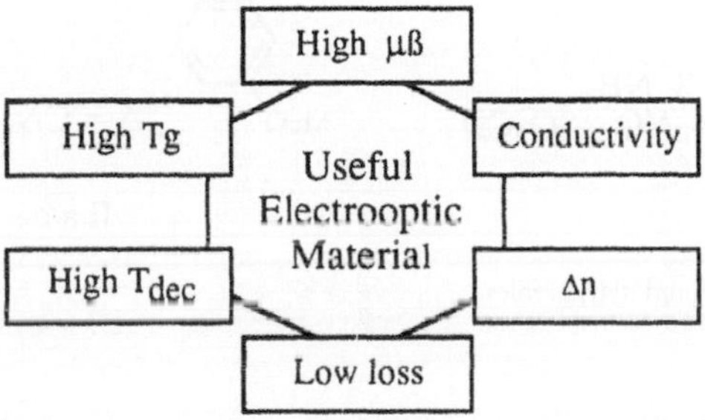

Figure 1. Property-performance criteria for electrooptic materials.

Amoco Ultradel 9000D® aromatic polyimides (Fig. 2, below) are a family of 7-butyrolactone (GBL) soluble, fully imidized, fluorinated polyimides developed for integrated optical applications. Thermal or photochemical cross-linking imparts a Tg approaching 400 °C and provides stability for poled polymer systems. Excellent optical transparencies have also been demonstrated in these materials.

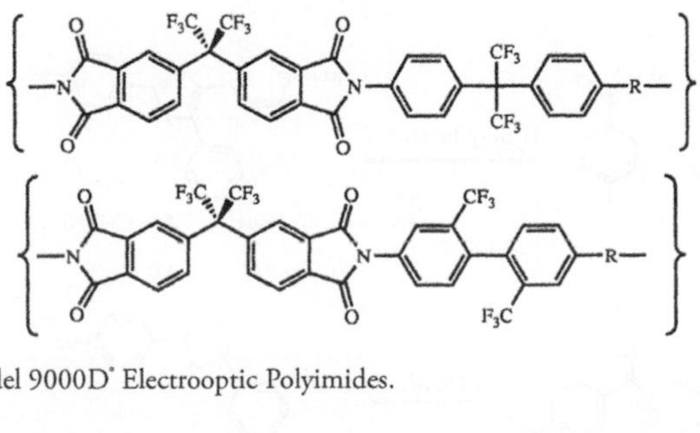

Figure 2. Amoco Ultradel 9000D® Electrooptic Polyimides.

Previously we reported results of characterization of Ultradel polyimides for electrooptic applications, as well as properties of these polymers heavily doped with donor-acceptor triaryl azole dyes (I) (Figure 3). In this study we have extended our investigations to examine novel diaryl oxazoles (II) and their properties in polyimide electrooptic films. Computational studies suggested that linear and nonlinear optical properties of diaryl oxazoles should be similar to their triaryl counterparts with the advantage of a higher nonlinearity to weight ratio. Thus, we undertook the synthesis and characterization of novel dyes II a-c.

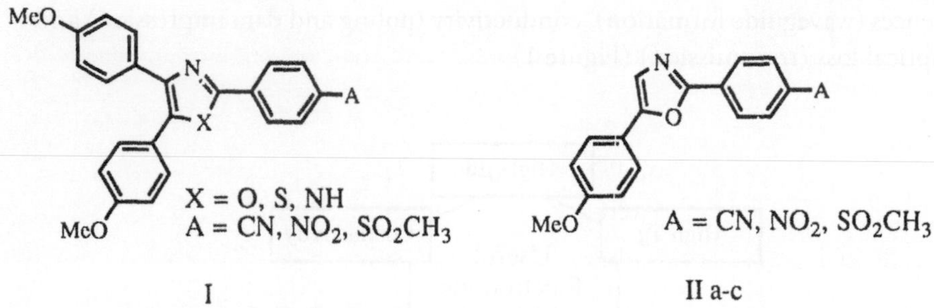

Figure 3. Donor-acceptor triaryl-and diaryl-azoles.

Results and Discussion

Triaryl oxazoles were prepared in multi-gram lots by condensation of an appropriate benzamide with anisoin in the presence of an acid catalyst (Figure 4.) A more elaborate route was developed to achieve the regioselectivity required of the diaryl oxazoles. For example, 4-cyanobenzoyl chloride was condensed with 2-amino-(4-methoxyphenyl)ethanone via a modified Schötten-Baumann procedure to give a 2-aza-l,4-butadione. Ring closure and dehydration were accomplished by refluxing the azadione in phosphorous oxychloride.

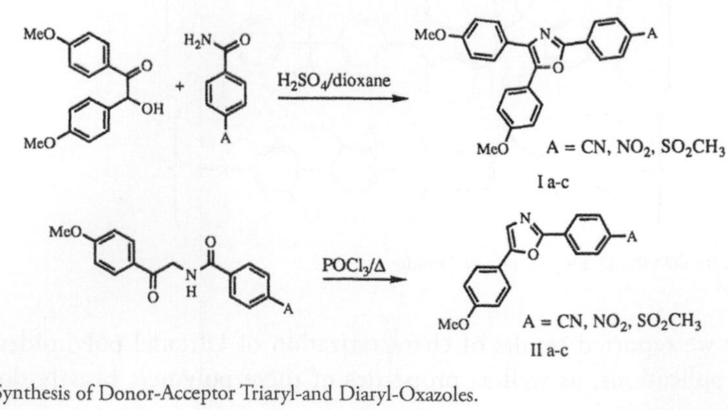

Figure 4. Synthesis of Donor-Acceptor Triaryl-and Diaryl-Oxazoles.

Compatibility with Ultradel 9000D® processing conditions demands dye thermal stability at 300 °C for one hour during the curing cycle, Thus, DSC studies were undertaken to evaluate the thermal stability of these dyes. Semi-preparative HPLC methods provided dye samples of high purity (>99%). Novel donor-acceptor oxazole dyes were assayed for thermal stability using sealed tube methods in order to eliminate concerns of dye sublimation during the DSC experiment. Samples of ~ 1 mg were sealed into 1.5 x 7.0 mm glass capillary tubes and inserted into aluminum holders designed to fit into a Perkin-Elmer DSC 7 and heated at 10 °C/min from 25-400 °C. The results from two experiments are shown in Figure 5.

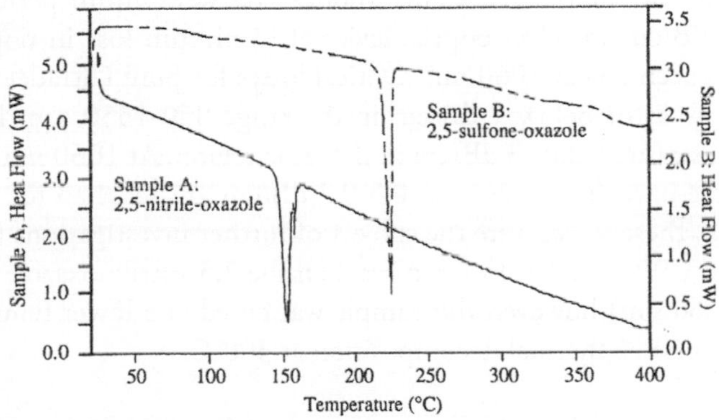

Figure 5. Sealed Tube DSC Spectra of Acceptor-Substituted Diaryl-Oxazoles (II-a,b).

From Table 1, below, no signs of thermal decomposition were observed below 400 °C for nitrile and sulfonyl substituted di-and triaryl oxazoles. Nitro compounds, in contrast, exhibited marked thermal decomposition exotherms with decomposition onset temperatures of 360-365 °C, conditions only marginally compatible with polyimide curing conditions. Nitro aromatics are also oxidants which may lead to long-term stability problems. As a result of these experiments nitrile and sulfonyl substituted diaryl oxazoles were selected for polymer doping studies.

Table 1. Summary of DSC Data for Donor-Acceptor Oxazoles.

Compound	$T_{m.p.}$ (°C)	$T_{dec., max.}$ (°C)	$T_{onset, dec.}$ (°C)
2,5-nitrile	155	>400	>400
2,5-nitro	167	385	361
2,5-sulfone	219	>400	>400
2,4,5-nitrile	183	>400	>400
2,4,5-nitro	207	396	364
2,4,5-sulfone	210	>400	>400

Photothermal deflection spectroscopy (PDS) provides a method for determining absorptive losses in thin films independent of scattering loss. PDS spectra of triaryl oxazoles I-a,b were previously reported. We now present similar results for 2,5-nitrile-oxazole (II-a) at varying concentrations of 5 to 20%-wt. DCM-doped polyimide was also examined for comparison, γ-Butyrolactone (GBL) solutions of polymer or polymer-dye combinations were spin coated onto infrasil quartz wafers and cured in a nitrogen-purged oven for 10 min at 100 °C, 30 min at 175 °C and 60 min at 300 °C. Sample thicknesses were measured on a Sloan Dektak profilometer and varied from 8 to 18 μm ± 1 μm. The results of PDS measurements for 2,5-nitrile-oxazole and DCM doped Ultradel 9000D® films are given in Figure 6.

Loss values for the undoped Ultradel 9000D® are well within performance criteria of < 1.5 dB/cm for electrooptic devices•l Minimum loss in doped and undoped samples occurs near 1060 nm. Optical losses for pure Ultradel 9000D® cured at 300°C are at or below 1 dB/cm in the range 850-1350 nm. For dye-doped samples, losses are below 3 dB/cm in the same region. At 1060 nm, loss for 20%-wt 2,5-nitrile-oxazole samples was 0.5-0.7 dB/cm greater than for the 5 or 10 %-wt samples. These losses were the subject of further investigations (vide infra). DCM doped U9000D® has lower losses than the 2,5-nitrile-oxazole samples (1.4 dB/cm at 1060 nm) however, this sample was cured at a lower temperature (250°C) due to excessive thermal decomposition at 300°C.

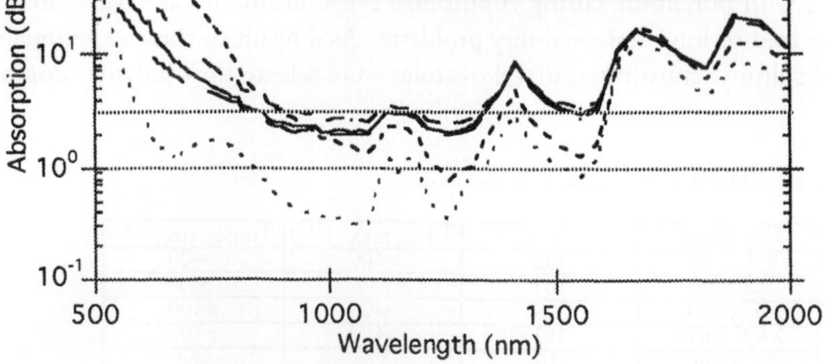

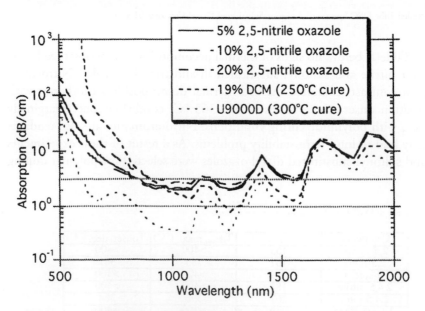

Figure 6. PDS Spectra of 2,5-Nitrile-Oxazole (II-a) Doped Ultradel 9000D®.

Table 2. Selected Loss Data for 2,5-Nitrile-Oxazole (II-a) Doped U9000D®, 300 °C Cure.

Sample	dB/cm_{826nm}	dB/cm_{1069nm}	dB/cm_{1320nm}
5%-wt	3.3	2.1	2.4
10%-wt	3.0	2.0	2.2
20%-wt	4.5	2.6	2.7
DCM (19%-wt)	5.4	1.4	1.0
U9000D®	1.2	0.32	1.0

Notes: DCM sample was cured at 250 °C. Loss for U9000D® at 830 nm was 0.7 dB/cm, as determined by waveguide loss spectroscopy.[2]

The origins of optical loss in dye-doped thin films of polyimides are not well understood. While the optical loss associated with the undoped Ultradel 9000D® is small (0.3 dB/cm at 1060 nm) as determined by PDS and waveguide loss spectroscopy, losses for oxazole-doped polyimides are greater (~ 1.6-1.8 dB/cm). Losses are similar at 5 and 10%-wt doping levels, with a jump of ~ 0.6 dB/cm at 1060 nm for the 20%-wt sample. The origin of this additional loss was the subject of further investigations.

We sought to examine the question of the added loss in 20%-wt 2,5-nitrile-oxazole by UV-Vis absorption studies. There is precedent for short-wavelength polymeric charge transfer absorptions in polyimides. We sought to examine the potential for dye-matrix interactions as the origin of the additional absorptions. Variation of solvent polarity was hypothesized to enhance the opportunity for dye-matrix charge transfer. For example broad, weak charge transfer from anisole to the cyanophenyl ring might be expected to give rise to long-wavelength absorptions.

UV-Vis absorption spectra for triaryl nitrile-oxazole were determined in solvents of varying polarity. Dilute solutions (~ 10 -5 M) of HPLC purified dyes were employed to eliminate concerns of spurious absorptions arising from dye-dye or dye-impurity interactions. Cell path lengths of 10.0 cm were used to maximize small absorption signals. Within experimental limits, no discernible charge-transfer absorptions were observed in any dye-solvent system in the region 600-900 nm. Minor solvatochromic effects were observed for absorption maxima.

DSC studies have demonstrated the excellent thermal stabilities of novel diaryl nitrile and sulfone dyes (II-a,c). In addition, dye-doped Ultradel 9000D® samples doped at 5 and 10%¬wt exhibit acceptable absorbance losses of ≤ 3 dB/cm at 820 and 1320 nm. Higher doping levels give rise to absorbance losses which cannot be explained solely on the basis of Beer's Law. It is clear that absorptive losses do not arise from the polymer itself. Our study of dye-solvent (dye-matrix) systems suggests that observed losses are not due to charge transfer interactions between dye and matrix (dye-solvent or dye-polymer).

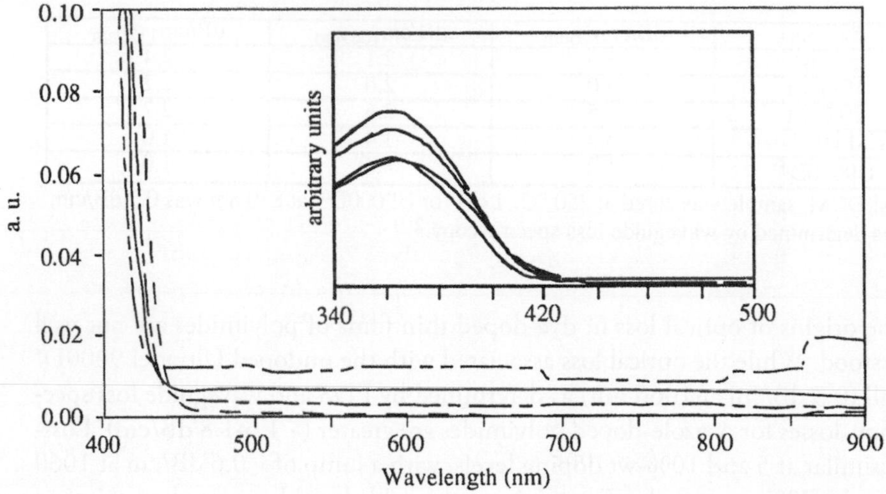

Figure 7. UV-Vis absorption spectra for nitrile-oxazole (I) in various solvents. Solvent (in descending order at 600 nm) benzene, ethyl benzoate, anisole and benzonitrile. Inset: solvent (in descending order at 360 rim) anisole, ethyl benzoate, benzonitrile and benzene.

Possible mechanisms of absorptive loss at high doping levels include dye aggregation and dye degradation during the curing process. PDS samples were free of apparent crystallinity as determined by optical microscopy. DSC samples examined after thermal cycling frequently revealed glassification upon cooling. These observations suggest the difficulty in ascertaining dye-polymer inhomogeneity.

Conclusions

DSC studies have demonstrated the excellent thermal stabilities of diaryl nitrile and sulfone oxazoles (II-a, c). In addition, dye-doped Ultradel 9000D® samples doped with dyes at 5 and 10%-wt 2,5-nitrile-oxazole (II-a) exhibit acceptable absorbance losses of < 3 dB/cm at 1060 nm. Higher doping levels of 2, 5-nitrile-oxazole give rise to unexpected absorbance losses. On the basis of this study, it seems apparent that these losses do not arise from either dye-matrix interactions or from the polymer itself. The distinct possibility remains that at high concentrations dye aggregation in cured polyimide films or dye degradation during the cure process may be the ultimate origin of these absorptive losses. Should this be the case, future systems may require enforced chromophore separation by copolymerization of the dye in order to attain losses of < 2 dB/cm.

Experimental

2-(4-Cyanophenyl)-4,5-Bis-(4-Methoxyphenyl)oxazole

A solution of 6.51 g (32.5 mmol) of anisoin and 9.30 g (34.1 mmol) of 4-bromobenzamide in 75 mL of 1,4-dioxane was treated with 0.5 mL of conc. sulfuric acid and heated at reflux for 18 h. Neutral workup provided a crude product which was purified by flash chromatography and recrystallized from ethyl acetate to give 2-(4-bromophenyl)-4,5-bis-(4-methoxyphenyl)oxazole, 8.78 g (62%) m.p.: 168-170 °C.

Next, a solution of the bromide (4.01 g, 9.20 mmol) and copper cyanide (4.12 g, 46.0 mmol) in N-methylpyrrolidinone (125 mL) was heated at 180 °C for 18 h. The mixture was poured over ice and the precipitated solids collected and dried. Flash chromatography (25% ethyl acetate-hexane), followed by recrystallization from ethyl acetate gave the desired product, 2.46 g (70%). m.p.: 184-186 °C.

The spectral data for 2-(4-cyanophenyl)-4,5-bis-(4-methoxyphenyl)oxazole were the following: FTIR (KBr) 3045, 3005, 2960, 2935, 2905, 2840, 2225 (CN), 1610, 1520, 1500, 1300, 1255, 1180, 1110, 1025, 965, 840 cm-1; 1H NMR (CDC13, 200 MHz) 8 8.20 (d, J = 8.7 Hz, 2 H, arom.), 7.73 (d, J = 8.7 Hz, 2 H, arom.), 7.59 (m, 4 H, arom.), 6.91 (m, 4 H, arom.), 3.83 (s, 6 H, OCH3)

MS calcd for C24H18N2O3: 382. Found: m/e 382.

UV-Vis: λmax(benzene) 362 nm, (εmax14600 L•mol-l).

(4-Methoxyphenyl)-2-Aminoethanone Hydrochloride

To a stirred solution of (4-methoxyphenyl)-2-bromoethanone (9.85 g, 43.0 mmol) in 300 mL of absolute ethanol was added 6.33 g (45.2 mmol) of hexamethylenetetraamine in one portion. After a brief induction period a white precipitate formed. After stirring at 25 °C for 8 h, 70 mL of a 2.5:1 solution of hydrochloric acid in water was added and the mixture stirred at 25 °C for an additional 8 h. The flask was stoppered and set in a freezer for several days and the crystalline solid was collected by suction filtration washing several times with cold ethanol to give 6.64 g (77%) of the amine hydrochloride, m.p. 204-206 °C, (dec).

The spectral data for (4-methoxyphenyl)-2-aminoethanone hydrochloride were the following: FTIR (KBr) 3000 (s), 1990, 1685 (CO), 1605, 1505, 1470, 1455, 1430, 1390, 1325, 1270, 1255, 1180, 1130, 1060, 1020, 970, 830, 810 cm-1; 1H NMR (CDCl3, 200 MHz) 5 7.86 (d, J = 8.9 Hz, 2 H, arom.), 6.89 (d, J = 8.9 Hz, 2 H, arom.), 4.05 (s, 2 H, CH2NH2), 3.82 (s, 3 H, OCH3).

2-(4-Cyanophenyl)-4-(4-Methoxyphenyl)-2-Aza-l, 4-Butadione

To 50 mL of an ice cold solution of 0.83 M sodium hydroxide was added 8.07 g (40.0 mmol) of (4-methoxyphenyl)-2-aminoethanone hydrochloride. After 30 min at 0 °C, a solution of 4-cyanobenzoyl chloride (6.51 g, 39.3 mmol) in 60 mL of tetrahydrofuran was added dropwise concomitantly with 50 mL of the sodium hydroxide solution over 1.5 h. After the addition was complete, the mixture was stirred for an additional 12 h. The reaction mixture was chilled and the precipated product was collected by filtration, dried and used without further purification, 6.01 g (52%), m.p. 190.0-193.5 °C.

The spectral data for 2-(4-cyanophenyl)-4-(4-methoxyphenyl)-2-aza-1,4-butadione were the following: FTIR (KBr) 3340, 3010, 2950, 2840, 2360, 2230, (CN), 1675, 1595, 1560, 1535, 1495, 1460, 1435, 1420, 1365, 1310, 1265, 1240, 1175, 1115, 1025, 1005, 995,900, 855, 868 cm-1; 1H NMR (CDC13, 200 MHz) 5 8.01 (d, J = 9.00 Hz, 2 H, anisyl), 7.99 (d, J = 8.63 Hz, 2 H, arom.), 7.78 (d, J = 8.63 Hz, 2 H, arom.), 7.45 (m, 1 H, NH), 7.00 (d, J = 9.00 Hz, 2 H, anisyl), 4.90 (d, J = 4.14 Hz, 2 H, CH2), 3.91 (s, 3 H, CH3).

2-(4-cyanophenyl)-5-(4-methoxyphenyl)oxazole

A solution of 6.01 g (20.4 mmol) of 2-(4-cyanophenyl)-4-(4-methoxyphenyl)-2-aza-l,4-butadione in 50 mL of $POCI_3$ was heated at reflux for 5 h. Excess $POC1_3$ was removed by vacuum distillation and the residual solids were suspended in water and collected by suction filtration to give 5.18 g of a crude product, m.p.: 156-158 °C. Recrystallization from acetonitrile provided 3.90 g (69%) of pure product, m.p. 151.5-153.0 °C as pale yellow needles.

The spectral data for 2-(4-cyanophenyl)-5-(4-methoxyphenyl)oxazole were the following: FTIR (KBr) 2995, 2840, 2225 (CN), 1610, 1575, 1500, 1485, 1440, 1410, 1395, 1280, 1255, 1175, 1130, 1115, 1060, 1025, 955, 840, 825, 810 cm-l; 1H NMR (CDC13, 200 MHz) 8 8.18 (d, J = 8.63 Hz, 2 H, arom.), 7.76 (d,J = 8.63 Hz, 2 H, arom.), 7.66 (d,J = 8.92 Hz, 2 H, anisyl), 7.39 (m, 1 H, NH), 6.99 (d, J = 8.92 Hz, 2 H, anisyl), 3.87 (s, 3 H, OCH3).

MS calcd for C17H12N2O2: 276.30. Found: m/e 276.

UV-Vis: λmax(benzene) 350 nm, (εmax27400 L.mol-1).

Acknowledgements

This work was performed, in part, at Sandia National Laboratories and was supported by U.S. Department of Energy under contract DE-AC04-94DP85000.

The authors wish to acknowledge Ken Singer and Tony Kowalczyk (Case Western Reserve University) for helpful discussions. The technical assistance of C. Allen (Amoco) is also acknowledged.

References

1. R. Lytel, G. F. Lipscomb, J. T. Kinney and E. S. Binkley, in Polymers for Lightwave and Intem'ated Optics, edited by L. A. Hornak, (Marcel Dekker, Inc., 1992) p. 433.
2. Amoco Chemicals Ultradel9000D®, Product Bulletin.
3. (a.) P. A. Cahill, C. H. Seager, M. B. Meinhardt, A. J. Beuhler, D. A. Wargowski, K. D. Singer, T. C. Kowalczyk, T. Z. Kosc, Proc. SPIE, 2025–07, 1993. In press. (b.) For examples of related work with triaryl azoles, see: C. R. Moylan, R. D. Miller, R. J. Twieg, K. M. Betterton, V. Y. Lee, T. J. Matray, C. Nguyen, Chem. Mater. 5, 1499 (1993).
4. K. Matsumoto and P. P. K. Ho, U.S. Patent No. 4322428 (30 March, 1982).
5. A. W. Ingersoll and S. H. Babcock, Org. Syn. Coll. Vol. II, 328–330 (1943).
6. M. A. Perez and J. M. Bermejo, J. Org. Chem. 58, 2628–2630 (1993).
7. L. F. Whiting, M. S. Labean, S. S. Eadie, Thermochemica Acta, 136, 231–245 (1988).
8. J. M. Salley, T. Miwa, C. W. Frank, in Materials Science of High Temperature Polymers for Microelectronics, edited by D. T. Grubb, I. Mita and D. Y. Yoon (Mater. Res. Soc. Proc. 227, Pittsburgh, PA, 1991) pp. 117–124.
9. C. Reichardt, Solvents and Solvent Effects in Organic Chemistry, 2nd ed. (VCH, New York, 1990).

CITATION
Meinhardt MB, Cahill PA, Seager CH, Beuhler AJ, Wargowski DA. Characterization of Thermally Stable Dye-Doped Polyimide Based Electrooptic Materials. Presented at the Fall Meeting of the Materials Research Society (MRS), Boston, MA, 29 Nov. - 3 Dec. 1993. Contracted by the NASA Scientific and Technical and Scientific Information (STI) Program. US Government publication.

Sol-Gel-Derived Silicafilms with Tailored Microstructures for Applications Requiring Organic Dyes

Monica N. Logan, S. Prabakar and C. Jeffrey Brinker

ABSTRACT

A three-step sol-gel process was developed to prepare organic dye-doped thin films with tailored porosity for applications in chemical sensing and optoelectronics. Varying the acid-and base-catalyzed hydrolysis steps of sols prepared from tetraethoxysilane with identical final H_2O/Si ratios, dilution factors and pH resulted in considerably different distributions of the silicate polymers in the sol (determined by ^{29}Si NMR) and considerably different structures for the polymer clusters (detennined by SAXS). During film formation these kinetic effects cause differences in the packing and collapse of the silicate network, leading to thin films with different refractive indices and volume fraction porosities. Under conditions where small pore-plugging species *were avoided, the porosities of as-deposited films could be varied by aging the sol prior to film deposition. This strategy, which relies on the growth and aggregation of fractal polymeric clusters, is compatible with the low temperature madnear neutral pH requirements of organic dyes.*

Introduction

Organic dye-doped thin films are being developed for applications in chemical sensing [1-3] and optoelectronics [4-7]. These applications require a rigid transparent matrix so the dye can exhibit a photochromic response, a low temperature synthesis technique to prevent decomposition of the dye, and depending on the intended application, the ability to control the thin film microstructure. Porous films are useful for chemical sensing while dense films are useful for applications requiring optical, dielectric or protective functions. Sol-gel-derived inorganic glass matrices exhibit superior chemical, thermal, mechanical and optical properties compared to organic polymer matrices, can be processed at low temperatures compatible with organic dye molecules, and permit tailoring the micrrostructttre of the matrix to the application.

Sol-gel processing offers numerous ways to control the microstructure of thin films. While there are many ways to affect the competition between phenomena that promote a more compact or a more porous structure in a dip-coated film, e.g., by changing the composition or dip-coating conditions [8-9], it is also possible to tailor the microstructure of films obtained from a single sol composition by changing the size or structure of the inorganic polymers in the sol [10-11]. Often the inorganic polymers constituting the sol are mass fractal objects, so they have two properties [12] we can exploit to tailor the microstructure of the resulting dip-coated films. First, the porosity of a mass fractal object increases with increasing size:

$$\text{Porosity} \sim \text{volume/mass} \sim r3/rD, \qquad (1)$$

where r = cluster size and D = mass fractal dimension ($0 < D < 3$). Second, for $D \geq 1.5$, the tendency for mass fractal objects to interpenetrate decreases with increasing size, preserving the porosity between clusters as the film is formed. Aging a sol allows the inorganic clusters to grow larger, and aging (at intermediate to basic pH) also causes the clusters to coarsen and become stiffer [13], so a film from an aged sol will have 1) more porous clusters, 2) less interpenetration of clusters as they are concentrated, and 3) less collapse of the clusters due to the capillary forces exerted during drying [8-10].

There are two complications to this simple method of varying film porosity by varying the age of the sol. First, the tendency for fractal objects to interpenetrate is a function of D as well as r. The tendency to interpenetrate is inversely related to the mean number of intersections M1,2 of two mass fractal objects of size r and mass fract_ddimension D placed in the stone region of space [12]:

$$M1,2 \propto r(2D -3). \qquad (2)$$

According to Equation 2, for D > 1.5, M1,2 increases with both increasing r and increasing D, leading to a reduced tendency for interpenetration. However, Equation 1 indicates that the porosity of individual clusters increases with decreasing D. To maximize porosity, an intermediate value of D is required that balances cluster porosity and cluster-cluster interpenetration: if D is too large, individual clusters are not very porous, whereas if D is too small, cluster interpenetration reduces the porosity between clusters. We will show that D is a function of the sequence and pH of the sol preparation steps. The second complication is that small inorganic clusters can "fill-in" the pores of the deposited film, masking the porosity created by aging a sol.

Experimental

A two- or three-step acid/base-catalyzed process was used to prepare sols identified as B2 [10, 14-15], AAB [11], or AAB(1/5), respectively. The first step was the same for all of the sols: tetraethoxysilane (TEOS), ethanol, water and HCI were mixed in the molar ratio 1:3.8:1.0:0.007, refluxed at 60°C for 90 rain and cooled to room temperature [151. This solution, referred to as stock solution, was used immediately or stored in a freezer at –20°C. For B2 the second hydrolysis step consisted of adding an aqueous solution of 0.05 M NH_4OH in additional ethanol, resulting in a final molar ratio of 1 TEOS:48 ethanol:3.7 H20:0.007 HCI:0.002 NH_4OH, and a final sol pH of 5.5 as estimated using colorimetric pH indicator strips (EM Science). For AAB the second step consisted of adding 1 M HCI diluted in ethanol, resulting in a H_2O/Si ratio = 2.5, and refluxing at 60°C for 60 rain. The third step consisted of adding an aqueous solution of 2 M NH_4OH diluted in ethanol, resulting in a final molar ratio of 1:48:3.7:0.028:0.05. For AAB(1/5) sols, the second step consisted of adding HC1, H_2O and ethanol and refluxing for 1 or 4 h, bringing the molar ratio to 1:20.6:2.5:0.0056. The third hydrolysis step consisted of adding 0.355 M NH_4OI-I in ethanol, resulting in a final molar ratio I of 1:30:3.7:0.0056:0.0078. All three sols had the same final H_2O /Si ratio and pH. The sols were aged in a Class-A (explosion-proof) oven at 50°C, and samples were removed at intervals up to the gel point or allowed to gel. The gelled samples were subjected to ultrasound to prepar sols of a consistency suitable for dip-coating. ^{29}Si NMR was employed to detennine the species distribution of the solutions after each step and after aging the sols. The effects of synthesis and aging conditions on the intermediate-scale sol structure (0.5 -30 nm) were determined by SAXS. Films were applied to Si substrates by dip-coating at rates varying from 10 to 25 cnVmin in a nitrogen atmosphere. After 20 rain drying under a heat lamp at 60°C, some films were supported on edge in quartz trays and fired in air at moderate temperatures (usually up to 400°C). Ellipsometry was used to detennine the refractive index and thickness of the as-deposited and fired films. TGA experiments were performed on bulk B2 and AAB gels that had been dried at 50°C.

Results and Discussion

Looking at film formation as the aggregation of non-interpenetrating fractal clusters, we expect that sol aging, which increases the size of the clusters, should increase film porosity and decrease refractive index. For films dip-coated from B2 sols, however, the expected effect did not appear until after heat treatment [10], as Table I shows. NMR studies [10, 16] of the B2 sols showed that about 5% Q^0 and 30% Q^1 species remain after aging. During dip-coating, these small, rather unreactive species fill in the pores of the network created by the non-interpenetrating clusters, so the as-deposited films show little variation in refractive index with sol age. TGA performed on a dried B2 gel [17] shows a 45% weight loss at 400°C owing to the removal of organics associated with these small unhydrolyzed or partially hydrolyzed species. Heating the films to 400°C reveals a range of porosity with sol age in the underlying network consistent with expectations from Equation 1.

Table I. Refractive indices and vol% porosities (calculated from the Lorentz-Lorenz relation [18]) for B2 and AAB films as a function of aging time normalized by the gelation time. Mass fractal dimension values are for sols aged for comparable normalized aging times.

B2 Sol:		As-deposited:		After 400°C:		AAB Sol:		As-deposited:		After 400°C:	
Normalized Aging Time	D	Refractive Index	Porosity (vol %)	Refractive Index	Porosity (vol %)	Normalized Aging Time	D	Refractive Index	Porosity (vol %)	Refractive Index	Porosity (vol %)
0.00	2.27	1.425	4.9	1.369	16.0	0.00	--	1.435	2.9	1.381	13.6
0.32	2.32	1.424	5.0	1.346	20.8	0.05	1.37	1.432	3.5	1.378	14.2
0.63	2.40	1.421	5.6	1.325	25.1	0.43	1.50	1.398	10.2	1.365	16.9
0.95	--	1.418	6.2	1.292	32.1	0.86	--	1.369	16.0	1.341	21.8
1.24	--	1.417	6.4	1.240	43.5	0.90	1.70	1.353	19.3	1.331	23.9

Because such a heat treatment is desla-uctiveto organic dyes, we added a second acid-catalyzed hydrolysis step (the AAB sol) to promote more complete reaction and to reduce the proportion of small, incompletely-hydrolyzed species. NMR studies of the AAB sols [16] showed that the second acid-catalyzed step promotes extensive hydrolysis and condensation: there were two broad peaks attributable to Q2 and Q3. The third (base-catalyzed) step promotes further condensation, and a lnagic angle spilming spectruln of an AAB gel 20 h after the third step shows 60% Q3 and 40% Q4 species. The TGA trace for a dried AAB gel [17] shows only about 20% weight loss at 400°C, and the as-deposited AAB films show the expected range in refractive index with sol age (Table I). Heating the AAB films to 400°C produces a small additional increase in porosity.

The values of the mass fractal dimension (obtained from SAXS [16]) show how the structure of the silicate polymers in the sol varies with the different routes these sols take to the same final dilution, H2O/Si ratio and pH. The mass fractal dilnensions are considerably different. We would expect AAB, which is right at the borderline of D = 1.5 for Equation 2, to exhibit different behavior than B2 with D = 2.3. However, Table I shows that both sols exhibit an increase in porosity with sol age (once the pore-plugging species are burned out of B2). The reason for this is the tradeoff between cluster porosity and cluster interpenetration. B2 clusters are less porous but pack less efficiently than AAB clusters. This tradeoff is illustrated schematically in Figure 1.

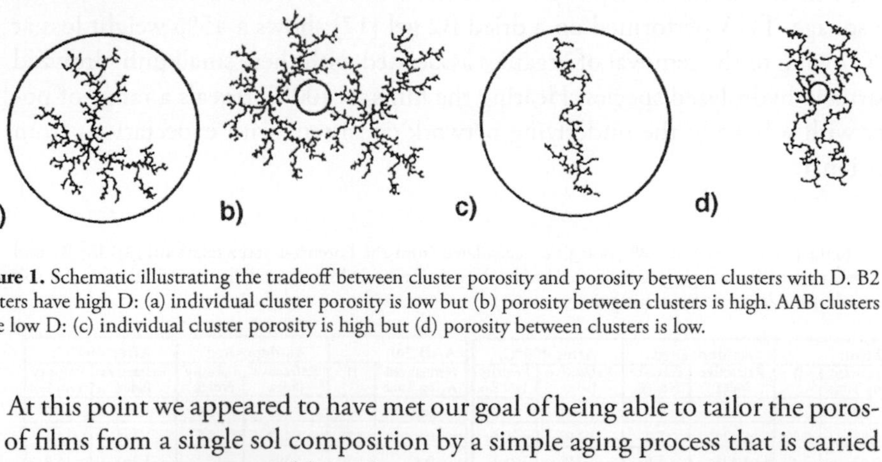

Figure 1. Schematic illustrating the tradeoff between cluster porosity and porosity between clusters with D. B2 clusters have high D: (a) individual cluster porosity is low but (b) porosity between clusters is high. AAB clusters have low D: (c) individual cluster porosity is high but (d) porosity between clusters is low.

At this point we appeared to have met our goal of being able to tailor the porosity of films from a single sol composition by a simple aging process that is carried out at temperatures compatible with organic dyes. The additional acid-catalyzed step eliminated the small pore-plugging species and the need for a heat treatment. The as-deposited AAB films showed a clear reduction in refractive index (increase in porosity) with aging, and the TGA results confinned that organic groups were greatly reduced in AAB. Unfortunately, some of the dried AAB gels were coated with a white powder. XRD analysis [19] confirmed our suspicion that the powder was NH4CI salt formed from the large amounts of HCI and NH4OH catalysts. XRD of the films showed that they were amorphous, but films that were dried at

50°C for the same period of time as the gels had loose white powder on the surface that looked crystalline under an optical microscope (30x).

Heating the films may be a simple way to remove the salt. Additional XRD studies showed that washing an AAB gel with ethanol eliminates the salt, so washing the films might also work. It would be better, however, if we found the irfinimum catalyst level that eliminates the small pore-plugging species. A new sol with less catalyst might also have D values that fall between those of B2 and AAB, and an intermediate value of D should produce more porous as-deposited films.

We prepared a series of sols from the stock solution with various fractions of the total moles of HCI in AAB. The base-catalyzed step was adjusted to produce a final pH of 5.5 for each sol. For sols with [HCI] = 1/4 or more of the [HCI] in AAB, XRD detected the presence of NH4CI in the dried gels. For sols with [HCI] = 1/10 of the [HCI] in AAB, the XRD was amorphous, the NMR showed mainly Q24 cyclics, and sols at pH 5.5 did not gel after 24 days at 50°C. For sols with [HCI] = 1/5 of the original AAB, the XRD was amorphous, and the sols gelled in less than a week. After the second acid-catalyzed step, which was refluxed for 1 h, the NMR showed 17% Q1, 62% Q2 and 21% Q3, and the Q1 species were more fully hydrolyzed than in B2 sols. After the base-catalyzed step, AAB(1/5) had 24% Q2, 49% Q3 and 27% Q4 species.

Preliminary refractive index data for AAB(1/5) show some dependence of refractive index on sol age, consistent with the trends observed for AAB, but a much stronger dependence on the substrate withdrawal rate. Normally, if the condensation rate is high, we would expect a faster withdrawal rate to produce more porous films. A fast withdrawal rate leads to a thicker film due to the balance between the viscous drag and the force of gravity [8-9]. Thicker films take longer to dry, so more condensation reactions can occur. This strengthens the film, making it more able to resist collapse due to capillary pressure during drying, so thicker films tend to be more porous as is generally observed for the AAB(1/5) series of films. We found that the withdrawal rate had no significant impact on the refractive index of B2 films [10], however. One possible explanation is that B2 has a large proportion of small, mainly unhydrolyzed species. (High resolution 29Si NMR studies indicate that about 75% of the Q1 species are unhydrolyzed dimer [20]). The small, unreactive, fully-ethoxylated clusters in B2 may interfere with condensation. Figure 2 shows the dependence of refractive index on withdrawal rate for AAB(1/5) as well as an upward trend for the refractive index at high ages that we cannot explain yet. Preliminary SAXS results indicate that we may not have the intermediate value of D that we hoped to achieve, so further adjustments to the process for AAB(1/5) may be needed.

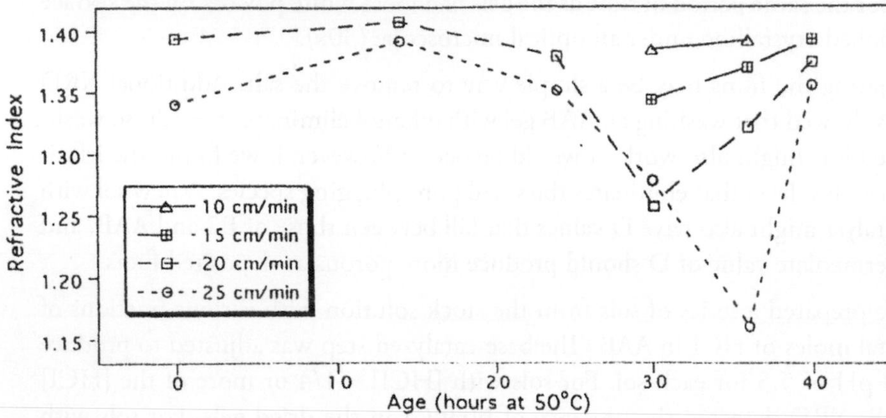

Figure 2. Preliminary refractive index data for the AAB(1/5) films.

What we have learned so far from B2, AAB and AAB(1/5) is that for sols with almost identical ingredients the sequence and timing of steps are important. This implies that kinetics determine the outcome. To test this, we prepared sols with compositions identical to AAB(1/5) but varying the order and tilning of steps. Figure 3 compares the NMR spectrum for an AAB(1/5)sol to the spectra of two sols with identical composition. One sol was made by reversing the order of the second and third steps; the other was made by adding all of the ingredients in one step. In both cases, we found that at the moderate H2O/Si ratio of 3.7, the differences from changing the order of the steps could not be erased, even after 15 h of refluxing.

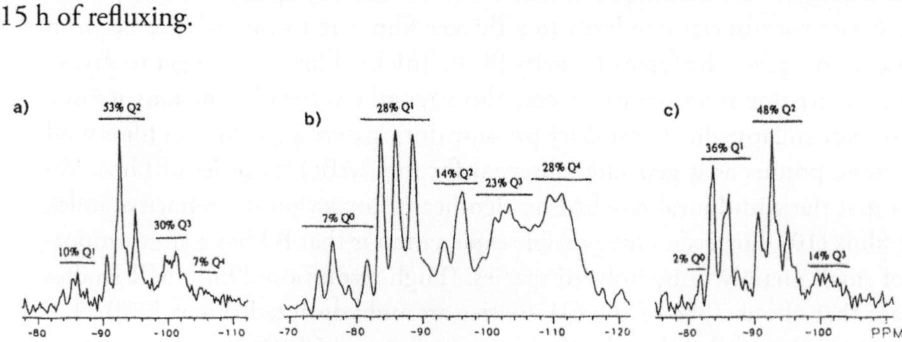

Figure 3. ^{29}Si NMR spectra of (a) an AAB(1/5) sol, (b) a sol with identical composition made by reversing the order of the second and third steps, and (c) a sol with identical composition made all in one step. The sols with differences in the timing and order of steps were refluxed for 15 h before the spectra were taken.

Next, we examined the effects of aging and refluxing between steps. We have found that stock solutions continue, to age during storage at -20°C. Figure 4 compares the NMR spectrum of a freshly-made stock solution to that of a two year old stock solution. Figures 4c and 4d show the corresponding spectra recorded after adding H2O and HCI in the second step of the AAB(1/5) process. In this case, at a lower H2O/Si ratio of 2.5, refluxing for 4 h eliminates differences arising from the different aging histories.

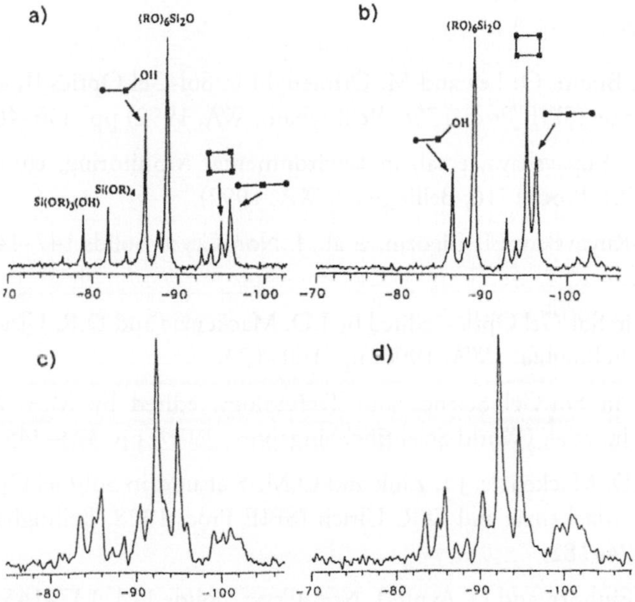

Figure 4. ^{29}Si NMR spectra for (a) freshly-made stock solution, (b) stock solution that was stored in a freezer for two years, and after 4 h refluxing of the second acid-catalyzed step of sols made from (c) the freshly-made stock, and (d) the two year old stock solution.

Summary

A multi-step hydrolysis procedure was used to produce sols with fractal inorganic clusters. The sequence and timing of the steps influence the distribution of the silicate polymers in the sol and the structure of the polymer clusters. At a low H_2O/Si ratio of 2.5, these kinetic differences can be overcome by extensive refluxing of the next step. However, differences between sols at moderate water ratios persist, even after long periods of refluxing. The profound differences in the species distributions of sols prepared with the same H_2O/Si ratio, pH and concentration, but with a different sequence of steps permit kinetic control over the final film microstructure. This ability to "tune" the refractive index by a simple aging process should allow the rational design of porous films for applications requiring organic dyes.

Acknowledgements

This work was performed at Sandia National Laboratories, supported by the U. S. Department of Energy under Contract # DE-AC04-94AL85000.

References

1. D. Avnir, S. Braun, O. Lev and M. Ottolenghi in Sol-Gel Optics II, edited by J.D. Mackenzie (SPIE Proc. 1758, Bellingham, WA, 1992) pp. 456–463.
2. O. Lev, B.I. Kuyavskaya, et al. in Environmental Monitoring, edited by T. Vo-Dink (SPIE Proc. 1716, Bellingham, WA, 1992).
3. B. Iosefzon-Kuyavskaya, I. Gigozin, et al., J. Non-Cryst. Solids 147–148, 808–812 (1992).
4. P.N. Prasad in Sol-Gel Optics, edited by J.D. Mackenzie and D.R. Ulrich (SPIE Proc. 1328, Bellingham, WA, 1990) pp. 168–173.
5. R. Reisfeld in Sol-Gel Science and Technology, edited by M.A. Aegerter, M. Jafelicci Jr., et al. (World Scientific, Singapore, 1989) pp. 323–345.
6. B. Dunn, J.D. Mackenzie, J.I. Zink and O.M. Stafsudd in Sol-Gel Optics, edited by J.D. Mackenzie and D.R. Ulrich (SPIE Proc. 1328, Bellingham, WA, 1990) pp. 174–182.
7. D. Levy, S. Einhorn and D. Avnir, J. Non-Cryst. Solids 113, 137–145 (1989).
8. C.J. Brinker, G.C. Frye, A.J. Hurd and C.S. Ashley, Thin Solid Films 201, 97–108 (1991).
9. C.J. Brinker, A.J. Hurd, G.C. Frye, P.R. Schunk and C.S. Ashley, J. of the Ceram. Soc. of Japan 99 (10), 862–877 (1991).
10. D.L. Logan, C.S. Ashley and C.J. Brinker in Better Ceramics Through Chemistry V, edited by M. Hampden-Smith, W.G. Klemperer and C.J. Brinker (Mat. Res. Soc. Proc. 271, Pittsburgh, PA, 1992) pp. 541–546.
11. D.L. Logan, C.S. Ashley, R.A. Assink and C.J. Brinker in Sol-Gel Optics II, edited by J.D. Mackenzie (SPIE Proc. 1758, Bellingham, WA, 1992) pp. 519–528.
12. B.B. Mandelbrot, The Fractal Geometry of Nature (W.H. Freeman, New York, 1983).
13. C.J. Brinker and G.W. Scherer, Sol-Gel Science (Academic Press, San Diego, CA, 1990) pp. 360–370 and 799–811.

14. C.J. Brinker, K.D. Keefer, D.W. Schaefer and C.S. Ashley, J. Non-Cryst. Solids 48 (1), 47–64 (1982).

15. C.J. Brinker, K.D. Keefer, et al., J. Non-Cryst. Solids 63, 45–59 (1984).

16. C.J. Brinker, N.K. Raman, D.L. Logan, R. Sehgal, T.L. Ward, S. Wallace and R.A. Assink, Polymer Preprints 34 (1), 240–241 (1993).

17. D.C. Goodnow and M.N. Logan, unpublished results.

18. M. Born and E. Wolf, Principles of Optics (Pergamon, New York, 1975) p. 87.

19. M. Gonzales and M.N. Logan, unpublished results.

20. D.H. Doughty, R.A. Assink and B.D. Kay in Silicon-Based Polymer Science: A Comprehensive Resource, edited by J.M. Zeigler and F.W.G. Fearon (Amer. Chem. Soc. Advances in Chemistry Series No. 224, 1990) pp. 241–250

CITATION

Logan MN, Prabakar S, and Brinker JC. Sol-Gel-Derived Silicafilms with Tailored Microstructures for Applications Requiring Organic Dyes. MRS Proceedings / Volume 346 / 1994. doi: http://dx.doi.org/10.1557/PROC-346-115. This work was performed at Sandia National Laboratory, supported by the US Department of Energy under contract # DE-AC04-94AL85000.

Synthesis and Analysis of Nickel Dithiolene Dyes in a Nematic Liquid Crystal Host

Irene Lippa

ABSTRACT

The Liquid Crystal Point Diffraction Interferometer (LCPDI) can be employed to evaluate the Omega Laser system for optimum firing capabilities. This device utilizes a nickel dithiolene infrared absorbing liquid crystal dye dissofved in a liquid crystal host medium (Merck E7). Three nickel dithiolene dyes were characterized for both their solubility in the E7 host and their infrared spectral absorption.

Project Goal

A family of infrared absorbing dyes for use in the LCPDI was synthesized and characterized. The LCPDI device assists with the diagnostics of the laser beam uniformity. The ideal dye for the LPDI should have (1) high optical absorbance at the desirable wavelength of the laser (λ_{max} 1054 nanometers); (2) high thermal stability to minimize loss of the dye chromophore to thermal degradation by the laser; (3) high solubility in the liquid crystal host to increase both dye loading and infrared optical density of the dye-host medium; and (4) a low impact on the degree of order (order parameter) of the host medium.

Previous work has shown the nickel dithiolene dye as a desirable candidate for the LCPDI. The nickel dithiolene has a unique aromatic ring structure, which imparts a relatively high infrared absorption and a high thermal and photochemical stability. With 2 alkoxyphenyl substitutions onto the nickel dithiolene core, these complexes possess optical absorbance bands at the desirable wavelength of the laser. However, these alkoxyphenyl substituted compounds have demonstrated limited solubility in the host liquid crystal medium. The potential of an alkylphenyl substituents to increase the solubility of the dithiolene dye in the host liquid crystal medium was the primary topic of this investigation.

In this project, p-butoxy-phenyl and p-nonoxy-phenyl substituted (-pC6H4-OC9H19 and -pC6H4-OC4H9) and dithiolene dyes were purified and characterized along with the total synthesis, characterization and purification of a novel p-butyl-phenyl (-pC6H4-C4H9) substituted dithiolene dye. The solubility of these three dyes was determined in the LCPDI liquid crystal host medium, and the absorption spectra were determined. The structure of the dithiolene dyes is shown in Figure 1.

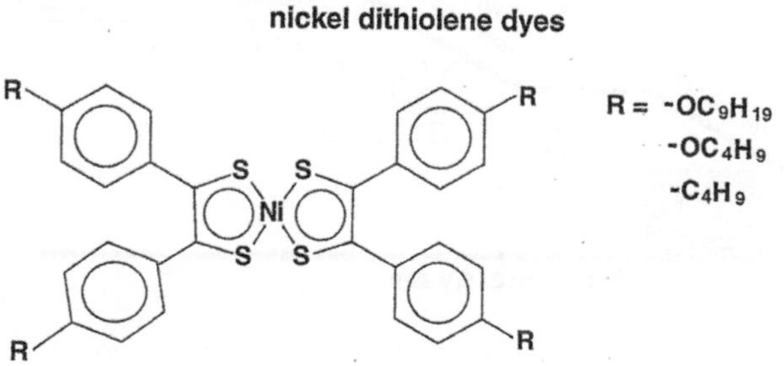

Figure 1

Liquid Crystals: Background

Liquid crystals are compounds that can exist in an intermediate mesophase between the solid and liquid phase.

A true crystalline solid has atoms or molecules fixed in position with no translational movement. These species are ordered in a regular repeating fashion in all three dimensions. A liquid has atoms or molecules with translational'movement in all directions, with no dimensional order. When the disorder of a liquid extends in all three dimensions, then true liquid is the isotropic phase. The phase transition from crystalline solid to liquid occurs at a sharp, distinct melting point. As melting occurs, all three dimensions of crystalline order are lost and the material enters the three dimensional disorder of the liquid phase.

Liquid crystals exist as a phase between the solid and liquid phase, showing a unique "double melting point." The liquid crystal can maintain partial order in one or two dimensions at an initial melting point. At this first melting point, one (or two) dimension(s) of order are lost, and the remaining dimensions of crystalline order are maintained. A second "melting" point occurs when the liquid crystal enters into the complete three dimensional disorder of a liquid. This "liquid crystalline" order that exists between the two melting points is called the mesomorphic phase. The mesomorphic phase exhibits anisotropic properties; i.e. different physical properties (optical and electrical) in different directions. These anisotropic properties are what make liquid crystals of such value for applications. The double melting point of a liquid crystal is shown in Figure 2.

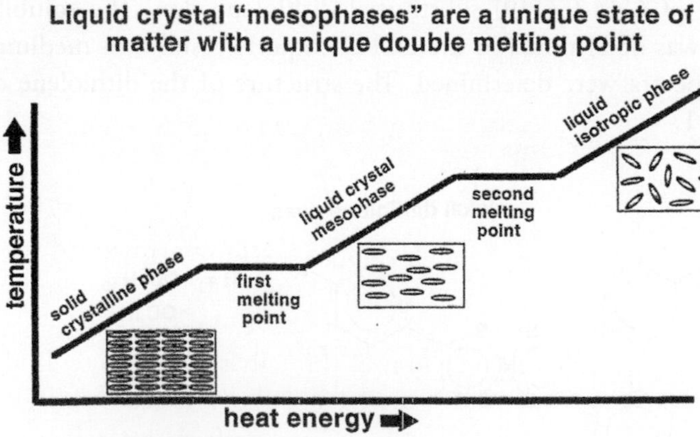

Figure 2

Liquid crystals are classified into four different groups according to the degree of ordering within the mesophase. The four groups from least to most highly ordered are nematic, cholesteric, smectic, and discotic. Nematic liquid crystals maintain, order in one dimension; here molecules are aligned with their long axes in a common direction in a one dimensional stacking pattern. Cholesteric, smectic and discotic liquid crystals maintain crystalline order in two (or more) dimensions; molecules are aligned in a common direction and in parallel planes (that can slip past each other) in a second dimension. The nematic, cholesteric and smectic liquid crystals are composed ojlinear (cigar-shaped) molecules. The discotic liquid crystal is composed of planar (disc-shaped) molecules arranged both face to face and end to end. The different liquid crystal structures are shown in Figure 3.

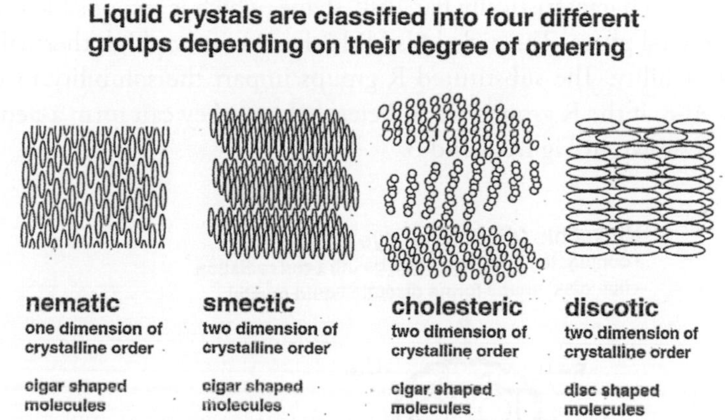

Figure 3

Merck E7TM is the liquid crystal host used in the LCPDI device. E7 is a nematic liquid crystal mixture.with a mesophase between -10°C and 60.5°C. The R-group substituents on the nickel dithiolene dye need to be selected for maximum solubility in the E7 host without compromising other important properties (absorption wavelength, optical density or thermal stability). Merck E7 is a room temperature eutectic mixture of cyanobiphenyl and cyanoterphenyl liquid crystal compounds; its components are shown in Figure 4.

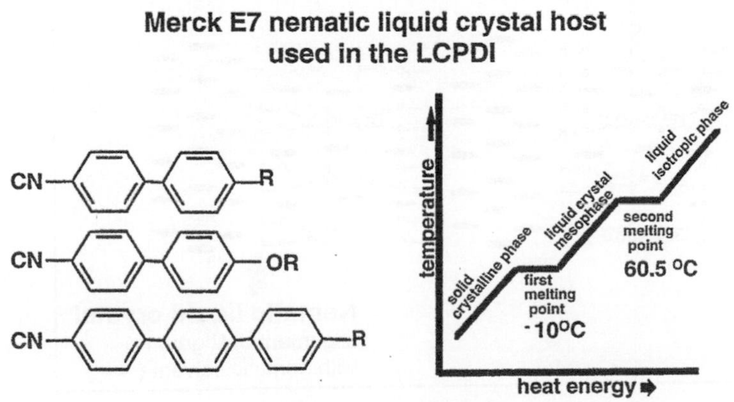

Figure 4

The substituted nickel dithiolene dyes are discotic liquid crystals dissolved in the E7 nematic host. There are two molecular "components" to the nickel dithiolene dye; the nickel bis (dithiolene) core and the substituted R groups. The nickel bis (ditholene) core is a unique conjugated transition metal aromatic complex which absorbs near infra-red radiation.

This group has a characteristically flat 'disk' shape, which is responsible for the discotic liquid crystal phase. The nickel bis (dithiolene) core has high thermal and photochemical stability. The substituted R groups impart the solubility in non-polar solvents. Also, if the R groups are sufficiently long, they can form a nematic liquid crystal, as shown in Figure 5 and 6.

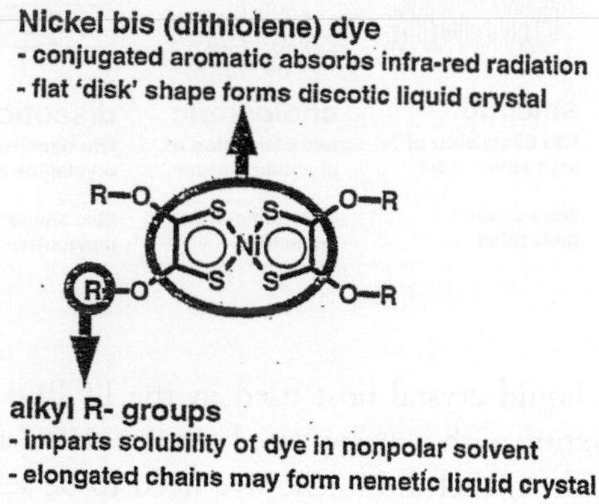
Nickel bis (dithiolene) dye
- conjugated aromatic absorbs infra-red radiation
- flat 'disk' shape forms discotic liquid crystal
alkyl R- groups
- imparts solubility of dye in nonpolar solvent
- elongated chains may form nemetic liquid crystal

Figure 5

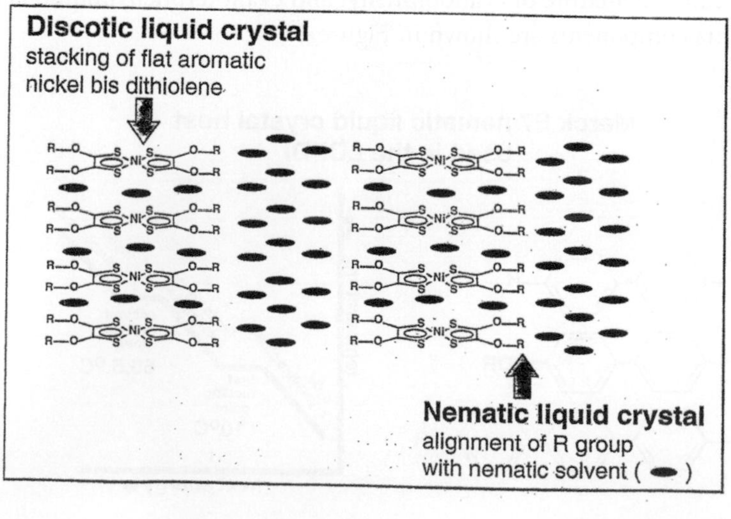
Discotic liquid crystal
stacking of flat aromatic
nickel bis dithiolene
Nematic liquid crystal
alignment of R group
with nematic solvent (▬)

Figure 6

Synthesis of Bis [1, 2 Di (4-n-Butyl Phenyl) Ethane 1, 2 Dithione] Nickel (0)

Synthesis of bis [1,2 di (4-n-butyl phenyl) ethane 1,2 dithione] nickel (0)

(1) **4-n-butyl phenyl magnesium bromide**

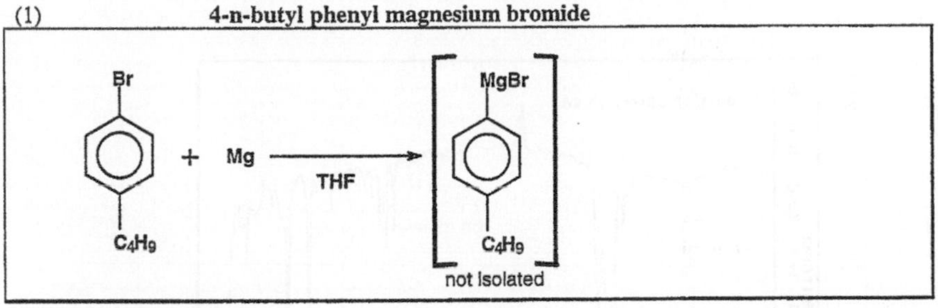

IMAGE (1) 4-n-butyl phenyl magnesium bromide

A three neck round bottomed flask was equipped with a dropping funnel, water cooled condenser, argon purge, electric heating mantle and magnetic stirrer. Sufficient tetrahydrofuran (THF) was dried over anhydrous MgSO4. The flask was charged with magnesium turnings and dry THF (20 ml) was added to the bottom of the three-neck round bottomed flask. 4-butyl-bromobenzene (18 g, 84 mmol) was added to the reaction dropwise and allowed to reflux for 12 hours. The resultant product was used in the next step without isolation or purification.

(2) **4,4'- di (n-butyl) stilbene**

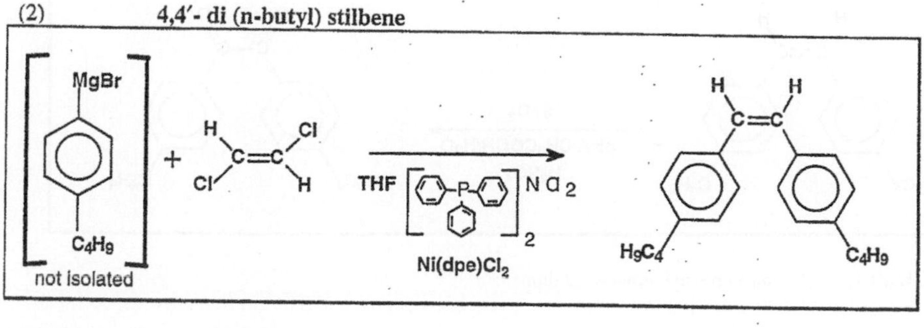

IMAGE (2) 4,4'-di (n-butyl) stilbene

The flask containing the 4-n-butylphenyl magnesium bromide from step I was cooled in an ice water bath, and 0.14 g (0.25 mmol) of dichloroethylene was added dropwise. The mixture was refluxed for 20 hours. The flask was cooled to room temperature and dilute aqueous Hel (l0%) is added. The resulting solid was collected, extracted with ethyl ether and dried over anhydrous NaSO4. The solvent was removed under vacuum using a rotary evaporator to give a crude mixture ofcis and transisomers of 4,4'-di (n-butyl) stilbene (16.4%). Recrystallization from ethyl acetate gave a slightly yellow solid in the trans form and a white crystalline solid in the-cis form (1.3 g trans, 2.23g cis) for an average 16.4% yield 12.2 theoretical. The IR spectrum of the cis-product is shown in Figure 7.

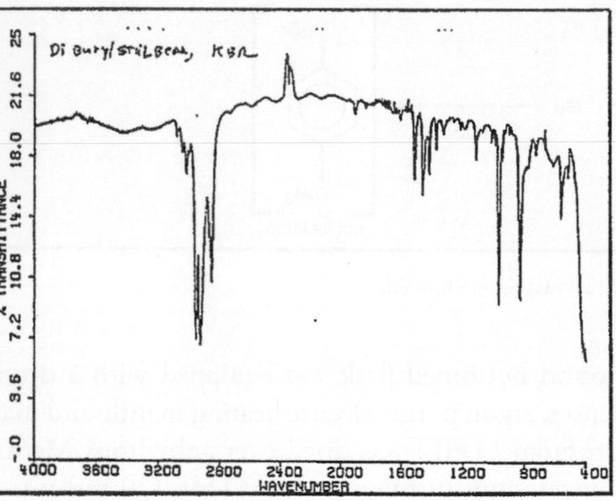

Figure 7. FTIR spectrum of 4,4'-di (n-butyl) stilbene

(3) **1,2 di(4-n-butyl phenyl)ethane 1,2 dione**

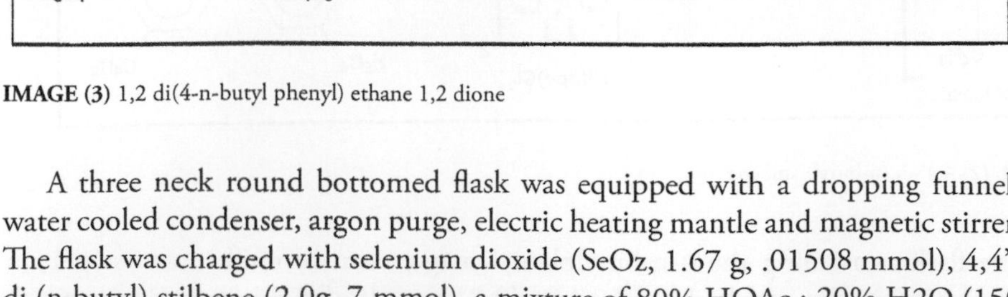

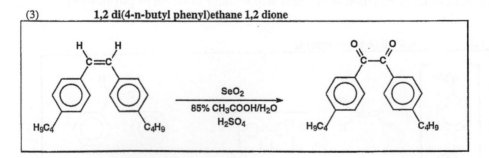

IMAGE (3) 1,2 di(4-n-butyl phenyl) ethane 1,2 dione

A three neck round bottomed flask was equipped with a dropping funnel, water cooled condenser, argon purge, electric heating mantle and magnetic stirrer. The flask was charged with selenium dioxide (SeOz, 1.67 g, .01508 mmol), 4,4'-di (n-butyl) stilbene (2.0g, 7 mmol), a mixture of 80% HOAc : 20% H2O (150 ml) and concentrated H2SO4 (1.5 ml). The reaction mixture was heated to reflux for 17 hours, insoluble solids were filtered, and the filtrate was extracted with ethyl ether. The ethyl ether extract was washed with saturated NaHCO3 solution until it was neutral. The extract was dried over MgSO4 and the solvent removed under vacuum to leave an oil in 80 % yield. The IR spectrum of the product, showing the characteristic c=o stretching at ~1700cm^{-1}, is given in Figure 8.

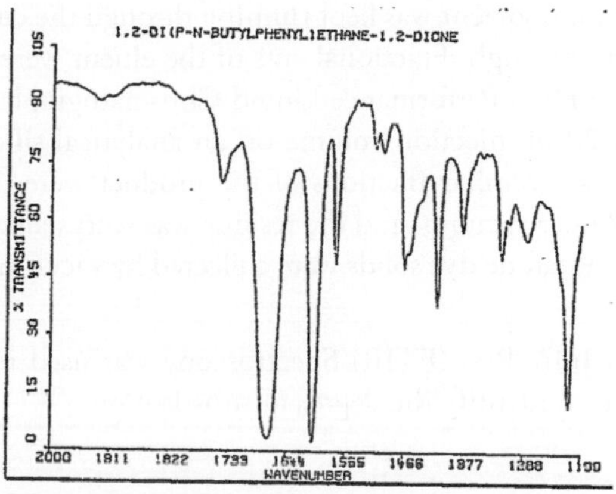

Figure 8. FTIR spectra of 1,2 di(4-n-butyl phenyl)ethane 1,2 dione

(4) **Bis [1,2 di (4-n-butyl phenyl) ethane 1,2 dithione] Nickel (0)**

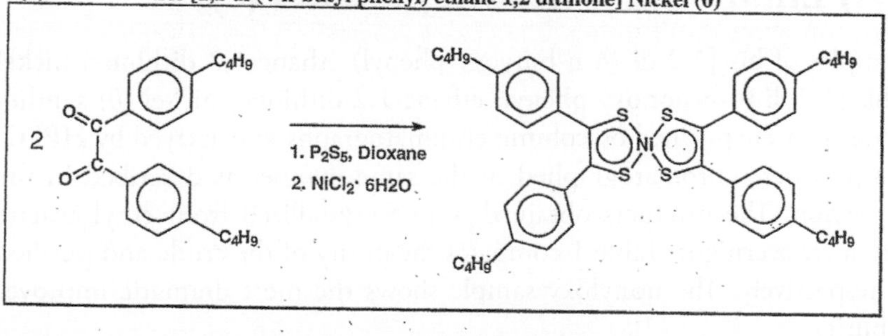

IMAGE (4) Bis [1,2 di (4-n-butyl henyl) ethane 1,2 dithione] Nickel (0)

A three neck round bottomed flask was equipped with a dropping funnel, water cooled condenser, argon purge, electric heating mantle and magnetic stirrer. The flask was charged with 1,2 di (p-n-butylphenyl)ethane 1,2 dione(1.8 g, 3.6 mmol), phosphorous pentasulfide (1.2 g 5., 5 mmol) and 30 dioxane. The mixture was refluxed for 5 hours. The reaction mixture was filtered hot to remove the unreacted phosphorous pentasulfide and the residue was washed with hot dioxane several times. The filtrate was returned to the cleaned round bottom flask, and nickel (II)chloride • hexahydrate (0.48 g, 2.0 mmol) in 10 ml of water was added. The reaction was refluxed for 2 hours, and then cooled in a ice water bath. The product, a crystalline black powder, was collected by filtration to give 1 g of the crude complex at 51 % yield.

The product was purified by liquid column chromatography, using a 1: 1 mixture of hexane and toluene as the eluent and a stationary phase of 5 μm porosity silica gel. The crude solids were dissolved into an aliquot of the solvent and added to the column. The eluting solvent was kept running through the column until all of the material had run through. Fractional cuts of the eluent were collected and assayed using a Hitachi High Performance Liquid Chromatograph (HPLC) with a UV-Vis detector. A 20 μl injection volume on an analytical silica gel column was used for the analysis. Similar fractions of the product were combined, the solvent was removed under vacuum, and the residue was recrystallized from ethyl acetate. The purified crystalline dye solids were collected by vacuum filtration and air dried.

Fourier Transform Infra-Red (FTIR) Spectroscopy was used to detect functional groups as well as to identify the degree of substitution.

Purification of Bis [1,2 Di (4-n-Butyoxy Phenyl) Ethane 1,2 Nickel (0) and Bis [1,2 Di (4-n-Nonoxy Phenyl) Ethane 1,2 Dithione]-Nickel (0)

Crude samples of bis [1,2 di (4-n-butyoxy phenyl) ethane 1,2 dithione]-nickel (0) and bis [1,2 di (4-n-nonoxy phenyl) ethane 1,2 dithione] nickel (0) synthesized previously were purified by column chromatography and assayed by HPLC. These samples were chromatographed in the same manner as described in the previous section. The products obtained were recrystallized from ethyl acatate and dried in air overnight. Table 1 compares the purity of the crude and purified samples, respectively. The nonyloxy sample shows the most dramatic improvement in purity.

Table 1

R-group name	R-group structure	Crude HPLC assay (area %)	Purified HPLC assay (area %)
p-nonoxy- phenyl	$-OC_9H_{19}$	46.0	90.0
p-butoxy-phenyl	$-OC_4H_9$	83.0	94.0
p-butyl-phenyl	$-C_4H_9$	98.5	99.4

Hot Stage Polarizing Microscopy

Phase transitions were characterized using a polarizing microscope with a hot stage attachment. The isotropic melting points are shown in Table 2.

Table 2

R-group name	R-group structure	second melting point (°C)
p-nonoxy- phenyl	$-OC_9H_{19}$	184.3-189.1
p-butoxy-phenyl	$-OC_4H_9$	246.3-248.7
p-butyl-phenyl	$-C_4H_9$	228.3-230.6

Spectroscopy

UV-Vis-near IR spectroscopy was used to find the location and strength of the electronic absorption maxima (λ_{max}). The point of interest for the LCPDI is 1054 nm, and the absorbance of all three dyes are in this vicinity. The absorbance characteristics of the C_4 and OC_4 dyes in the liquid crystal host over the 800-1600 nm region are shown in Figure 8. The λ_{max} for all three dyes is given in Table 3.

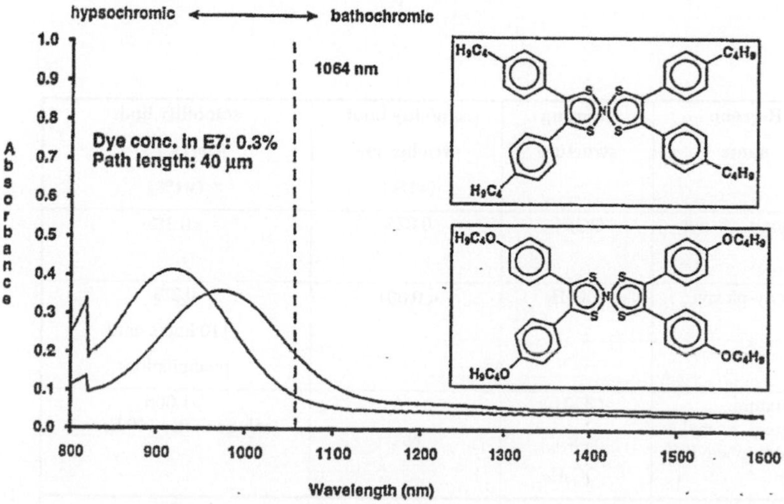

Figure 8a. The enhanced solubility of the C4 metal complex is compromised by the hypsochromic shift of its peak absorbance

Table 3

R-group name	R-group structure	λ max cyclohexane (nm)
p-nonoxy- phenyl	$-OC_9H_{19}$	912
p-butoxy-phenyl	$-OC_4H_9$	910
p-butyl-phenyl	$-C_4H_9$	870

Solubility

Solubility limits of each dye were determined in both cyclohexane and Merck E7 liquid crystal. Samples of each dye were mixed in a series of weight percents (0.3 wt % to 1wt %) in E7 to make a total volume of 2 ml. The dyes were dissolved in LC host by heating to100°C and stirringseveral hours. Upon cooling, each sample was filtered through a 0.5 μm Teflon membrane filter to remove undissolved dye and insoluble particles. An additional set of samples containing a 1: 1 ratio of the OC_4 and OC_9 dyes was also prepared over the same range of concentrations.

All samples were checked periodically both visually and by microscopic inspection for signs of dye precipitation. The solubility results in E7 are given in Table 4 and Figure 9.

Table 4

R-group name	R-group structure	solubility limit cyclohexane (wt%)	solubility limit Merck E7™ (wt%)
p-nonoxy- phenyl	$-OC_9H_{19}$	0.025	<0.3%
p-butoxy-phenyl	$-OC_4H_9$	< 0.001	0.3% (110 hours until precipitation)
1:1 mixture p-nonoxy- phenyl & p-butoxy-phenyl	$-OC_9H_{19}$ & $-OC_4H_9$	NA	>1.000 (solution after 140 hours)
p-butyl-phenyl	$-C_4H_9$	0.050	0.3% (solution after 140 hours)

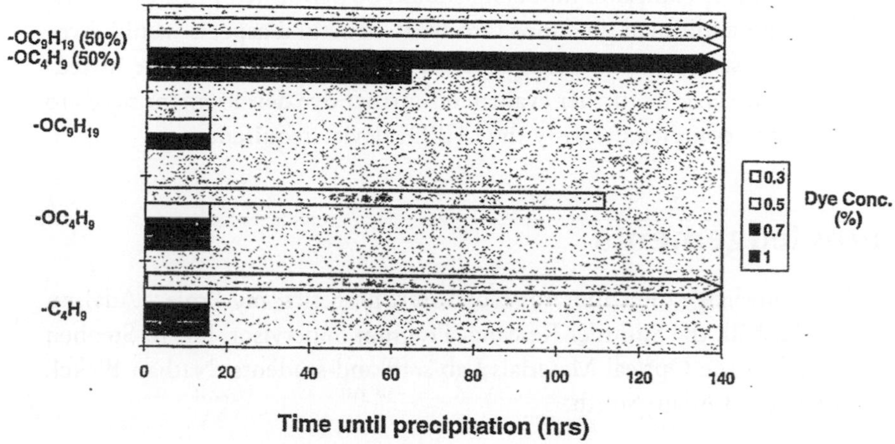

Figure 9

At the time of this writing, the lower concentration of 1: 1 OC4/OC9 dye mixture in E7 was still in solution and its solubility limit was larger than the rest of the dyes; however, an actual number of hours until precipitation has yet to be detertnined. The 0.3%, 0.5%, and 0.7% weight percent dye mixtures had not precipitated after 140 hours. The -C4H9/E7 mixture was also stable at 0.3%. This shows us the high solubility rate of the -C4H9 as well the important discovery that the mixing of dyes with similar structures can increase the solubility limit.

Summary

A new alkyl phenyl-substituted nickel dithiolene was successfully synthesized and its solubility was evaluated in a nematic LC host. The solubility of this dye in the same host was benchmarked against existing alkoxy-substituted analogs. Both the solubility and the spectral properties of nickel dithiolenes are affected by relatively minor changes in terminal functional groups. Although the only differences between the $-OC_4$, $-C_4$ and $-OC_9$ were different length chains and lack of oxygen present, there was a dramatic effect on both the solubility limits and the λ_{max} absorption. The new $-C_4$ nickel dithiolene shows enhanced solubility in liquid crystal, but a hypsochromic shift of its peak absorbance compromises its efficiency at 1054 nm. Substantial improvements in dye solubility can be made by using mixtures of dyes with similar structures. The net gain in solubility limit of mixed dyes in E7 has made a substantial improvement over the solubility limits of either dye by itself.

In the future, more computer modeling should be performed prior to synthesis in order to give additional guidance into what structural aspects would favor desirable characteristics. The long term stability of both the -C4 and the mixed-dye systems also needs to be further evaluated. Further dye development needs to be done in order to completely satisfy the LCPDI device requirements.

Acknowledgements

I gratefully acknowledge the Laboratory for LaserEnergetics; my Project Advisor, Kenneth L. Marshall, the Summer Research Program Supervisor, Dr. R. Stephen Craxton, as well as the Optical Materials Lab staff and students, Nathan Bickel, Joann Starowitz, and Adam Smith.

References

1. Mueller-Westerhoff, U. The Syntnesis of Dithiolene Dyes With Strong Near-IR Absorption. Tetrahedron, Vol.47, No6. pp909–932, 1991, Pergamon Press plc.
2. Ohta, K. Discotic Liquid Crystals of Transition Metal Complexes. 4: 1 Novel Discotic Liquid Crystals Obtained from Substituted Bis(dithiolene)nickel Complexes by a New Method. Mol. Cryst. Liq. Cryst.,1987 Gordon and Breach Science Publishers S.A., Vol 147, pp15–24.

CITATION

Lippa I. Synthesis and Analysis of Nickel Dithiolene Dyes in a Nematic Liquid Crystal Host. Mar 1999; 27 p; DOE/SF/19460--299-PT.6; Contract Fc03-92sf19460. US Government publication.

Synthesis of a Photoresponsive Polymer and Its Incorporation into an Organic Superlattice

Alfred M. Morales, James R. McElhanon, Phillip J. Cole and Chris J. Rondeau

ABSTRACT

The synthesis of a photoswitchable polymer by grafting an azobenzene dye to methacrylate followed by polymerization is presented. The azobenzene dye undergoes a trans-cis photoisomerization that causes a persistent change in the refractive index of cast polymer films. This novel polymer was incorporated into superlattices prepared by spin casting and the optical activity of the polymer was maintained. A modified coextruder that allows the rapid production of soft matter superlattices was designed and fabricated.

Introduction

Since the invention of the laser, the use of light in information technology has increased tremendously. However, of the three major components of information technology, namely signal transmission, storage, and manipulation, most progress has been made in the area of signal transmission by developing high quality optical materials such as doped silica and claddings for use in optical fibers (1). High speed all-optical computing has not been fully realized mainly because of the lack of materials with specific properties that allow for more sophisticated light storage and for manipulation of the amplitude, frequency, phase, or direction of light without having to either convert light into electrical signals or having to mechanically modify the signal.

Historically, two concepts in modern materials science have been particularly successful in generating new material properties. First, novel physical properties have been obtained by arranging nanometer thick semiconducting inorganic layers into superlattices. By stacking what are essentially 2D inorganic layers, interesting optical, electrical, and magnetic properties can be generated through band gap engineering, quantum confinement, and tailored coupling. An impressive application of this idea is the design, fabrication, and characterization of the quantum cascade laser at Bell Labs (2).

The second enabling concept in materials science is the realization that modern organic chemistry enables the physical and chemical properties of soft matter (polymers, oligomers, and organic molecules in general) to be manipulated much more easily than the properties of inorganic compounds. Soft matter is no longer thought of as a merely insulating, compliant material: chemists and materials scientists can now endow soft matter with electrical conductivity, optical activity, superconductivity, magnetism, and many other properties. In fact, this precise control of the properties of materials has sparked a huge worldwide race to fabricate efficient and inexpensive organic light emitting materials (3). By contrast, the chemical manipulations that can be carried out in solid state inorganic nanolayers are often limited by a variety of reasons including equilibrium thermodynamics and lattice matching considerations.

In this report we document our accomplishments combining these two approaches and create new materials that can be used to manipulate light. We investigate the application of organic synthesis to create polymeric materials that undergo a change in their refractive indexes when stimulated by an ultraviolet (UV) trigger signal. We also explore techniques to incorporate these polymeric materials into superlattices and thus create novel reflecting materials. Figure 1 summarizes our strategy.

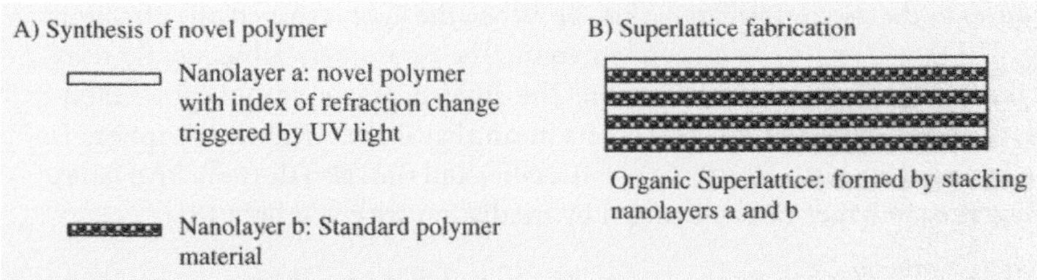

Figure 1. Schematic representation of our strategy for creating new optical materials. (a) a novel polymer that changes its index of refraction when triggered by a UV light signal is synthesized and then (b) combined with another polymer to form a superlattice.

Specifically, our strategy for creating new optical materials will build upon the work reported by Weber et al. (4) on birefringent polymer nanolayer superlattices with enhanced reflectivities. These superlattices consisted of alternating nanolayers of isotropic polymethyl methacrylate (PMMA) and either birefringent polyester or birefringent syndiotactic polystyrene. The reflectivities of the p- and s-polarized light from the superlattices is determined by the indexes of refraction, the thickness of the layers, and the angle of incidence. Thus, by combining nanolayers with different indexes of refraction, the value of the reflection coefficient at a specific angle at the nanolayer interfaces could be predictably controlled. By building nanolayers of specific thicknesses, the center wavelength and the bandwidth of a high reflectivity band versus the angle of incidence can be determined. However, as reported these superlattices are passive materials: once fabricated, the optical properties cannot be changed in response to the reflected light or to a trigger signal. Our strategy in this project is to modify and tailor, using organic chemistry, the composition of one of the nanolayers to include photoresponsive molecules that could change the indexes of refraction of that set of nanolayers when triggered by an external signal. This change in refractive indexes will change the overall reflectivity and the value of the Brewster angle thereby enabling optical switching.

Experimental

Synthesis of Photoresponsive Polymers

Our synthetic efforts were focused on grafting azobenzene dye (ABD) chromophores into methacrylate polymers. As reported in the literature, ABD's change their conformation when irradiated with UV light (5). ABD's exist as two isomers, a rod-like thermally stable trans isomer and a bent, metastable cis isomer (6,7). When the trans isomer is irradiated with light of the appropriate wavelength it isomerizes to the cis-conformation (Fig. 2). When the light is turned off, a fraction of the ABD's will retain the cis configuration. The cis-isomers subsequently relax back into the trans isomer configuration. The differences in electronic polarizability between the trans and the cis isomers mean that if the ABD chromophore is grafted to a polymer, the electronic polarizability and thus also the refractive index of the grafted polymer can be changed by irradiation with UV light (8).

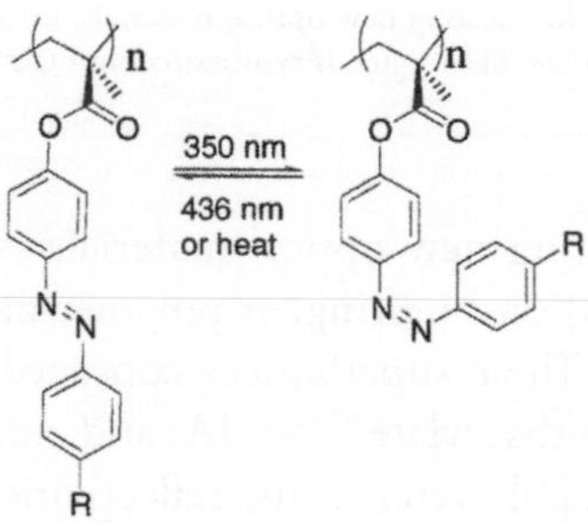

Reaction 1

1 R=H
2 R=NO_2

Figure 2. Expected photoinduced isomerization of ABD grafted onto PMMA .

Two types of ABD-grafted PMMA were initially prepared with different polar and electronic substituents (Fig. 2). ABD-grafted polymer 1 contains no substituent in the azobenzene para position while polymer 2 contains a para nitro group. Potentially, two synthetic routes are possible to generate ABD-grafted PMMA. In approach PMMA is first synthesized and then the ABD is grafted onto the polymer chains. In another approach, ABD-grafted methylmethacrylate monomer units are first synthesized and then polymerized in a subsequent step.

We generated ABD-grafted PMMA by first synthesizing grafted monomer units and thus ensuring minimal steric effects on the final yield. 1and 2 were respectively prepared by reacting 2-methacryloyl chloride (3) with 4-(phenylazo) phenol(4) or 4-(4- nitrophenylazo)pheno1(5) in the presence of triethyl amine in THF (reaction 2) at room temperature followed by AIBN catalyzed free radical polymerization (reaction 3) in refluxing THF. The final product was isolated by dripping the THF solution into cold petroleum ether. ABD-grafted PMMA 1 and 2 were obtained in yields of 85% and 15% respectively. Low yield of 2 was likely due to formation of insoluble high molecular weight precipitates.

As a control, a non-polymerizable azobenzene model compound (9) similar to monomer 6 was prepared. 9 was synthesized in 95% yield through reaction of 4-(phenylazo)phenol (4) with isobutyryl chloride (8) in the presence of triethyl amine in THF (reaction 4).

Reaction progress was monitored using 1H-NMR and the optical activity of the ABD-grafted PMMA, MMA, and of model compound 9 was verified by collecting UV-vis absorbance spectra in solution. All compounds were initially prepared in approximately 5 g batches. Approximately 50 g batches of compound 1 was later prepared with no significant procedural modifications.

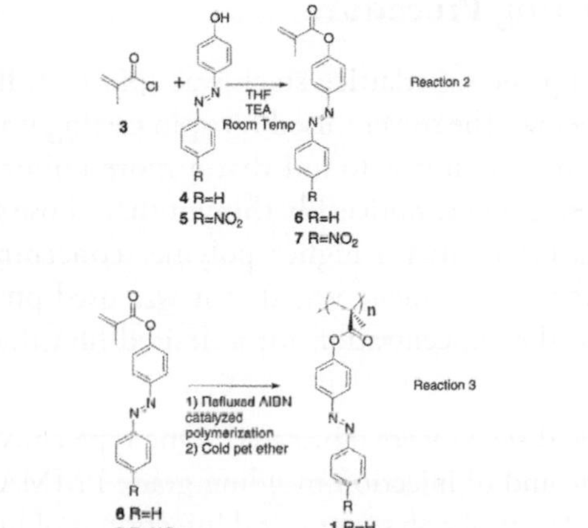

Synthetic Schemes: Organic synthesis reactions useG to generate ABD-grafted compounds. Reactions 2–4

Physical Characterization of ABD-Grafted Polymers

The optical response of the ABD-grafted PMMA was verified by collecting UV-visible (UV-vis) absorbance spectra of methylene chloride solutions. Since the desired novel optical materials will be used in the solid state, the optical activity of the ABD-grafted PMMA was verified by taking UV-vis absorbance spectra of solid thin films spin cast onto quartz slides. The persistent change in the index of refraction for each grafted PMMA was quantified by taking ellipsometric readings on solid thin films spin cast onto silicon wafers. The thermal stability of the most promising polymer, 1, was checked using thermal gravimetric analysis.

Superlattice Fabrication

Two routes were explored to fabricate organic superlattices: a modified spin casting procedure and multilayer coextrusion. Spin casting was selected as a relatively quick, low tech way to select the most promising photoresponsive polymer.

Modified Spin Casting Procedure

Each layer that makes up the superlattice stack was spin cast, floated, and then collected as described below. The solvent used for spin casting was toluene. Given the lower volatility of toluene, it was found that a more uniform film could be obtained. However, these films are noticeably thinner than those spun with higher volatility solvents. This meant that a higher polymer concentration had to be used. Another reason for using toluene was that it was used previously in creating polystyrene films, and a concentration for a desired film thickness had been predetermined.

Two types of 8-layered stacks were fabricated. One type consisted of alternating layers of polystyrene and of injection molding grade PMMA. The other type consisted of alternating layers of polystyrene and injection molding grade PMMA mixed with 1. The ratio of the PMMA to 1 was 2:1 respectively. All of the films were spun from a solution of 10%weight fraction in toluene (the solution for the PMMN/1 film contained 0.3338 of 1, 0.666g of PMMA, and 90g of toluene). To prepare the toluene solutions, the polymers were stirred into the toluene and the container was then placed into a sonicator for approximately 2 hours and then placed into a 40 °C oven for a minimum of 24 hours. The solutions were then filtered through 5μm and 1 μm diameter membranes.

The procedure for spin-coating was as follows: A glass slide was thoroughly washed with toluene and then dried with an inert gas (argon). The slide was then placed into an ozone cleaner for 5 minutes. The slide was put on the spin coater vacuum chuck and flooded with toluene. It was then spun at 1500rpm for approximately 60 seconds until it was dry. Finally, the solution with the polymer was added to the slide by means of a glass pipette (which was cleaned with toluene prior to use) and the slide spun. The parameters for the spin coating steps were 1500rpm for 120 seconds.

After the film was spun on the glass slide, a scalpel was used to scrape the sides of the slide so that the film would come off more easily. A large container to hold deionized (D.I.) water for floating the film off the glass slide was cleaned with toluene and dried with argon. The cleaning process was repeated for each film that was floated. Once the container was cleaned and filled with D.I. water, the glass slide was placed into the water at a shallow angle of attack. The film would separate from the slide and float to the free surface of the water. The floating released film was carefully scooped out onto a wafer holding the stack of previously produced films.

The initial film on each of the 8-layered samples was the polystyrene film (10% concentration). This initial film was spun directly onto a silicon wafer, not a glass slide. The second layer was the PMMN/1 film. Each film thereafter alternates between the polystyrene and PMMN/1 film. After a new layer was added, the wafer was air dried vertically to remove any trapped water.

Multilayer Coextrusion

The purpose of multilayer coextrusion is to bring together two or more polymers into a superlattice structure with continuous layers of controlled thickness. Once the coextrusion system is set up, superlattice production will be fast and inexpensive. However, considerable time and resources must be invested to design, fabricate, and test the coextrusion system. The funding and time frame of this project only allowed for the design and fabrication of the coextrusion system. Figure 3 diagrams the coextrusion system built at Sandia. Details on the principles of coextrusion and on the design of the Sandia coextruder will be presented later.

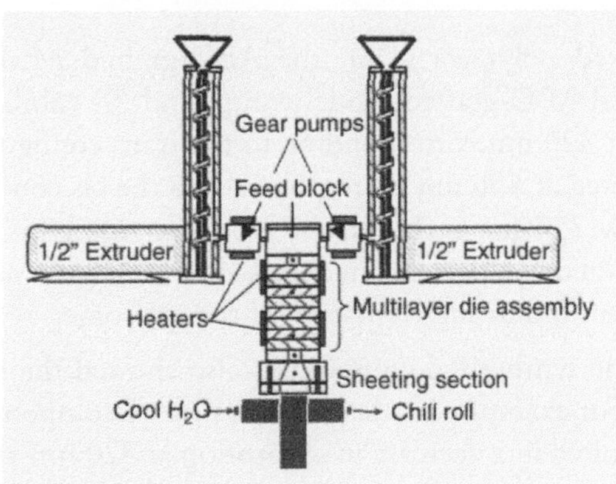

Figure 3. Diagram of coextrusion system.

Results and Discussion

Physical Properties of ABD-Grafted PMMA

All the ABD-grafted PMMA showed optical response in methylene chloride solutions (Fig. 4).

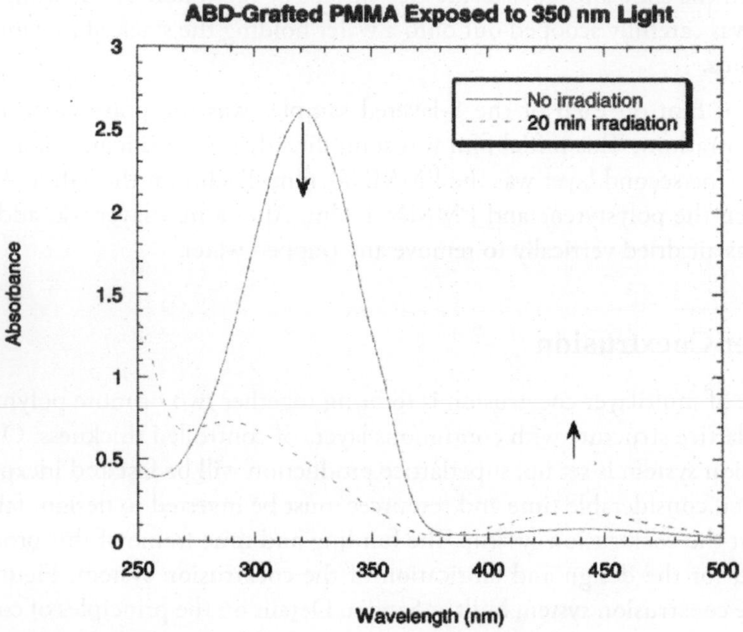

Figure 4. Absorbance spectra of 1 before and after irradiation with 350 nm light.

All the ABD-grafted polymers (1,2), the ABD-grafted MMA monomers (6,7), and the saturated ABD-grafted model compound (9) exhibited an absorbance peak centered at 320 nm corresponding to the trans configuration and an absorbance peak centered at 436 nm corresponding to the cis configuration (the absorbance peak below 250 nm is due to the aromatic rings) (5-7). Irradiation with 350 nm light resulted in a decrease in absorbance at 320 nm and an increase in absorbance at 436 nm indicating formation of the cis isomer.

Thin films of all the synthesized compounds also showed the expected photochemical behavior. For example, irradiation of a film of compound 1with 350 nm light over time resulted in a decrease in absorption at 320 nm and an increase in absorption at 436 nm, again indicating formation of the cis isomer (Fig. 5a). After 15 minutes of irradiation, a photostationary state was reached and no further changes in absorption were observed. The lower relative change in absorption at 436 nm is due to the cis isomer having a lower extinction coefficient than the trans isomer at 320 nm. Exposure of the film to a visible lamp resulted in complete isomerization back to the trans form within 2.5 hours (Fig. 5b). The same transformation occurred within minutes when the cis film was exposed to sunlight. The thermal isomerization from cis to trans was also monitored in the dark at room temperature and the first order rate constant of 4.6 x 10-6 s-l was obtained. This rate constant is consistent with other ABD-grafted polymers reported in the literature (9).

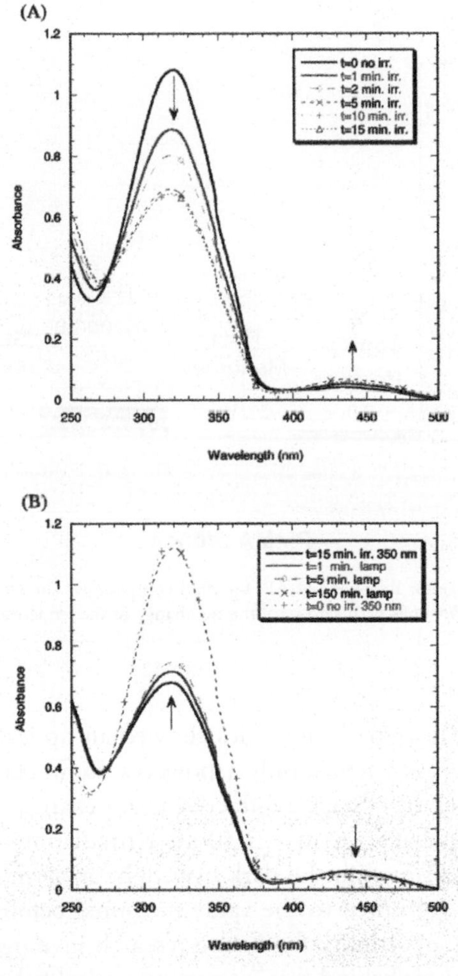

Figure 5. (A) Trans to cis isomerization of a film of ABD-grafted PMMA 1 and, (B) cis to trans isomerization of 1 with visible light.

Ellipsometric measurements (633 nm probe light at a 70" incidence angle) of the persistent change in the indexes of refraction after irradiating the samples with 350 nm light were carried out on films spun from methylene chloride solutions of 75% PMMA/25% ABD-grafted compound. The films were deposited onto silicon wafers whose ellipsometric constants where measured a priori and accounted for. Figure 6 summarizes the observed index changes.

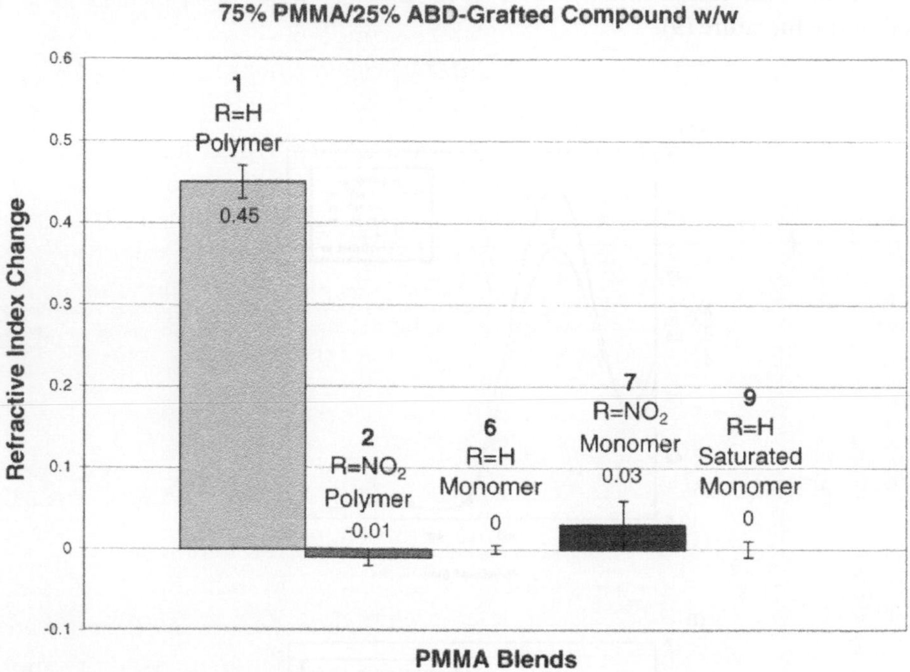

Figure 6. Refractive index change for thin films of ABD-grafted compounds. The refractive index change is plotted in absolute refractive index units with zero meaning no change in the refractive index after irradiation with 350nm light.

Although all the ABD-grafted molecules showed an optical response in the solid state absorbance measurements, only 1showed a useful persistent change in refractive index. 1 had on the average a refractive index change of 0.45, changing from η = 2.85before irradiation to η = 3.30 after irradiation. After irradiation with 350nm light, 1 was stored in the dark overnight at room temperature and its refractive index value returned to the value measured before irradiation. We believe that there is an opportunity for future research investigating the fundamental physical mechanisms behind this interesting optical behavior.

Another important property for the eventual incorporation of 1 into a coextruded superlattice is its thermal stability. Coextrusion can potentially expose the extruded polymer to temperatures as high as 230°C for less than a minute in a nitrogen atmosphere. In order to test for thermal and oxidative robustness, compound 1 was heated at a rate of 1°C/min in air in a thermal gravimetric analysis instrument. As seen in Figure 7, the compound exhibited negligible mass change below 250°C and the same photoinduced refractive index change was measured in film samples after heating to 230 °C.

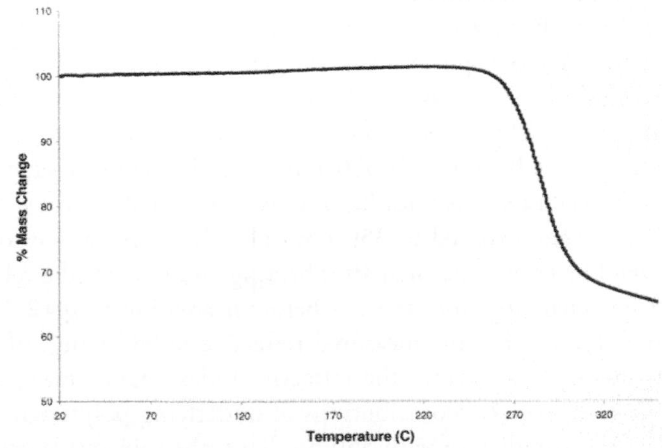

Figure 7. Thermal gravimetric analysis of compound 1 taken at a heating rate of 1°C/min in air.

Incorporation of ABD-Grafted PMMA into Superlattice

After determining that compound 1 had a large photoinduced index of refraction change in the solid state and that it had enough thermal stability for incorporation into extruded superlattices, a large 50 g batch of the compound was synthesized for eventual use in the coextrusion system. We also demonstrated the incorporation of 1 into a superlattice fabricated via the modified spin casting procedure (Figure 8).

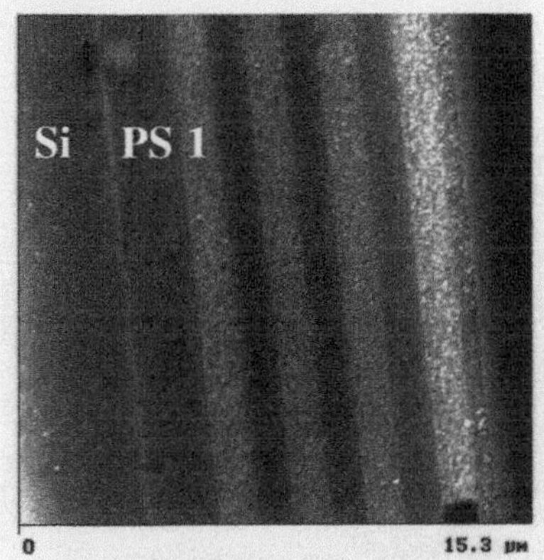

Figure 8. Atomic force microscopy cross-sectional scan of spun cast superlattice. Si: silicon substrate; PS: polystyrene layer; 1:ABD-grafted PMMA compound 1.

The layers fabricated by the modified spin casting procedure were approximately lpm in thickness. Both a spin cast control superlattice containing polystyrene and pure PMMA and a spin cast superlattice containing polystyrene and 2: 1 PMMAcompound 1were heated to 200 °C for 2 minutes and the refractive indexes measured with an ellipsometer (633 nm probe light at a 70° incidence angle) before and after irradiation with 350 nm light. The control superlattice maintained its layer structured after heating but as expected showed no change in its refractive index when exposed to 350 nm light. The superlattice containing 1 also maintained its layer structured after heating and it showed a refractive index change of 0.15, changing from η = 2.1 before irradiation to η =2.25 after irradiation. We hypothesize that the measured refractive index change of 1 in a superlattice configuration differs from the refractive index change measured on a thin film because of the optical contributions of underlying polystyrene layers in the superlattice. After irradiation with 350 nm light, the spin cast superlattice containing 1 was stored in the dark overnight at room temperature and its refractive index value returned to the value measured before irradiation. Thus, we have shown that compound 1 is thermally stable, can form thermally stable superlattices in combination with polystyrene, and has a reversible change in its refractive index that can be triggered by irradiation with UV light.

Progress in Multilayer Coextrusion

Figure 9 shows the configuration of the multilayer coextruder built as part of this project.

Figure 9. Sandia's multilayer coextrusion system.

In multilayer coextrusion, a different polymer is fed into each extruder, which melts and pressurizes the polymer for further processing. The output of each extruder enters a gear pump. The gear pumps arecritical to multilayer fabrication for two reasons. First, as the flight of the extruder screw passes the extruder outlet it creates a pressure pulse in the molten polymer stream. During layer multiplication a pressure pulse can lead to instabilities in the flow field and a break-up of the layered structure. The gear pump isolates the pressure pulse, removing a significant source of flow instability. Second, there is a combination of drag and pressure-driven flow within the extruder. Thus, for a given extruder screw rotation rate, the output of molten polymer will depend on the viscoelastic properties of each polymer. The implication is that to match the thickness of each layer of the two polymer streams, the flow rate must be calibrated for each polymer under the extrusion conditions. The intermeshing teeth of the gears within the gear pumps have a fixed volume. Therefore, the gear pumps deliver a precise volume of molten polymer, independent of the polymer properties.

Thc formation of the multilayered structure begins by bringing the output of the two gear pumps together in a feedblock. As shown in Figure 10, this creates a bilayer that is "multiplied" to achieve a many-layered product. The multiplication method is simple, and was patented by Dow Chemical in 1969 (10). In the 1990's the technology was licensed to the 3M Company, primarily finding use as reflective materials in light pipes. In each multiplication element the flow is cut by a vertical piece in the flow field. The left half of the molten stream is sent througha channel to the lower portion of the multiplication element and spread back to the full channel width. Similarly, the right half is sent to the upper portion of the element and re-spread, such that at the end of the multiplication element the number of layers has doubled. Stacking these multipliers together causes the number of layers to increase as 2n+1, in which n is the number of multiplication elements.

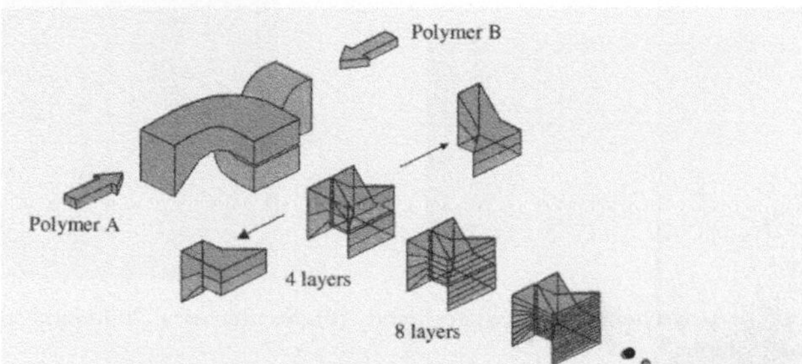

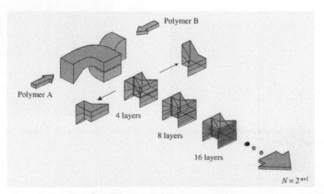

Figure 10. Schematic representation of multilayer creation via coextrusion.

After the desired number of multiplication elements, a sheeting section is placed in the flow field. The sheeting section is designed to vertically compress and horizontally stretch the multilayered structure to produce a sheet or film. The molten film exits the multilayer coextrusion line and enters a chill roll take-up device. The water-cooled steel rollers freeze the multilayered structure, preventing it from deforming.

There are many advantages to producing multilayered structures using this method. First, the polymers need only be melt-processable (i.e. dissolution in solvent is not required). Second, the film can be produced in near-infinite length, provided polymer is fed to the extruders. Third, the thickness of the film is determined by the thickness of the sheeting die exit and the take-up speed, while the film thickness and the number of multipliers control the thickness of each layer. Thus, ten multipliers and a sheeting die with an exit thickness of 1 mm produce a film with each layer approximately 0.4 microns thick. Fourth, the layers do not mix significantly, due to the high viscosity of the polymers and the low flow rates. For example, estimating the Reynolds number using typical coextrusion parameters (melt viscosity of l000 Pa-s, density of 1000 kg/m3, flow rate of 1x10-6 m3/s, channel diameter of 0.125 m) gives l x 10-5, which is orders of magnitude below the laminar-to-turbulent flow transition. Thus, structures such as those shown in Figure 11a (32-layer sample of polystyrene alternating with polypropylene, produced at the University of Minnesota) or Figure 11b (2000-layer sample of polycarbonate alternating with poly(methy1 methacrylate) produced at the Dow Chemical Company) can be produced. Figure 1la is an optical image, in which each layer is approximately 25 microns thick. The AFM image in Figure 1 lb shows layers with a thickness of approximately 200 nm, which produce a final layered structure with mirror-like properties.

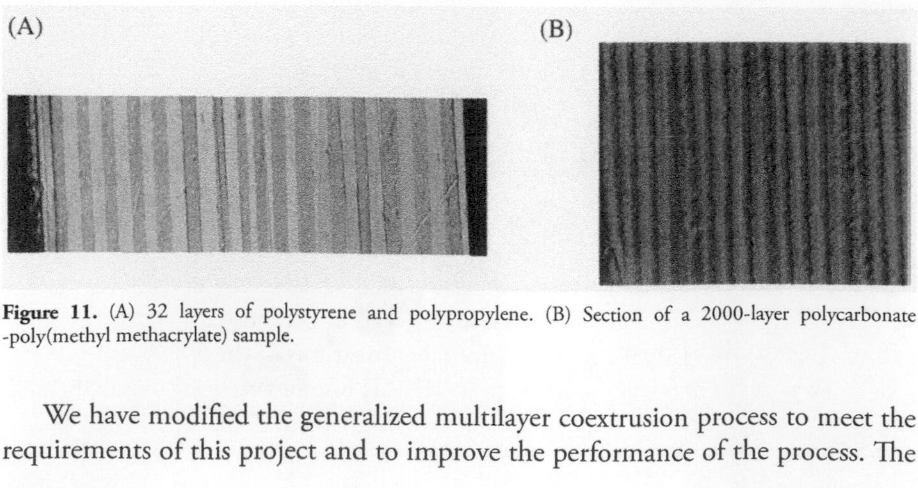

Figure 11. (A) 32 layers of polystyrene and polypropylene. (B) Section of a 2000-layer polycarbonate -poly(methyl methacrylate) sample.

We have modified the generalized multilayer coextrusion process to meet the requirements of this project and to improve the performance of the process. The common feed screw in each extruder has been replaced with a modified mixing screw. In addition to providing the shear flow and pressure necessary to melt and extrude the polymer, these modified screws have elements that produce extensional flow and should lead to enhanced mixing. This is critical if we areto blend polymers synthesized at Sandia California with other commercial materials.

Additionally, the multiplication process has been altered. Inhomogeneities in the velocity profile are produced in each multiplication element. If the multiplication elements are stacked sequentially, the vertical cut of the layered sample will transect an inhomogeneous flow. This leads to large variations in the thickness of each layer. Two modifications were made to the design that are expected to significantly reduce this effect. First, an open channel is added after each multiplier to re-equilibrate the velocity profile. Second, the orientation of the multipliers is reversed after each multiplication. The multiplication shown in Figure 10 has a clockwise orientation. We have placed a multiplier with a counterclockwise (right section flows down, left section flows up) orientation after to further balance the polymer flow field.

Lastly, each multiplication increases the instabilities in the flow field and the likelihood that the layers will either break-up or have non-uniform thickness. Increasing the number of layers produced in the feedblock can reduce the number of layer multiplications required. The Sandia feedblock has been designed to separate the incoming extruder flow into four layers. Thus, eight total layers exit the feedblock. Since the flow division is accomplished before the layered structure is produced, we expect that there will be no negative impact on the layer uniformity or flow homogeneity.

The Sandia coextruder system was assembled at the end of this project. Additional funding from other sources will be needed to demonstrate and optimize the coextrusion of superlattices and to measure the optical performance of such materials. We expect the new coextrusion system along with polymer 1 to be the starting point for the synthesis of novel optical materials that will ultimately result in new optical devices and applications.

Conclusions

We have successfully synthesized a photoresponsive polymer by grafting an azobenzene dye to MMA followed by polymerization. The azobenzene dye undergoes a trans-cis photoisomerization that causes a persistent change in the refractive index of cast polymer films. This novel polymer was incorporated into superlattices prepared by spin casting and the optical activity of the polymer was maintained. A modified coextruder that allows the rapid production of soft matter superlattices was designed and fabricated.

References

1. K. Booth, S. Hill, The Essence of Optoelectronics, Prentice Hall, London, p. 4 (1998).
2. G. Scamarcio et al., "High Power Infrared (8-Micrometer Wavelength) Superlattice Lasers," Science 276,773 (1997).
3. J. H. Schon et al., "An Organic Solid State Injection Laser," Science 289,599 (2000).
4. M. F. Weber et al., "Giant Birefringent Optics in Multilayer Polymer Mirrors," Science 287,2451 (2000).
5. T. Yamamoto et al., "Holographic gratings and holographic image storage via photochemical phase transitions of azobenzene polymer liquid crystal films," J. Mater. Chem. 10,337 (2000).
6. P. O. Astrand et al., "Five-membered rings as diazo components in optical storage devices: an ab initio investigation of the lowest singlet excitation energies," Chem Phys Letters 325, 115 (2000).
7. N. K. Viswanathan et al., "Surface relief structures on azo polymer films," J. Mater. Chem. 9, 1941 (1999).
8. W. D. Callister Jr, Materials Science and Engineering, An Introduction, New York, p. 698 (1997).
9. L. X. Liao et al., "Photochromic dendrimers containing six azobenzenes," Macromolecules 35,319 (2002).
10. W.J. Schrenk, T. Alfrey Jr, "Some physical properties of multilayer films," Polym. Eng. Sci. 9, 393-399 (1969).

CITATION

Morales AM, McElhanon JR, Cole PJ and Rondeau CJ. Synthesis of a Photoresponsive Polymer and Its Incorporation into an Organic Superlattice. Work performed by Sandia National Laboratories for the US Department of Energy. DOE contract # AC04-94AL85000. January 2009. US Government publication.

An Efficient Drug Delivery Vehicle for Botulism Countermeasure

Peng Zhang, Radharaman Ray, Bal Ram Singh, Dan Li, Michael Adler and Prabhati Ray

ABSTRACT

Background

Botulinum neurotoxin (BoNT) is the most potent poison known to mankind. Currently no antidote is available to rescue poisoned synapses. An effective medical countermeasure strategy would require developing a drug that could rescue poisoned neuromuscular synapses and include its efficient delivery specifically to poisoned presynaptic nerve terminals. Here we report a drug delivery strategy that could directly deliver toxin inhibitors into the intoxicated nerve terminal cytosol.

Results

A targeted delivery vehicle was developed for intracellular transport of emerging botulinum neurotoxin antagonists. The drug delivery vehicle consisted of the non-toxic recombinant heavy chain of botulinum neurotoxin-A coupled to a 10-kDa amino dextran via the heterobifunctional linker 3-(2-pyridylthio)-propionyl hydrazide. The heavy chain served to target botulinum neurotoxin-sensitive cells and promote internalization of the complex, while the dextran served as a platform to deliver model therapeutic molecules to the targeted neurons. Our results indicated that the drug delivery vehicle entry into neurons was via BoNT-A receptor mediated endocytosis. Once internalized into neurons, the drug carrier component separated from the drug delivery vehicle in a fashion similar to the separation of the BoNT-A light chain from the holotoxin. This drug delivery vehicle could be used to deliver BoNT-A antidotes into BoNT-A intoxicated cultured mouse spinal cord cells.

Conclusion

An effective BoNT-based drug delivery vehicle can be used to directly deliver toxin inhibitors into intoxicated nerve terminal cytosol. This approach can potentially be utilized for targeted drug delivery to treat other neuronal and neuromuscular disorders. This report also provides new knowledge of endocytosis and exocytosis as well as of BoNT trafficking.

Background

Botulinum neurotoxins (BoNTs) are produced by the anaerobic Clostridium botulinum species of bacteria and are the cause of botulism, a life-threatening neuroparalytic disease. They are extremely potent food poisons, with a mouse LD50 of 0.1 ng/kg for type A [1,2]. Aerosol exposure of BoNTs does not occur naturally, but could be attempted by bioterrorists to achieve a widespread effect. It has been estimated that a single gram of crystalline toxin, evenly dispersed and inhaled, could kill more than one million people [2].

BoNTs are large proteins with a molecular weight of 150 kDa. They are produced as a complex containing the neurotoxins and associated proteins [3]. They are synthesized as inactive single chain protoxins and are activated by protease nicking to form a dichain molecule (a 50 kDa light chain (LC) and a 100 kDa heavy chain (HC)) linked through a disulfide bond [4]. The HC is responsible for binding to the target nerve cells (through its C-terminus) and translocating the LC into the cell cytoplasm (through its N-terminus) [5,6].

Inside the neuronal cytosol, the LC acts as a Zn^{2+}-endopeptidase against specific intracellular protein targets present either on the plasma membrane or on the synaptic vesicle, and inhibits neurotransmitter release by disabling the exocytotic docking/fusion machinery [5,6]. BoNTs catalyze proteolysis of specific proteins of the soluble NSF attachment protein receptor (SNARE) complex that have been implicated in the exocytotic machinery [5,7]. BoNT/A,/C, and /E cleave a 25 kDa synaptosomal associated protein (SNAP-25).

Current therapy for botulism involves respiratory supportive care and the administration of antitoxin. The antitoxin could be the currently available equine BoNT antibodies or potentially more effective recombinant multivalent antibodies. However, only a few antitoxins, which must be administered before toxins reach the nerve cells, are available. Thus, the therapeutic window for using an antitoxin is short. Once the syndrome is developed, the antitoxin is less effective since it cannot penetrate the nerve cell to neutralize the toxin. The flaccid muscle paralysis caused by BoNT/A lasts for several months [8]. Therefore, patients who have already developed the syndrome must be put under respiratory intensive care during paralysis [1,2,9]. Should a bioterrorist attack occur, public health crisis could arise due to the lack of effective antidotes against botulism, especially in the absence of reliable presymptomatic diagnostics.

For relief from BoNT-mediated paralysis, it is important to rescue the poisoned nerve cells through restoration of the neurotransmitter release process. While drugs have been designed to block the BoNT endopeptidase activity, which is believed to be responsible for the inhibition of neurotransmitter release, delivery of the drugs specifically to the poisoned nerve terminals remains a major hurdle. Therapeutic targeting is important for two main reasons: (a) delivering an effective high concentration of the therapeutic compound to the site of toxicity, i.e., nerve terminals for botulism, and (b) minimizing systemic toxicity, if any, due to treatment compounds. At present, some examples of the proposed pharmacological antidotes for BoNT poisoning are a protease inhibitor, a phospholipase A2 activator or a modulator of intracellular free $Ca2+$ concentration. Since all of these parameters are involved in normal body functions, a systemic therapeutic approach is inadvisable due to potential toxicity concerns.

Therefore, we developed a drug delivery vehicle (DDV) comprising the non-toxic recombinant heavy chain of BoNT-A coupled to a 10-kDa amino dextran via the heterobifunctional linker 3-(2-pyridylthio)-propionyl hydrazide. The heavy chain served to target botulinum neurotoxin-sensitive cells and promote internalization of the complex, while the dextran served as a platform to deliver model therapeutic molecules to the targeted cells.

Results

Structure of DDV

Initially we designed a DDV utilizing the recombinant BoNT/A heavy chain (rHC), which is known to specifically bind to the presynaptic nerve terminals and be internalized via endocytosis. The DDV construct was a modification of that developed by Goodnough et al. [10] consisting of a targeting molecule, Cy3 labeled purified (from the holotoxin) HC linked by a disulfide bond to a drug simulant, which was Oregon green 488 (OG488) labeled 10 kDa dextran (Fig. 1). The DDV structure was used for further experiments. To our knowledge, this is the first experimental demonstration of a prospective therapeutic approach to treat botulism in a relevant peripheral neuronal model combined with a feasible targeted drug delivery technology.

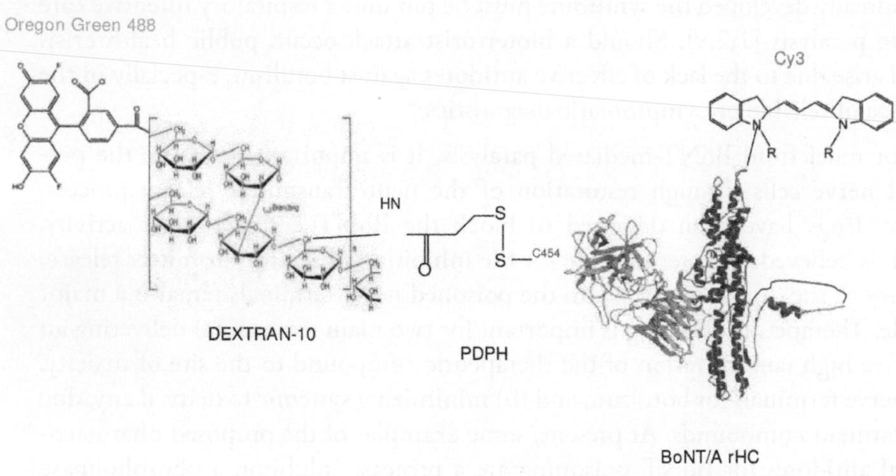

Figure 1. Schematic representation of the DDV for transport of BoNT/A antagonists. The schematic representation of the DDV construct without drug. The PDPH linker is bound to one of four possible cysteine (C) sulfhydryl groups on the BoNT/A rHC. It is attached to C454, which normally participates in the disulfide linkage with the LC. Cy3 and Oregon green 488 are bound to O-amino groups of lysine in the rHC and dextran, respectively. The dextran is conjugated to the rHC by a C-N bond in one of the glucose residues. In a functional DDV, multiple drug molecules may be attached to dextran carrier.

rHC was a Safe DDV Component for Delivery of BoNT Antidotes

To exclude any possible toxicity of the rHC component in our DDV construct, we compared the inhibition of 80 mM K^+ stimulated [^{3}H]glycine release due to increasing concentrations of rHC or native BoNT/A holotoxin by the assay described in Methods. In these experiments, the results obtained using the particular batch of toxin showed that BoNT/A was quite toxic, as expected, with an IC50 (toxin concentration to cause 50% inhibition of neuroexocytosis in untreated control cells) of approximately <1 pM and a total inhibition at ~0.1 nM.

However, a much higher concentration of rHC, up to 200 nM, did not show any inhibition of [^{3}H]glycine release under the same assay conditions (Fig. 2).

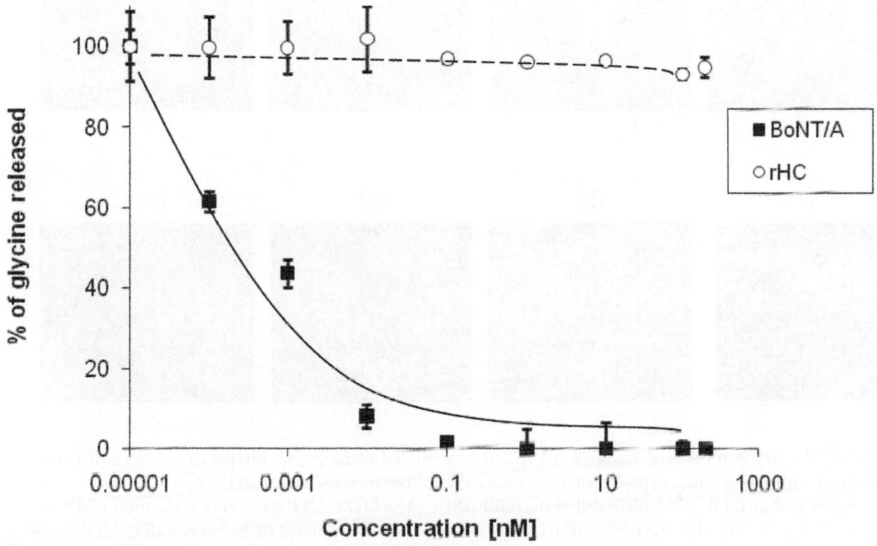

Figure 2. rHC was a safe DDV component to delivery antidotes. Primary cultures of mouse spinal cord were exposed to BoNT/A (∎) or rHC (o) respectively in indicated concentrations for 16 hrs. Potassium-evoked glycine release was measured. Results expressed as percentage glycine release compared with untreated control. Data points are the mean (± SD) of three separate experiments each determined in triplicate.

DDV Entry into Neurons via BoNT/A Receptor Mediated Endocytosis

The experimental design was to mimic a therapeutic application of the DDV strategy to treat individuals poisoned with BoNT/A and exhibiting clinical symptoms of botulism. Since the targeted DDV approach is based on the premise of a selective entry of DDV into presynaptic nerve terminals via BoNT/A receptor mediated endocytosis, we demonstrated by competition experiments that the uptake of the DDV-Mas-7 was via BoNT/A receptors. In these experiments, 3-week old spinal cord cultures were exposed for 16 hours to DDV (200 nM) alone or to DDV plus a 1-, 3-, or 10-fold excess of unlabeled rHC (Fig. 3, A1-A4) or BoNT/A holotoxin (Fig. 3, B1-B4) added to cultures simultaneously. As seen in Fig. 3, in the absence of rHC or BoNT/A, DDV uptake and dextran separation were as expected; however, a 10-fold excess of rHC or BoNT/A holotoxin completely blocked the uptake of DDV. These results suggested that DDV entry into neurons occurred by the same route as used by BoNT/A.

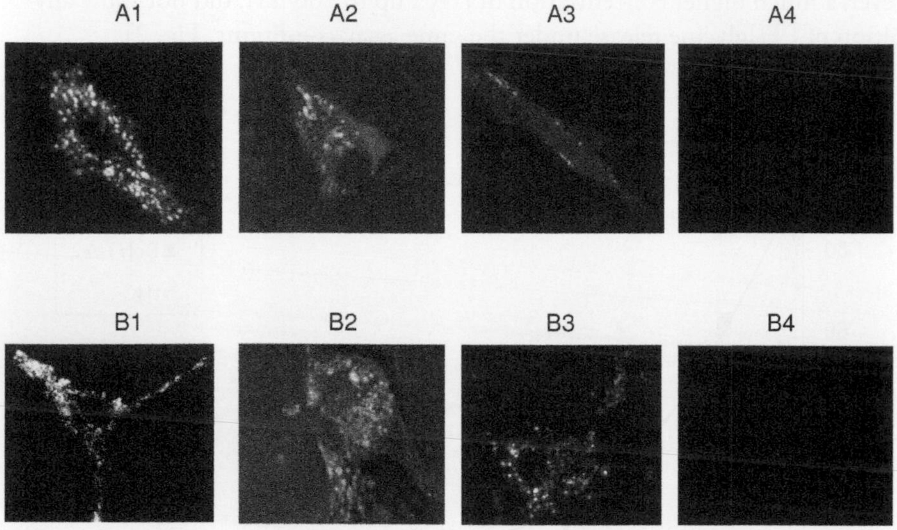

Figure 3. DDV entry into neurons via BoNT/A receptor mediated endocytosis. Images in rows A and B were obtained from triplicate cultures exposed for 16 h under the following conditions: (A1) DDV (200 nM) in the absence of rHC; (A2) DDV (200 nM) and rHC (200 nM); (A3) DDV (200 nM) and rHC (600 nM); and (A4) DDV (200 nM) and rHC (2 μM); (B1) DDV (200 nM) in the absence of BoNT/A; (B2) DDV (200 nM) and BoNT/A (200 nM); (B3) DDV (200 nM) and BoNT/A (600 nM); and (B4) DDV (200 nM) and BoNT/A (2 μM). DDV and rHC or BoNT/A were added simultaneously. Note the progressive reductions in fluorescence with increasing concentrations of BoNT/A or rHC. Micrographs were obtained on a Bio-Rad 2000 laser microscope confocal microscope using a 100× oil immersion objective. The fluorescence colors of labelled molecules are: red-rHC; green-OG488-dextran.

Drug Carrier could be Separated from DDV in a Fashion BoNT/A LC Dissociates from the HC in the Holotoxin

To determine the efficacy of delivering the therapeutic compound, we studied the separation of the drug carrier from DDV. Confocal microscopy was used to detect the separation of the DDV components. Spinal cord neurons were treated for 16 hours with 200 nM labeled DDV at 37°C. The staining pattern of unseparated DDV was orange (red plus green labeling) and punctate (Fig. 4, D1, D2 and 4, D3). The punctate nature of the staining suggested clustering of DDV in vesicles. The images shown in Fig. 4 highlight the presence of released drug carrier (green) in the particles present in the nerve terminal cytosol. Inclusion of endosome staining (blue) indicated that the DDV was intra-endosomal (Purple, which was red plus blue staining, Fig. 4, E1, E2 and 4, E3) as expected for material transported by BoNT HC. It indicated that the separation of the drug carrier from DDV was in a fashion similar to the dissociation of BoNT/A LC from the HC in the holotoxin.

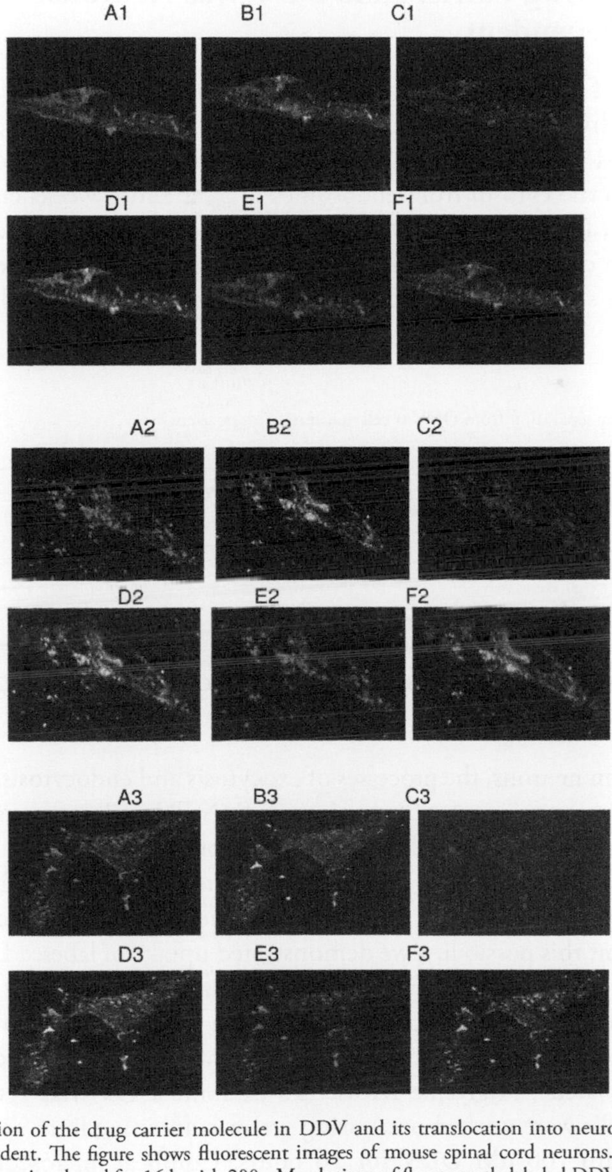

Figure 4. Separation of the drug carrier molecule in DDV and its translocation into neuronal cytosol is cell maturation dependent. The figure shows fluorescent images of mouse spinal cord neurons. Cells at different stages of culture were incubated for 16 h with 200 nM solutions of fluorescently labeled DDV, and then labeled with anti-endosome antibody as described under methods. Confocal images shown are as follows: A1-A3, red-rHC; B1-B3, green-OG488-dextran; C1-C3, bright blue-Alexa 633-endosomes; D1-D3, overlay of red and green showing either co-localization (orange) or separation of rHC and dextran; E1-E3, overlay of red and blue showing either the localization (magenta) of rHC in the endosomes or its release into the cytosol, if any; F1-F3, overlay of green and blue showing either localization (light blue or greenish blue) of dextran in the endosomes or its release into the cytosol. The numerical suffixes as in A1, A2 and A3 indicate culture age, i.e., one- (top panels), two- (middle panels) or three-week (bottom panels) old. The results clearly demonstrated that the rHC component of the DDV remained localized in the endosomes, while the OG488-dextran separated from the DDV and migrated into the cytosol.

Separation of Drug Carrier from DDV was Neuronal Maturation-Dependent

Different stages of spinal cell culture growth were used to evaluate the efficacy of separation of drug carrier from DDV. Confocal image analysis revealed that about 20, 32 and 40% of the drug carrier component separated from DDV and diffused into the cytosol from endosomes in 1, 2 and 3 weeks culture, respectively (Fig. 4, D1, D2 and 4, D3; table 1). These results indicated that the separation of the drug carrier from DDV is neuronal maturation-dependent. Furthermore, the separations of DDV components occur in a time-dependent manner.

Table 1. The drug carrier separation from DDV is cell maturation-dependent

Cell growth period	I week	2 weeks	3 weeks
Drug carrier separation rate (%)	20 ± 3	32 ± 5	40 ± 4

The Processes of Exocytosis and Endocytosis are not Tightly Coupled

It is possible that in neurons, the processes of exocytosis and endocytosis are tightly coupled, i.e., interruption of exocytosis, as in BoNT/A poisoning, might halt endocytosis as well. If true, the DDV approach as presented here would not be a feasible drug delivery system in BoNT poisoned neurons because the uptake of DDV, via endocytosis, could be blocked as a sequel of exocytosis blockade by BoNT. To discount this possibility, we demonstrated uptake of labeled DDV (red fluorescence), of which Oregon green 488 was omitted and and only Cy3 was used, in spinal cord neurons previously exposed to a high concentration (1 nM) of Alexa 488-labeled BoNT-A (green florescence); 1 nM BoNT-A had completely blocked K^+-stimulated $[^3H]$glycine release. To examine DDV-Mas-7 uptake in these cells, the cells were washed once using warm culture medium and reincubated at 37°C with 100 nM DDV for 16 hours. Confocal microscopy results indicated that both BoNT/A and DDV were taken up in the same cell pool, but localized in separate population of endosomes (Fig. 5), demonstrating internalization of DDV via endocytosis into BoNT/A poisoned neurons. This suggested that the exocytosis and endocytosis are not tightly coupled in BoNT/A poisoned neurons.

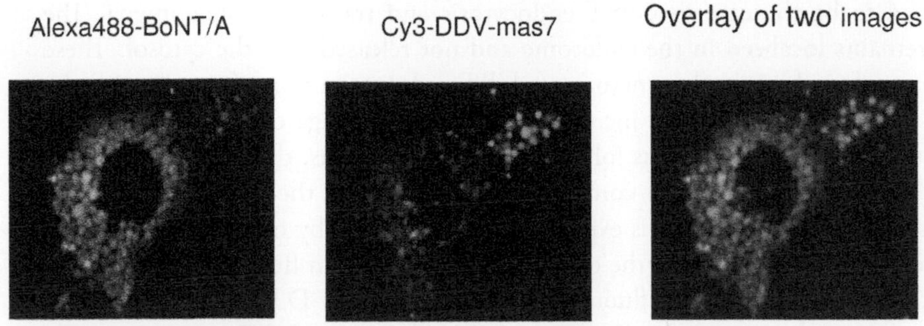

Figure 5. Exocytosis and endocytosis are not tightly coupled in BoNT/A poisoned neurons. (a) BoNT/A was labeled with Alexa 488 (green fluorescence). Three week-old cultured mouse spinal cord neurons were incubated with a culture medium containing 1 nM of Alexa 488-BoNT/A for 8 hours at 37°C. Excess Alexa 488-BoNT/A was removed by washing 3 times with fresh culture medium. The cells were incubated with fresh medium for 1 hour and then incubated in medium containing 100 nM Cy3 (red fluorescence) labeled DDV for 16 hours (b). (c) Overlay of red and green showing both BoNT/A (green) and DDV (red) were taken up in the same cell pool, but localized in separate population of endosomes. Fluorescent images were analyzed by Bio-Rad 2000 laser confocol microscope. Alexa 488 was excited at 488 nm line of an argon laser and detected with a 530-nm cutoff filter; Cy3 was excited at 543 nm line of an argon laser and detected with a 565-nm cutoff filter.

Discussion

Primary cultures of spinal cord represent a convenient and sensitive system to study mechanisms of neurotransmitters release [11,12]. Internalized neurotransmitters by spinal cord neurons in culture were released quantitatively in response to depolarization and Ca2+. This release is inhibited by tetanus toxin and botulinum neurotoxins in a concentration- and time-dependent manner [13-17]. Therefore, this system serves as a suitable model to examine the efficacy of prospective BoNT countermeasures. Sheridan and Adler indicated that the evoked release of neurotransmitters, notably glycine, in this system was time-dependently increased [18]. In our studies, there was a pronounced time-dependent increase of the drug carrier separation from DDV, which paralleled an enhancement of transmitter release.

We postulate that the ability of drug carrier to separate from DDV in our spinal cord cell culture model is dependent upon neuronal maturation at increasing age of the cultures. Regarding DDV uptake and processing in neurons, the results in Fig 4 clearly demonstrated the following facts. The red (Cy3-rHC) (A1-A3) and the blue (Alexa 633-endosome) (C1-C3) fluorescence signals were strongly present in neurons at any stage of development, however, were always in a distinct punctate localization. Moreover, the red and the blue signals were always co-localized (E1-E3) and the red signals never diffused into the cytosol. These observations indicated that the DDV was readily internalized into neurons apparently via endocytosis and the rHC component remained in the endosomes as generally believed to be the case in BoNT endocytosis and trafficking phenomena. The HC remains localized in the endosome and not released into the cytosol. These micrographs exhibit both punctuate and diffused distribution of signals with the level of diffusion apparently increasing with increased age of cultures. This observation may be explained as follows: in mature neurons, the DDV is taken up into endosomes, the dextran component separates from the rHC and then gets released into the cytosol. This explanation is supported by our results presented in the micrographs showing the overlays of red and green fluorescence (D1-D3) and those of green and blue fluorescence (F1-F3). In D1-D3, overlap of red and green generated orange indicating intact DDV; the released OG488-dextran was green. It should be noted that the orange was punctate suggesting endosomal localization, whereas the green was diffuse suggesting the release of OG488-dextran component of the DDV into the cytosol, which was enhanced with increasing age of cultures. Using the images in D1-D3, we calculated the separation and release rates of OG488-dextran from the DDV by utilizing the Bio-Rad AutoDeblur and AutoVisualize software to quantitate fluorescence intensity. The separation/release rates were expressed as 100% (total in images) minus percentage of co-localization rate (Table 1). In F1-F3, overlap of bright blue and green generated light or greenish blue indicating OG488-dextran remaining in the endosomes; the released OG-dextran was green. On examining the D1-D3, E1-E3 and F1-F3 micrographs, it is interesting to note that intact DDV (D1-D3), rHC (E1-E3) and unreleased dextran all seem to be localized in the same endosomal pool; this provides additional support to our proposed mechanism of DDV uptake and processing. Relevant to this proposition, most significant was our demonstration that in neurons, the DDV function that required an efficient endocytotic mechanism was developmentally regulated, i.e., neuronal maturation-dependent. Moreover, our results showed that the efficiency in neuroexocytosis was also a function of mature neurons.

To demonstrate the feasibility of delivering a therapeutic compound via the DDV, we examined the separation of the prototype drug carrier dextran from DDV. The results indicated that the drug carrier components were satisfactorily separated from DDV and diffused into cytosol. As described in the results section, the targeting component of the DDV, i.e., rHC was nontoxic. Therefore, the DDV approach presented here may be a physiologically compatible and a feasible targeted drug delivery method to counteract botulism.

Although it was believed to be the case, our results provided the first experimental evidence that the HC of BoNT/A upon internalization into neurons remains localized in the endosomes and thus, does not participate in the cytosolic mechanism of BoNT/A toxicity. Very meaningful was the fact that BoNT/A poisoned neurons, which were totally incapable of stimulated exocytosis, could still incorporate the HC via endocytosis. This was apparently due to new or spare receptors available for HC molecule binding on the plasma membrane. Furthermore, the exocytosis and endocytosis phenomena in neurons may not necessarily be coupled tightly. Neale et al. also reported that BoNT/A blocked synaptic vesicle exocytosis but not endocytosis at nerve terminal [20].

In conclusion, this report provides new knowledge of endocytosis and exocytosis, as well as of BoNT trafficking and action. Notably, application of this DDV approach to antagonize botulism is not necessarily limited to the neuronal targeting of BoNT as shown here, but also should be useful for delivery of other prospective antidotes, such as protease inhibitors to protect the vesicle fusion proteins as applicable. Finally, the success in the DDV strategy against botulism shown here may open new avenues in developing technologies to treat other neurological disorders that require a targeted delivery of therapeutics to affected neurons or tissues. Before the actual studies are conducted, we can only speculate on a possible route of administration of the DDV as a therapeutic in a patient. Oral or inhalation routes are inadvisable. The oral administration may result in DDV degradation in the gastrointestinal system. The Inhalation administration may result in a slower DDV absorption and may also require a high DDV concentration, which could possibly trigger a cell-mediated immune response. Based on these considerations, we propose the intravenous route which should rapidly achieve a high DDV level in the circulation for an effective drug delivery into BoNT poisoned nerve terminals.

Conclusion

An effective botulinum neurotoxin-based drug delivery vehicle can be used to directly deliver toxin inhibitors into the intoxicated nerve terminal cytosol. The concept may possibly be utilized for drug delivery for other neuronal and neuromuscular disorders. Besides a BoNT therapeutic approach, this report also provides new fundamental knowledge of endocytosis and exocytosis as well as of BoNT trafficking in neurons.

Methods

Construct a Model Drug Conjugated Drug Delivery Vehicle

We designed a DDV utilizing recombinant BoNT/A heavy chain (rHC), which is known to specifically bind to the presynaptic nerve terminals and be internalized via endocytosis. The DDV construct was a modification of that developed by

Goodnough et al., [10]. The DDV consisted of a targeting molecule, Cy3 labeled rHC linked by a disulfide bond at Cys454 of rHC to a drug simulant, Oregon green 488 (OG488) labeled 10 kDa dextran (Fig. 1). The DDV construct was soluble in aqueous medium and was stable under our experimental conditions, i.e., at 37°C.

Spinal Cord Cultures

Timed pregnant C57BL/6NCR mice were obtained from the Frederick Cancer Research and Development Center (Frederick, MD). Research was conducted in compliance with the Animal Welfare Act and other federal statutes and regulations relating to animals and experiments involving animals and adheres to principles stated in the Guide for the Care and Use of Laboratory Animals, NRC Publication, 1996 edition. Spinal cords were removed from fetal mice at gestation day 13. Cells were dissociated with trypsin and plated in collagen-coated 4 well coverslips or 35 mm diameter 6-well culture plates at a density of 105 cells/cm2. Cells were grown in Eagle's Minimum Essential Medium with 5% heat-inactivated horse serum and a nutrient supplement (N3) at 37°C in 90% air/10% CO2. Cell cultures were treated with 54 mM 5-fluoro-2-deoxyuridine and 140 mM uridine from day 5-9 after plating to inhibit glial proliferation. Cultures were fed 1-2 times per week and were used for experiments at 1 to 3 weeks after plating.

^{3}[H]glycine Release Assay

^{3}[H]glycine release was determined by a modification of the method described by Williamson et al. [19]. Spinal cord cells were incubated at 37°C for 30 min in HEPES-buffered saline (HBS) containing 2 mCi/ml 3 [H]glycine. The cells were washed briefly with Ca2+-free HBS and incubated sequentially for 7 min in each of the following modified HBS solutions: 5 mM KCl/0 mM Ca2+, 80 mM KCl/2 mM Ca2+ and 5 mM KCl/0 mM Ca2+. Each incubation solution was collected, and the radioactivity was determined by scintillation counting.

Uptake of DDV by Spinal Cord Neurons and Release of Dextran

Cells were exposed to DDV, Cy3-labeled rHC, or Oregon green 488-labeled dextran in growth medium for 16 h at a concentration of 200 nM at 37°C. Cells were subsequently washed three times with growth medium and fixed overnight using

2% paraformaldehyde. The coverslips containing fixed cells were mounted between a glass slide and glass coverslip and viewed on a Bio-Rad 2000 laser confocal microscope. Oregon green 488 was excited at 488 nm and read through a 515-nm cutoff filter. Cy3 was excited at 543 nm and read through a 565-nm cutoff filter. To minimize photobleaching, Slowfade Light was added to the mounting medium. Micrographs were obtained using a Bio-Rad laser confocal microscope with a 100× oil immersion objective. Images were collected with Bio-Rad software. Co-localization of rHC and dextran and separation rates of dextran from DDV were then calculated by utilizing the Bio-Rad AutoDeblur and AutoVisualize software to quantitate fluorescence intensity. Separation rates were expressed as % total minus % co-localization.

Determination of rHC Localization

Experimental procedures were similar to those described above under "Uptake of DDV by spinal cord neurons and release of dextran," except that after fixing with 2% paraformaldehyde, cells were subsequently washed three times with D-PBS and permeabilized with 0.2% TritonX-100 for 10 min in room temperature (RT). Cells were blocked with 4% BSA in D-PBS at 4°C for 1 h and subsequently incubated with goat anti-EEA1 antibody at RT for 1 hr. After washed five times with D-PBS, cells were incubated with a secondary antibody (donkey anti goat-Alexa 633) at RT for 0.5 h. Cells were subsequently washed three times with D-PBS. The coverslips containing fixed cells were mounted between a glass slide and glass coverslip and viewed on a Bio-Rad 2000 laser confocal microscope. Excition and emmission for Cy3 and OG-488 were as stated above. Alexa 633 was excited at 632 nm and detected with a 649 nm cutoff filter. Micrographs were obtained using a Bio-Rad laser confocal microscope with a 100× oil immersion objective.

List of Abbreviations

BoNT: Botulinum neurotoxin; LC: light chain; HC: heavy chain; DDV: drug delivery vehicle.

Competing Interests

The authors declare no competing financial interests. The opinions or assertions contained herein are the private views of the author, and are not to be construed as official, or as reflecting true views of the Department of the Army or the Department of Defense.

Authors' Contributions

PR and RR conceived the project, guided research directions, and edited the manuscript. RR provided special advice on neurobiological issues. PR procured the funding and served as the principal investigator being responsible for overall supervision and all reporting requirements. PZ performed the experiments, analyzed data, drafted and revised the manuscript. BRS supervised all synthetic chemistry work and provided the products to include all unlabeled and labelled botulinum toxin heavy chain, drug simulant and the delivery vehicle molecules. BRS also served as an expert consultant for the project and provided advice on experiments. DL performed some initial experiments and analyzed data. MA served as an expert consultant for the project and edited the manuscript. All authors read and approved the final manuscript.

Acknowledgements

This work was supported by a grant from the Defence Threat Reduction Agency - Joint Science and Technology Office, Medical S&T Division (to PR) and partly by DARPA GRANT W911NF-07-1-0623 (to BRS).

References

1. Greenfield RA, Brown BR, Hutchins JB, Iandolo JJ, Jackson R, Slater LN, Bronze MS: Microbiological, biological, and chemical weapons of warfare and terrorism. Am J Med Sci 2002, 323:326–340.
2. Arnon SS, Schechter R, Inglesby TV, Henderson DA, Bartlett JG, Ascher MS, Eitzen E, Fine AD, Hauer J, Layton M, Lillibridge S, Osterholm MT, O'Toole T, Parker G, Perl TM, Russell PK, Swerdlow DL, Tonat K: Botulinum toxin as a biological weapon: medical and public health management. J Am Med Assoc 2001, 285:1059–1070.
3. Cai S, Sarkar HK, Singh BR: Enhancement of the endopeptidase activity of botulinum neurotoxin by its associated proteins and dithiothreitol. Biochemistry 1999, 38:6903–6910.

4. Inoue K, Fujinaga Y, Watanabe T, Ohyama T, Takeshi K, Moriishi K, Nakajima H, Inoue K, Oguma K: Molecular composition of Clostridium botulinum type A progenitor toxins. Infect Immun 1996, 64:1589–1594.

5. Singh BR: Intimate details of the most poisonous poison. Nature Struct Biol 2000, 7:617–619.

6. Li L, Singh BR: Structure-function relationship of clostridial neurotoxins. J Toxicol-Toxin Reviews 1999, 8:95–112.

7. Montecucco C, Schiavo G: Structure and function of tetanus and botulinum neurotoxins. Quart Rev Biophys 1995, 28:423–472.

8. Cherington M: Clinical spectrum of botulism. Muscle Nerve 1998, 21:701–710.

9. Rosenbloom M, Leikin JB, Vogel SN, Chaudry ZA: Biological and chemical agents: a brief synopsis. Am J Therapeutics 2002, 9:5–14.

10. Goodnough MC, Oyler G, Fishman PS, Johnson EA, Neale EA, Keller JE, Tepp WH, Clark M, Hartz S, Adler M: Development of a delivery vehicle for intracellular transport of botulinum neurotoxin antagonists. FEBS Letters 2002, 513:163–168.

11. Daniels-Holgate PU, Dolly JO: Productive and non-productive binding of botulinum neurotoxin A to motor nerve endings are distinguished by its heavy chain. J Neurosci Research 1996, 44(3):263–271.

12. Simpson LL: Identification of the major steps in botulinum toxin action. Annu Rev Pharmacol and Toxicol 2004, 44:167–193.

13. Bergey GK, Bigalke H, Nelson PG: Differential effects of tetanus toxin on inhibitory and excitatory synaptic transmission in mammalian spinal cord neurons in culture: a presynaptic locus of action for tetanus toxin. J Neurophysiol 1987, 57(1):121–131.

14. Hall YH, Chaddock JA, Moulsdale HJ, Kirby ER, Alexander FC, Marks JD, Foster KA: Novel application of an in vitro technique to the detection and quantification of botulinum neurotoxin antibodies. J Immunol Methods 2004, 288(1-2):55–60.

15. Williamson LC, Fitzgerald SC, Neale EA: Differential effects of tetanus toxin on inhibitory and excitatory neurotransmitter release from mammalian spinal cord cells in culture. J Neurochem 1992, 59(6):2148–2157.

16. Keller JE, Cai F, Neale EA: Uptake of botulinum neurotoxin into cultured neurons. Biochemistry 2004, 43(2):526–532.

17. Sheridan RE, Smith TJ, Adler M: Primary cell culture for evaluation of botulinum neurotoxin antagonists. Toxicon 2005, 45(3):377–382.

18. Sheridan RE, Adler M: Growth factor dependent cholinergic function and survival in primary mouse spinal cord cultures. Life Sciences 2006, 79:591–595.

19. Williamson LC, Halpern JL, Montecucco C, Brown JE, Neale EA: Clostridial Neurotoxins and Substrate Proteolysis in Intact Neurons. J Biol Chem 1996, 271:7694–7699.

20. Neale EA, Bowers LM, Jia M, Bateman KE, Williamson LC: Botulinum Neurotoxin A Blocks Synaptic Vesicle Exocytosis but Not Endocytosis at the Nerve Terminal. J Cell Biol 1999, 147:1249–1260.

CITATION

Zhang P, Ray R, Singh BR, Li D, Adler M, and Ray P. An Efficient Drug Delivery Vehicle for Botulism Countermeasure. BMC Pharmacology 2009, 9:12. doi:10.1186/1471-2210-9-12.

Estimation of Synthetic Accessibility Score of Drug-Like Molecules Based on Molecular Complexity and Fragment Contributions

Peter Ertl and Ansgar Schuffenhauer

ABSTRACT

Background

A method to estimate ease of synthesis (synthetic accessibility) of drug-like molecules is needed in many areas of the drug discovery process. The development and validation of such a method that is able to characterize molecule synthetic accessibility as a score between 1 (easy to make) and 10 (very difficult to make) is described in this article.

Results

The method for estimation of the synthetic accessibility score (SAscore) described here is based on a combination of fragment contributions and a complexity penalty. Fragment contributions have been calculated based on the analysis of one million representative molecules from PubChem and therefore one can say that they capture historical synthetic knowledge stored in this database. The molecular complexity score takes into account the presence of non-standard structural features, such as large rings, non-standard ring fusions, stereocomplexity and molecule size. The method has been validated by comparing calculated SAscores with ease of synthesis as estimated by experienced medicinal chemists for a set of 40 molecules. The agreement between calculated and manually estimated synthetic accessibility is very good with $r^2 = 0.89$.

Conclusion

A novel method to estimate synthetic accessibility of molecules has been developed. This method uses historical synthetic knowledge obtained by analyzing information from millions of already synthesized chemicals and considers also molecule complexity. The method is sufficiently fast and provides results consistent with estimation of ease of synthesis by experienced medicinal chemists. The calculated SAscore may be used to support various drug discovery processes where a large number of molecules needs to be ranked based on their synthetic accessibility, for example when purchasing samples for screening, selecting hits from high-throughput screening for follow-up, or ranking molecules generated by various de novo design approaches.

Background

The assessment of synthetic accessibility (SA) of a lead candidate is a task which plays a role in lead discovery regardless of the method the lead candidate has been identified with. In the case of a de novo designed molecule the experimental validation of its activity requires synthesis of the compound. In the case of experimental or virtual screening exploration of the SAR around the hit, synthetic access to the chemotype is required as well. The more difficult the synthesis of the lead candidate is, the more time and resources are needed for the exploration of this particular area of chemical space. Lead candidates are normally prioritized according to criteria such as drug-likeness [1,2], natural-product likeness [3], predicted activity or freedom to operate with respect to intellectual property. Since sooner or later in the drug discovery process the candidates will be ranked, or even eliminated by their synthetic accessibility, it is desirable to include this aspect into the prioritization of compounds early on. When compounds are purchased from off-the-shelf catalogues in order to augment the screening library, compounds likely to fail later on because of problems with their synthetic tractability may be removed already at this stage. Also in the selection of follow-up candidates from large primary screening results, prioritization by synthetic accessibility can ensure that compounds chosen for validation in dose-response experiments are less likely to be later rejected based on problems with their synthesis. In these two cases, the compounds that are to be validated exist, which means that chemical synthesis must in principle be feasible despite possible complications. When chemical structures are constructed during the de novo design process, one cannot take for granted that the chemical synthesis of such compounds is feasible at all. Therefore it is even more important to estimate whether these compounds can be synthesized with reasonable effort. While experienced chemists are able to estimate synthetic accessibility of individual compounds, performing this estimation for large numbers of compounds requires computational methods.

Several computational approaches to assess synthetic accessibility of molecules exist [4]. They may be roughly divided into two groups: complexity-based and retrosynthetic-based. Complexity-based methods use sets of rules to estimate complexity of target structures (features like presence of spiro-rings, non-standard ring fusions, or large number of stereocenters) which is then directly related to SA. The second group of methods is based on the full retrosynthetic approach when the complete synthetic tree leading to the molecules needs to be processed. Such a procedure is quite time consuming, because the size of the synthetic tree grows exponentially with the number of required steps. Additionally, retrosynthetic methods rely on reaction databases as well as lists of available reagents, which both need to be kept up-to-date. This high requirement on maintenance is probably one of the reasons why methods for estimation of SA based on the retrosynthetic approach have been developed mainly by large academic teams (for example group of Prof. Gasteiger at Erlangen University with the WODCA system [5] or group of Prof. Johnson at Leeds University with the SPROUT/CAESA program [6]).

The major problem when developing methods for estimation of SA is the validation of results. It is not straightforward to extract synthetic complexity out of the protocol describing molecule synthesis. While the overall yield over the sequence of synthetic steps gives some information, this depends also on the effort which has been undertaken to optimize the process; and if only low amounts are needed for initial experiments, then a non-optimal synthesis is tolerable. Another possible measure of synthetic accessibility of a molecule could be its price in catalogues of chemical providers. The price, however, depends on too many factors not related to SA (for example novelty of the reagent, demand, packaging, marketing issues) to be relied on as an objective measure of SA. We were not able to get any reasonable correlation between normalised catalogue price and various structural descriptors for a large set of reagents. The total cost of production of pharmaceutical substances, where the whole process is highly optimized concerning the cost of goods and manufacturing expenses, would be probably the most useful parameter in this respect, but unfortunately this type of data is one of the most guarded secrets in the pharmaceutical industry.

Therefore currently the only way to assess the performance of the calculated synthetic accessibility score is to rely on a ranking done by experienced medicinal chemists.

Several studies focused on performance of chemists in ranking molecules or estimating their synthetic accessibility. In the work of Takaoka et al. [7] 5 chemists ranked 3980 molecules according to their ease of synthesis into three categories: easy, possible and hard. Correlation coefficients between scores assigned by various chemists were in the range 0.40 to 0.56 with an average 0.46. The authors concluded, however, that the models based on the average of chemist estimations may be useful for classification of molecules. Baber and Feher [4] described an experiment where 8 medicinal chemists scored 100 drug-like compounds according to their ease of synthesis. The mean absolute error in chemists' estimations was around 10%, for some compounds, however, there was a variation of up to 70%. In the study of Lajiness at al. [8], 13 chemists reviewed sets of 2000 diverse compounds containing also a common set of 250 compounds, with the goal of removing those that are unacceptable for any reason (too complex, having too complicated synthesis, unsuitable for launching a drug discovery campaign etc): the objective was to see the consistency of chemists in picking "bad" molecules. The study has shown that chemists are not very consistent in their rejection of compounds: only 24% of the compounds rejected by one chemist were also rejected by another. Boda et al. [9] asked 5 chemists to rank 100 molecules selected randomly from the Journal of Medicinal Chemistry according to their ease of synthesis. The chemists seemed to agree on synthetic accessibility for very simple and quite complex molecules; in the middle range, however, larger divergence was observed. The agreement among chemists was acceptable with correlation coefficients in the range 0.73 – 0.84. The ranks entered by chemists have been then used to train the synthetic accessibility score function described in the publication.

All these studies indicate that even experienced chemists differ in their estimations of ease of synthesis. This, of course, is nothing surprising. Chemists have different backgrounds, different areas of research (medicinal chemists, natural product chemists, chemists working in combinatorial synthesis, etc.) or experience based on projects they have been working on. Therefore, to use ranks assigned by chemists as a measure of SA, a consensus score based on several estimations is required. The situation is additionally complicated by the fact that the ease of synthesis for a particular molecule is not a constant. It evolves within time as a consequence of introduction of new synthetic methods and availability of new reagents and building blocks. For example, an introduction of methods like carbon-carbon coupling reactions, sophisticated organometallic catalysts or use of enzymes in organic synthesis allows currently relatively easy synthesis of molecules, which would be very difficult to make just a decade ago [10].

Calculation of Synthetic Accessibility Score

The goal of the present study was to develop a method for estimation of SA which could be used in various drug discovery activities. The fact that the method should be able to process very large numbers of molecules (several millions when making a selection from large commercial catalogues or processing virtual libraries), as well as a decision not to rely on comprehensive databases of reactions and reagents (with the related maintenance hurdle) clearly favored implementation of a method based on molecular complexity. Pure complexity-based approaches, however, have known deficiencies: they do not take into account easy availability of complex reagents, which allows us to introduce some complex features to molecules relatively easily [6], neither the fact that some simple reactions can produce quite complex structures (condensation reactions, cycloadditions, various cyclizations). To account for this deficit of a pure complexity-based approach we have decided to implement a method which would be a compromise between fast complexity-based, and resource-intensive full retrosynthetic approaches. In addition to several standard rules identifying known synthetically problematic molecular features, we wanted to capture also the "synthetic chemistry knowledgebase" by analyzing common substructures in a very large number of already synthesized molecules. For this purpose, a representative subset of molecules from the PubChem database [11] was used. PubChem contains currently 37 million unique molecules including common drugs and agrochemicals, structures extracted from patents, and large numbers of samples from numerous compound providers. One million molecules representatively selected from PubChem served as a training set to identify common (and therefore one can assume also easy to make) structural features. Our approach is similar to those presented by Boda and Johnson [12] who based their estimation of molecular complexity on a set of simple fragments collected from a database of drug-like molecules. Our fragment approach differs, however, in using different types of fragments, as well as by a different method to calculate fragment contributions.

The synthetic accessibility score – SAscore in our approximation is calculated as a combination of two components:

$$\text{SAscore} = \text{fragmentScore} - \text{complexityPenalty}$$

The fragmentScore, as already mentioned, was introduced to capture the "historical synthetic knowledge" by analyzing common structural features in a large number of already synthesized molecules. The score is calculated as a sum of contributions of all fragments in the molecule divided by the number of fragments in this molecule. The database of fragment contributions has been generated by statistical analysis of substructures in the PubChem collection as described in the following section.

934,046 representative molecules from the PubChem database were fragmented. Extended connectivity fragments (ECFC_4# fragments) as implemented in Pipeline Pilot [13] were used. This type of fragment includes a central atom, as well as several levels of neighbors connected to the central atom by one to three bonds. The type of atoms in the last level is not specified and they all are marked as "star" atoms only. To illustrate this procedure the fragments generated for Aspirin are shown in Figure 1 together with their frequency of occurrence. The size of fragments we used has been chosen intentionally to be quite large to capture also information about rings, and relative positions of multiple substitution points on the rings (information which is very important for estimation of SA [12]).

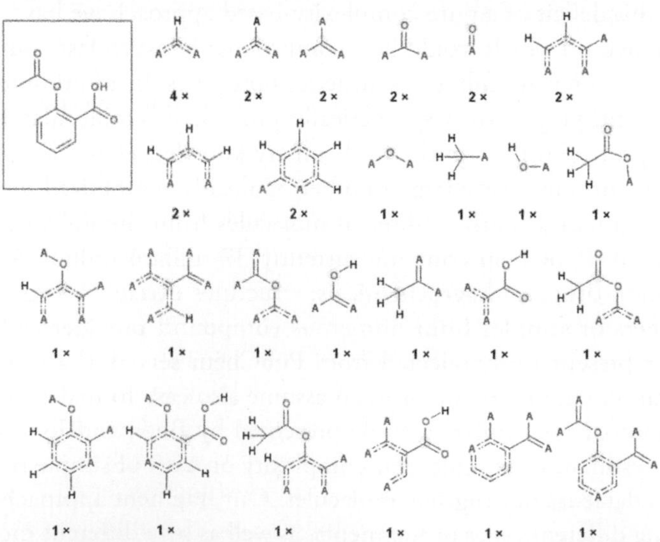

Figure 1. Substructures obtained by fragmentation of Aspirin, the "A" represents any non-hydrogen atom, "dashed" double bond indicates an aromatic bond, number below the fragment indicates the count of this substructure in the molecule and the yellow circle marks the central atom of the fragment.

Altogether 605,864 different fragment types have been obtained by fragmenting the PubChem structures. Most of them (51%), however are singletons (present only once in the whole set). Only a relatively small number of fragments, namely 3759 (0.62%), are frequent (i.e. present more than 1000-times in the database). The most common fragments are shown in Figure 2.

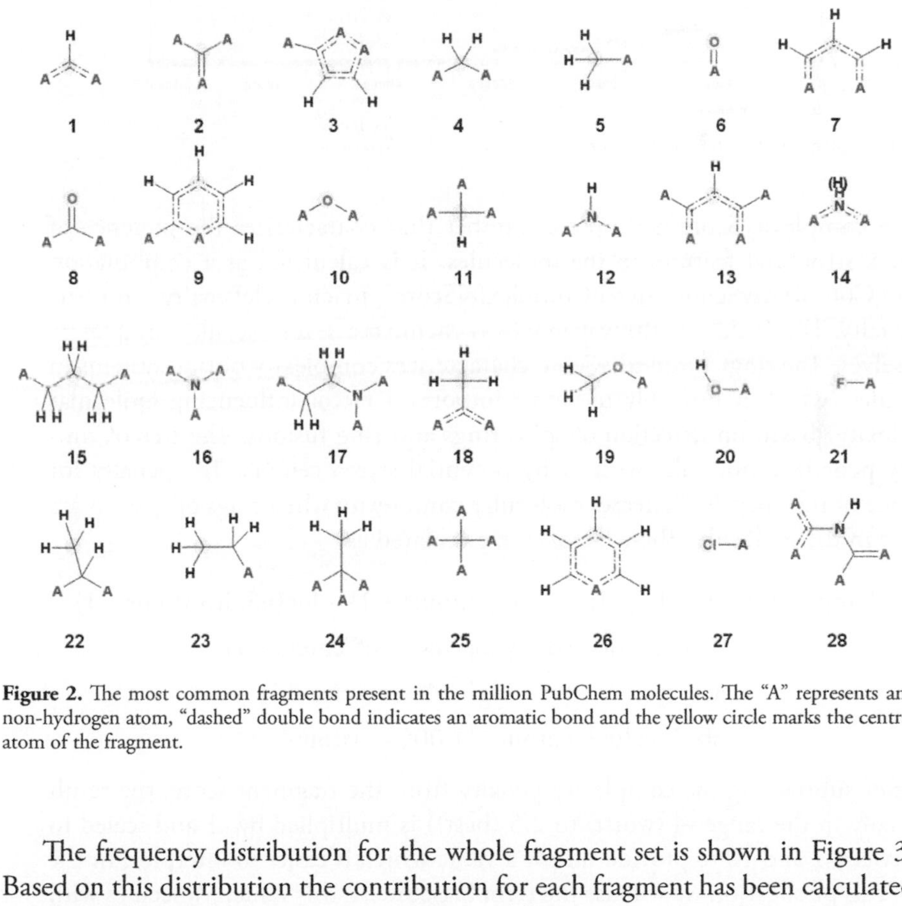

Figure 2. The most common fragments present in the million PubChem molecules. The "A" represents any non-hydrogen atom, "dashed" double bond indicates an aromatic bond and the yellow circle marks the central atom of the fragment.

The frequency distribution for the whole fragment set is shown in Figure 3. Based on this distribution the contribution for each fragment has been calculated as a logarithm of the ratio between the actual fragment count and the number of fragments forming 80% of all fragments in the database. As a result the frequent fragments have positive scores and less frequent fragments have negative scores. The whole approach is based on a simple assumption that the fragment frequency is related to their synthetic accessibility – substructures which are easy to prepare are present in molecules quite often, those which are difficult to synthesize or are unstable are rare.

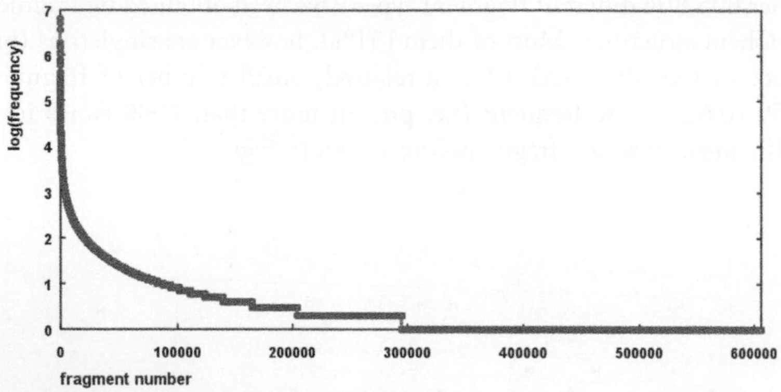

Figure 3. Frequency distribution of fragments.

The complexityScore is simply a number that characterizes the presence of complex structural features in the molecules. It is calculated as a combination of ringComplexityScore, stereoComplexityScore, macrocyclePenalty and the sizePenalty. This is close to the way in which chemists assess molecular complexity themselves. The ringComplexityScore characterizes complexity of ring systems in molecules (which is probably the most important factor influencing molecular complexity) based on detection of spiro rings and ring fusions. The stereoComplexity penalizes molecules with many potential stereo centers. The penalty for presence of macrocycles increases molecular complexity when rings of size > 8 are present in the molecule. These factors are calculated as:

$$\text{ringComplexityScore} = \log(\text{nRingBridgeAtoms} + 1) + \log(\text{nSpiroAtoms} + 1)$$

$$\text{stereoComplexityScore} = \log(\text{nStereoCenters} + 1)$$

$$\text{macrocyclePenalty} = \log(\text{nMacrocycles} + 1)$$

$$\text{sizePenalty} = \text{natoms}^{**}1.005 - \text{natoms}$$

After subtracting the complexity penalty from the fragment score, the result (normally in the range -4 (worst) to 2.5 (best)) is multiplied by -1 and scaled to be between 1 and 10 to provide simply a value which is easier to interpret. In the rest of this publication we will use the term SAscore for this value. Molecules with the high SAscore (say, above 6, based on the distribution of SAscore shown in the Fig. 4) are difficult to synthesize, whereas, molecules with the low SAscore values are easily synthetically accessible. We did not make any attempts to find optimal weights of complexity and fragment contributions (as was done for example in the study [9] where these weights have been optimized to fit the ranks assigned by chemists). Our dataset (40 molecules) was relatively small and any optimization of parameters would probably lead to overfitting. The complexity parameters act in this sense rather as "indicator variables" increasing the SAscore for molecules containing synthetically problematic features.

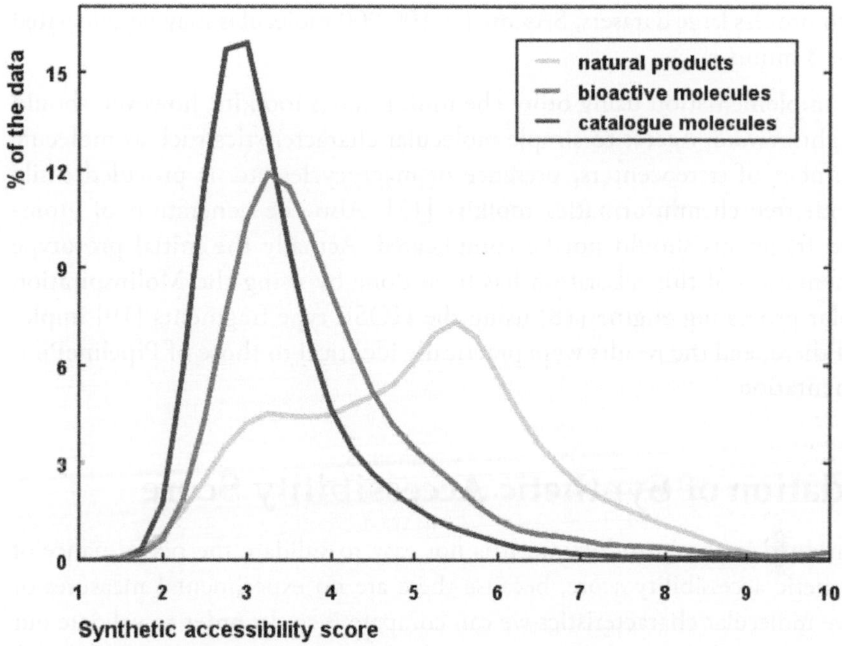

Figure 4. Distribution of SAscore for natural products, bioactive molecules and molecules from catalogues.

To illustrate the performance of the new SAscore, its distribution for 100,000 synthetic molecules from catalogues of commercial compound providers (not used in the training process), 100,000 bioactive molecules randomly selected from the WDI [14] and MDDR [15] databases and 100,000 natural products from the Dictionary of Natural Products [16] is shown in Figure 4. The graph is consistent with the common presumption that natural products are much more difficult to synthesize than "standard" organic molecules. Bioactive molecules have their SAscore somewhere in the middle between these two sets. This graph should provide some feeling about the meaning of the score and how it is distributed in different molecular data sets.

SAscore Implementation

To make the SAscore as broadly available as possible at Novartis, we implemented the algorithm in the Pipeline Pilot environment [13]. PipelinePilot protocols are used routinely at Novartis to support various drug discovery activities. The heart of the calculation protocol is the "SAscore Calculator" component, which is a custom component written in PERL, where the actual calculation of the score described in the previous section is implemented. The speed of the protocol is sufficient to process large datasets; SAscore for 100,000 molecules may be calculated in about 3 minutes.

The implementation using other cheminformatics toolkits, however, should be straightforward. Access to simple molecular characteristics such as molecule size, number of stereocenters, presence of macrocycles etc. is provided easily by several free cheminformatics toolkits [17]. Also the generation of atom-centered fragments should not be complicated. Actually the initial prototype implementation of this algorithm has been done by using the Molinspiration molecular processing engine [18] using the HOSE type fragments [19] implemented there, and the results were practically identical to those of PipelinePilot implementation.

Validation of Synthetic Accessibility Score

As mentioned in the introduction, it is not easy to validate the performance of the synthetic accessibility score, because there are no experimental measures or objective molecular characteristics we can compare it to. In order to validate our algorithm we decided therefore to compare the SAscore with an "ease of synthesis" ranking assigned by synthetic chemists. For this purpose 40 molecules were selected randomly from the PubChem database in such a way that the whole range from small to large molecules was covered. Stereochemistry information was discarded, because it is not used directly in the calculation protocol. The number of possible stereoisomers, however, is used in the generation of the complexity part of the score, so the stereocomplexity is captured in this way. Nine Novartis chemists, with long experience in various medicinal chemistry projects, were asked to rank these molecules. To make the ranking process easy, a simple web interface was prepared where all 40 molecules were displayed along with a menu which allowed selection of a score between 1 and 10. The "chemist score" we use in the rest of this article is simply an average of 9 scores entered by chemists (Figure 5). We want to stress here that this validation experiment was performed only after the development and implementation of the SAscore protocol had been completely finished, and therefore the results could not be used in any way to "tune" the calculation algorithm.

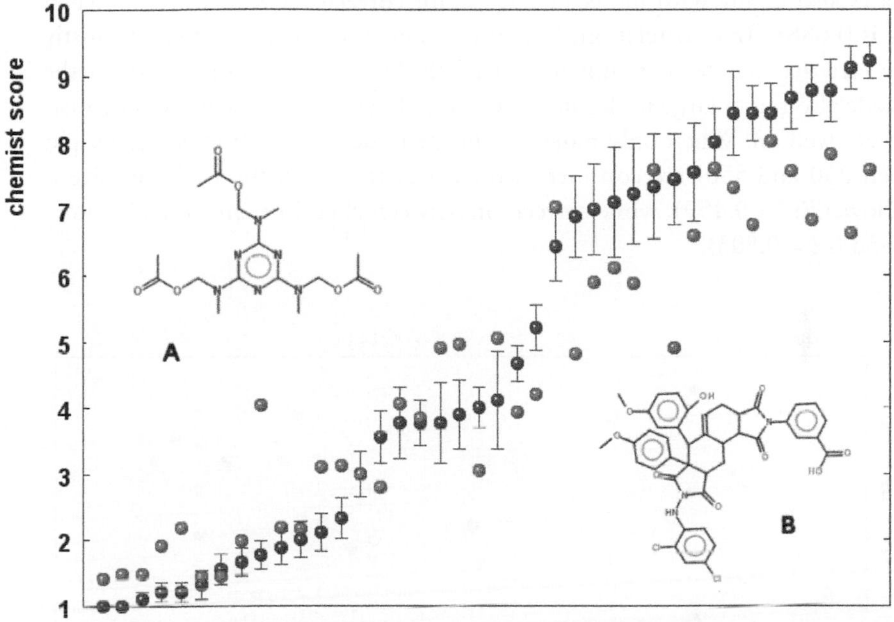

Figure 5. Average of chemist ranks for 40 test molecules (blue) compared with the computed SAscore (red). Error bars on blue points indicate standard error of mean of estimations by 9 chemists.

The agreement among chemists in their rankings is quite good, the r2 ranges between 0.450 and 0.892 with the average r2 for all "chemist pairs" being 0.718. For a few molecules, however, scores by some chemists differ by 6 or more ranks, and for 7 molecules out of 40 the standard deviation is above 2. Average standard deviation for all 40 molecules is 1.23 and average standard error of mean (shown for all molecules in Figure 5 as error bars) is 0.41. The chemists seem to agree on scores for very simple and very complex molecules better than for structures in the middle region (as mentioned already in [9]). Our results are consistent with the outcome of previous studies, indicating that in order to use ranking by chemists as a reference, one has to use the average of several estimations that smoothes somehow the high individual variation.

The correlation between calculated SAscore and the average of chemist ranks is shown in Figure 6. The agreement between these two values is very good, with r2 = 0.890 (which means that 89% of variation in the synthetic accessibility as seen by chemists is explained by the SAscore), standard deviation is 0.742. When the SAscore is separated into its two components, the complexityScore provides also very good correlation with the chemist rank (r2 = 0.872), while the fragmentScore correlates with the chemist score with r2 = 0.628. The chemist score correlates also highly with molecule size (r2 for correlation with the number of atoms is 0.688). This correlation, however, is somehow artificial, caused mostly by the fact that our set contains several relatively small molecules, and on the other side also rather large molecules, about which, as already discussed, chemists agree very well on. When only molecules in the middle range (molecular weight between 250 and 550) are considered, the correlation with the molecule size is much lower, (r2 = 0.459), while correlation between chemist score and SAscore is still good (r2 = 0.803).

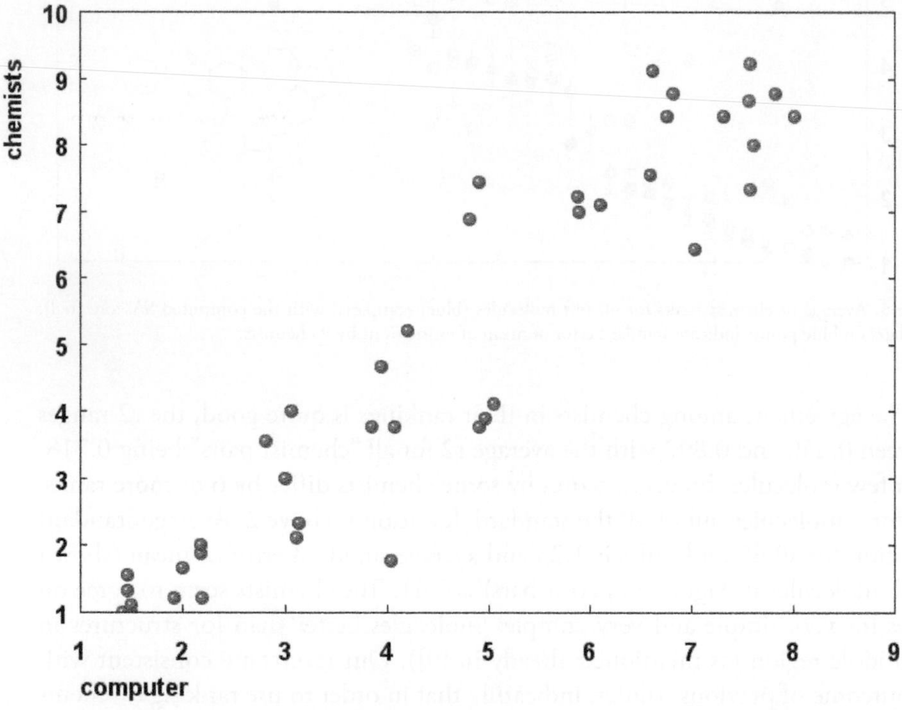

Figure 6. Correlation of calculated SAscore and average chemist estimation for 40 molecules (r^2 = 0.890).

We are, of course, aware of the fact that our set containing only 40 molecules is not large enough to draw too general conclusions from the results. The sole purpose of this exercise was to check whether the calculated SAscore correlates with assessment of ease of synthesis by chemists. The data presented here clearly indicates good support for validity of SAscore with both its components (molecular complexity and fragment score) being important for its good performance.

Discussion

Particularly large differences between chemists and computers could be seen for molecules A and B, both shown in Figure 5. A is a highly symmetrical molecule, which makes synthesis easier, but this factor is not considered when computing the SAscore. We plan therefore to introduce recognition of molecule symmetry in the next version of our SAscore.

Another example where the chemist score and SAscore differ significantly is structure B. In this case chemists overrate the complexity of synthesis. On a first look, the molecule with a central scaffold consisting of 4 fused aliphatic rings indeed looks large and complex. When checking PubChem, however, more than 39,000 molecules with this particular central scaffold can be found. This system may be actually easily synthesized by a sequence of Diels-Alder reactions from simple starting materials [20]. This example nicely illustrates how the fragment score can recognize easy to make substructures even without the necessity to rely on the reaction databases.

In order to get a better understanding of the fragment contributions in the SAscore method, it is helpful to study the most common fragments depicted in Figure 2. They can be grouped into three general groups. The first group consists of frequent side chains. Fragments 5 (methyl), 19 (methoxy), 20 (hydroxy), 21 (fluoro), 23 (ethyl) and 27 (chloro) belong into this category. Fragments 18 and 22 encode a methyl group in a specific environment: attached to an aromatic ring and to an aliphatic carbon. Fragment 26 describes a 6-membered aromatic ring with maximally one substituent and maximally one heteroatom, which must be identical with the substitution site. With the exception of the relatively rare pyridinium group, the simple phenyl group shows this pattern, and therefore 26 can be also counted as a typical side chain fragment. These side chain fragments are also among the most frequent substituents identified in [21]. It is worth noting that many simple, mono-substituted 5-ring hetero-aromatics often used as side chains, such as thiophene, furane, or pyrrole, share fragment 3 regardless of whether substitution is in position 2 or 3. These side-chains are typically available for all types of building blocks and, with the exception of the hydroxyl group, do not generally interfere with most chemical linkage reactions used in parallel synthesis.

Another group of fragments is directly related to these typical linkage reactions. Each molecule synthesized with one of the linkage reactions listed in Table 1 contains at least two of the most frequent fragments as shown in Figure 2. The comparison of the most common fragments with the RECAP bond cleavage types (Figure 2 in [22]) shows that of the 11 RECAP cleavage types only "olefin," "quarternary nitrogen" and "aromatic nitrogen – aliphatic carbon" are not represented by one of the fragments shown in Figure 2. When comparing the cleavage types with the most common fragments, it is noteworthy that the cleaved bond is not always included in the fragments listed in Figure 2. Fragments containing these bonds are often characteristic for one linkage reaction related to one cleavage type, whereas the most common fragments are those which cover more than a single linkage reaction; for example fragment 8 representing carbonyl groups in general or the even more generic fragment 2 describing a carbon atom with one double and two single bonds to non-H atoms.

Table 1. Relation between common linkage reactions and most common fragments shown in Figure 2.

Linkage Reaction	Fragment(s)
Amide bond formation or Urea formation	**2, 6, 8, 12** (from primary amine) or **16** (from secondary amine), **28** (only if aniline)
Sulfonamide formation	**6, 12** (from primary amine) or **16** (from secondary amine)
Ester formation	**2, 6, 8, 10**
Reductive amination	**4** (the CH_2 group from the aldehyde carbon), **12** (from primary amine) or **16** (from secondary amine)

A third group of the most common fragments generally represent frequent structural features. Fragments 1, 3, 7, 9, 13 highlight the prevalence of aromatic rings in the space of easily accessible chemistry. Fragment 14 represents any aromatic nitrogen. Usage of piperazine as a linker is represented beside the fragments listed in Table 1 and also by presence of fragment 17.

Conclusion

A novel methodology to calculate synthetic accessibility score of drug-like molecules has been developed. The method is based on the combination of molecule complexity and fragment contributions obtained by analyzing structures of a million already synthesized chemicals, and in this way captures also historical synthetic knowledge. The method provides good reliability and is sufficiently fast to process very large molecular collections. The performance of the SAscore has been validated by comparing it with the "ease of synthesis" ranks estimated by experienced medicinal chemists, with very good agreement between these two values (r^2 = 0.890). The application area of the SAscore is to rank large collections of molecules, for example to prioritize molecules when purchasing samples for screening, support decisions in hitlist triaging or rank de novo generated structures.

Despite the good performance of the SAscore documented above, we are well aware also of limitations of this method. The SAscore cannot compete with more sophisticated approaches for estimation of synthetic accessibility which reconstruct the full synthetic path, in cases when the results are critical, for example when making decision about selection of a development compound from several candidates. And the ultimate measure for assessing synthetic accessibility of complex organic molecules still remains to be a cumulative experience of skilled medicinal chemists.

Competing Interests

The authors declare that they have no competing interests.

Authors' Contributions

PE (http://peter-ertl.com webcite) developed the SAscore method. AS contributed to the development and discussion and provided indispensable contribution to the development of PipelinePilot SAscore protocol.

Acknowledgements

The authors want to thank nine Novartis chemists who were willing to rank the test molecules and in this way helped to validate the score, as well as to Richard Lewis for critically reading the manuscript and for helpful comments.

References

1. Clark DE, Pickett SE: Computational methods for the prediction of 'drug-likeness.' Drug Discov Today 2000, 5:49–58.
2. Ertl P, Rohde B, Selzer P: Fast calculation of molecular polar surface area as a sum of fragment-based contributions and its application to the prediction of drug transport properties. J Med Chem 2000, 43:3714–3717.
3. Ertl P, Roggo S, Schuffenhauer A: Natural Product-likeness Score and Its Application for Prioritization of Compound Libraries. J Chem Inf Model 2008, 48:68–74.
4. Baber JC, Feher M: Predicting Synthetic Accessibility: Application in Drug Discovery and Development. Mini Rev Med Chem 2004, 4:681–692.
5. Gasteiger J, Ihlenfeldt WD: Computer-Assisted Planning of Organic Syntheses: The Second Generation of Programs. Angew Chem Int Ed Eng 1996, 34:2613–2633.

6. Gillet VJ, Myatt G, Zsoldos Z, Johnson AP: SPROUT, HIPPO and CAESA: Tools for de novo structure generation and estimation of synthetic accessibility. Perspect Drug Disc Design 1995, 3:34–50.
7. Takaoka Y, Endo Y, Yamanobe S, Kakinuma H, Okubo T, Shimazaki Y, Ota T, Sumiya S, Yoshikawa K: Development of a Method for Evaluating Drug-Likeness and Ease of Synthesis Using a Data Set in Which Compounds Are Assigned Scores Based on Chemists' Intuition. J Chem Inf Comput Sci 2003, 43:1269–1275.
8. Lajiness MS, Maggiora GM, Shanmugasundaram V: Assessment of the Consistency of Medicinal Chemists in Reviewing Sets of Compounds. J Med Chem 2004, 47:4891–4896.
9. Boda K, Seidel T, Gasteiger J: Structure and Reaction based Evaluation of Synthetic Accessibility. J Comput Aided Mol Des 2007, 21:311–325.
10. Kündig P: The Future of Organic Synthesis. Science 2006, 314:430–431.
11. The PubChem Database [http://pubchem.ncbi.nlm.nih.gov/]
12. Boda K, Johnson AP: Molecular Complexity Analysis of de Novo Designed Ligands. J Med Chem 2006, 49:5869–5879.
13. Pipeline Pilot, version 6.0 [http://www.accelrys.com] Accelrys, Inc., San Diego, CA, USA.
14. WDI – Derwent World Drug Index, version 2007.04 [http://www.thomsonscientific.com/]
15. MDDR – MDL Drug Data Report, version 2007.2 [http://www.prous.com/product/electron/mddr.html]
16. CRC Dictionary of Natural Products, v 15.1 [http://www.crcpress.com/] CRC Press; 2006.
17. Ertl P, Jelfs S: Designing Drugs on the Internet? Free Web Tools and Services Supporting Medicinal Chemistry. Curr Top Med Chem 2007, 7:1491–1501.
18. mib – Molinspiration molecular processing engine, version 2007.10, Molinspiration Cheminformatics, Slovensky Grob, Slovak Republic [http://www.molinspiration.com].
19. Bremser W: HOSE – A Novel Substructure Code. Anal Chim Acta 1978, 103:355–365.
20. Kanemasa S, Sakoh H, Wada E, Tsuge O: Diene-transmissive Diels-Alder Reaction Using 2-Ethoxy-3-methylene-1,4-pentadiene and 2-(2-Bromo-1-ethoxyethyl)1,3-butadiene. Bulletin of the Chemical Society of Japan 1985, 58:3312–3319.

21. Ertl P: Cheminformatics Analysis of Organic Substituents: Identification of the Most Common Substituents, Calculation of Substituent Properties, and Automatic Identification of Drug-like Bioisosteric Groups. J Chem Inf Comput Sci 2003, 43:374–380.

22. Lewell XQ, Judd DB, Watson SP, Hann MM: RECAP – Retrosynthetic Combinatorial Analysis Procedure: A Powerful New Technique for Identifying Privileged Molecular Fragments with Useful Applications in Combinatorial Chemistry. J Chem Inf Comput Sci 1998, 38:511–522.

CITATION

Ertl P and Schuffenhauer A. Estimation of Synthetic Accessibility Score of Drug-Like Molecules Based on Molecular Complexity and Fragment Contributions. Journal of Cheminformatics. 2009; 1: 8. doi: 10.1186/1758-2946-1-8.

Rational Mutagenesis to Support Structure-Based Drug Design: MAPKAP Kinase 2 as a Case Study

Maria A. Argiriadi, Silvino Sousa, David Banach, Douglas Marcotte, Tao Xiang, Medha J. Tomlinson, Megan Demers, Christopher Harris, Silvia Kwak, Jennifer Hardman, Margaret Pietras, Lisa Quinn, Jennifer DiMauro, Baofu Ni, John Mankovich, David W. Borhani, Robert V. Talanian and Ramkrishna Sadhukhan

ABSTRACT

Background

Structure-based drug design (SBDD) can provide valuable guidance to drug discovery programs. Robust construct design and expression, protein

purification and characterization, protein crystallization, and high-resolution diffraction are all needed for rapid, iterative inhibitor design. We describe here robust methods to support SBDD on an oral anti-cytokine drug target, human MAPKAP kinase 2 (MK2). Our goal was to obtain useful diffraction data with a large number of chemically diverse lead compounds. Although MK2 structures and structural methods have been reported previously, reproducibility was low and improved methods were needed.

Results

Our construct design strategy had four tactics: N- and C-terminal variations; entropy-reducing surface mutations; activation loop deletions; and pseudoactivation mutations. Generic, high-throughput methods for cloning and expression were coupled with automated liquid dispensing for the rapid testing of crystallization conditions with minimal sample requirements. Initial results led to development of a novel, customized robotic crystallization screen that yielded MK2/inhibitor complex crystals under many conditions in seven crystal forms. In all, 44 MK2 constructs were generated, ~500 crystals were tested for diffraction, and ~30 structures were determined, delivering high-impact structural data to support our MK2 drug design effort.

Conclusion

Key lessons included setting reasonable criteria for construct performance and prioritization, a willingness to design and use customized crystallization screens, and, crucially, initiation of high-throughput construct exploration very early in the drug discovery process.

Background

Structure-based drug design (SBDD) can be an effective contributor to the identification and optimization of drug candidates by providing a structural rationale for the design of improved compounds. Protein crystallization, structure determination, and the rapid determination of multiple protein/ligand complexes can be expensive and time-consuming. Major variables include protein construct design, mutations, and post-translational modifications, the nature of protein impurities (chemical or conformational), the choice of ligands or even proteins for co-crystallization, and the crystallization conditions themselves. These variables represent an enormous matrix of experimental possibilities that is difficult or impossible to explore systematically. Despite these challenges, the availability of structural information at the preliminary stages of a drug discovery program is critical to maximize impact. Therefore, efficient methods developments, techniques and strategies to deliver structures early in a project are clearly needed.

MAPKAP kinase 2 (MK2) plays a key role in the production of pro-inflammatory cytokines such as TNF-α. MK2 is activated by the mitogen-activated protein (MAP) kinase p38 [1-3]. Activated MK2 phosphorylates a number of target proteins in immune cells resulting in cytokine production and cellular proliferation and activation. Mice lacking MK2 are healthy and fertile, but they fail to increase production of pro-inflammatory cytokines such as TNF-α, IL-6, and IFN-γ [4] in response to stimuli such as lipopolysaccharide. MK2 knockout mice are resistant to the development of collagen-induced arthritis, a model for human rheumatoid arthritis [5]. The catalytic activity of MK2 is required to mount the pro-inflammatory response [6]. These and related studies have attracted attention to MK2 as a target for the design of therapeutic treatments for rheumatoid arthritis and other TNF-α-driven diseases.

Although the data supporting MK2 as a promising drug target have been available for nearly ten years, to our knowledge there are no MK2 inhibitors in clinical development. Many companies have initiated MK2 projects, but little success has been reported. Anecdotally, a common problem has been that high-throughput screening for lead MK2 inhibitors has been unproductive. We believe SBDD targeting MK2 could help address this issue. Yet, despite several reports of MK2 crystal structures at moderate (2.7–3.8 Å) resolution [7-10], the routine production of well-diffracting MK2 crystals bound to compounds of diverse structure remains difficult. More robust methods are needed to enable efficient SBDD.

The domain structure of MK2 may contribute to these difficulties [2]. Its proline-rich N-terminal domain (residues 1–65) is unique, having no counterpart in other MAP kinases [3]. This domain binds c-ABL Src homology 3 domain in vitro [11]. The sequence of the kinase domain (66–327) identifies MK2 as a Ser/Thr kinase family member. MK2 exhibits low homology to other Ser/Thr kinases, however, with the exception of the close homologs MK3 and MK4. The regulatory domain at the C-terminus (328–400) contains an autoinhibitory α-helix [7,8] followed by nuclear export signal (NES) and nuclear localization signal (NLS) sequences [2,3,12-15]. The NES and NLS are essential for MK2 complex formation with p38 and subsequent translocation to the nucleus. Deletion of the entire MK2 regulatory domain results in a marked increase in catalytic activity [15]. There are three critical phosphorylation sites on MK2: Thr222, Ser272, and Thr334 [16,17]. Phosphorylation at these residues activates MK2 by causing a conformational change in the C-terminal regulatory α-helix: on kinase activation, the helix displaces from the kinase surface and thereby allows substrates to bind [7,9,16].

Construct design is known to be a critical factor in producing large quantities of soluble protein and reproducible crystals. For example, it can be difficult to predict precisely domain boundaries and to identify the surface residues of globular proteins, alteration of which might enhance solubility or crystallization. Altering or deleting features such as surface hydrophobicity, post-translational modifications, side-chain flexibility, secondary structural elements, or even entire domains can dramatically modulate protein physical characteristics, especially solubility. Protein solubility also depends on details such as the expression vector, host cell, culture conditions, and protein fusion partner used. To increase crystallization robustness and improve crystal diffraction, we thought such wide-ranging approaches needed to be explored with MK2.

Here we report the optimization of several steps in human MK2 structure determination: rapid and systematic exploration of construct design and expression screening; high-throughput protein purification; and wide screen crystallization with customized factorial grids. Our methods expand upon those of Malawski et al. [18], who examined only N- and C-terminal truncations of the MK2 catalytic domain. We took advantage of the fact that MK2 is one of the few kinases that expresses well in Escherichia coli, which facilitates high-throughput construct design and production, to explore the effect of not only truncations but also two kinds of surface mutations and several internal deletions.

Our strategy had four components: First, the N- and C-termini of the protein were varied. Second, surface-exposed lysine and glutamate residues with high conformational entropy [19] were mutated to drive novel crystallization contacts and thereby enhance crystallization. Third, internal flexible regions were deleted, again to foster novel crystal forms. Fourth, phosphorylation sites [16,17] were altered to provide homogenous MK2 rather than a heterogeneous mixture of unactivated (no phosphorylation) and activated (partial or full phosphorylation) forms.

We implemented a high-throughput, parallel approach to enable construct production, expression, and purification of all mutants within a short time, nearly all of which expressed well and were tested in customized, kinase-specific robotic crystallization screens. The methodological improvements implemented here enabled the screening of 44 MK2 constructs, resulting in seven crystal forms, diffraction testing of ~500 crystals, and high-resolution data collection and structure determination of ~30 MK2/inhibitor complexes.

Results and Discussion

We initiated MK2 crystallographic studies with a construct (Table 1) comprising part of the proline-rich domain, the kinase and C-terminal regulatory domains, and a point mutation introduced to abolish kinase activity. MK2(36–401, K93R) disrupts the highly conserved catalytic lysine residue; the catalytically-inactive

K93R mutation was described previously [6]. We began with an inactive construct because use of inactive kinases proved critical to our obtaining homogenous protein suitable for crystallography on several earlier projects. It quickly became apparent, however, that these MK2 constructs were not suitable for structural studies, due to both low expression levels and relative insolubility, likely due in part to the proline-rich segment. We switched to a construct that had been used to determine the first reported MK2 crystal structure, MK2(47–400) [7]. Although protein expression and behavior improved, suitable crystals were not forthcoming, despite success in another laboratory. We believed that a new, more robust approach was clearly needed.

Table 1. Representative MK2 expression constructs.

Rationale	Constructs
N- and C-terminal variations	MK2(36–400, K93R) MK2(41–364) MK2(47–357) MK2(47–366)
Entropy-reducing surface mutations	MK2(41–364, K64A) MK2(41–364, K343A, E344A) MK2(47–366, K84A) MK2(47–366, E88A, K89A)
Activation loop deletions	MK2(41–364, Δ(L220-G238)) MK2(47–366, Δ(L220-G238))
Pseudoactivation mutations	MK2(41–364, T222E) MK2(41–364, T334E) MK2(41–364, T222E, T334E) MK2(47–366, T222E) MK2(47–366, T334E) MK2(47–366, T222E, T334E)

The complete list of MK2 constructs is available in Additional File 1 Table S7.

Construct Design Strategy

We thus implemented a broad, four-point MK2 construct design strategy coupled with the high-throughput production and testing of multiple crystallographic constructs. First, we sought to parlay the domain organization of MK2 into a series of N- and C-terminal truncation mutants that either incorporated multiple domains of the intact protein, or defined the kinase domain more precisely than a single construct would. Our approach was significantly more extensive than that of Malawski et al. [18]. Second, we mutated surface-exposed lysine and glutamate residues to alanine, to reduce the high conformational entropy of these residues [19]. This approach has been used successfully in a variety of contexts [20]. Third, the flexible internal activation loop of MK2 was deleted. Fourth, phosphorylation sites were altered to provide homogenous MK2 rather than a heterogeneous mixture of unactivated (no phosphorylation) and activated (partial or full phosphorylation) forms. A representative subset of the constructs we used is shown in Table 1; the complete list of constructs is also available. Figure 1 illustrates the location of all mutation sites mapped onto the MK2 three-dimensional structure.

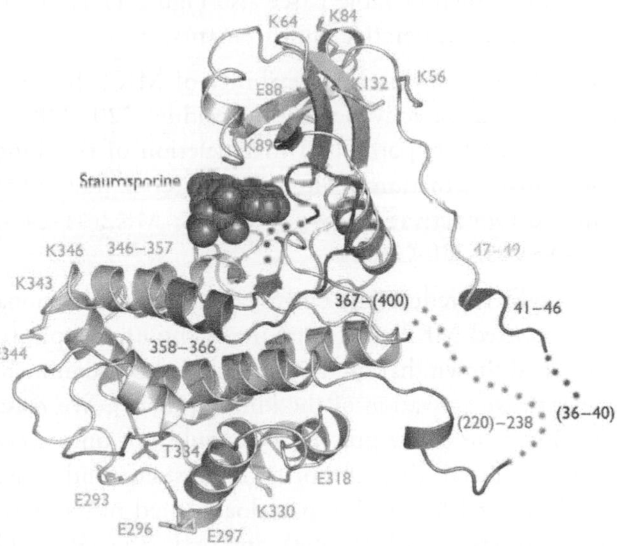

Figure 1. MK2 structure and location of mutagenesis sites. MK2 is represented as a ribbon diagram, a composite of Protein Data Bank entries 1KWP and 1NXK. The mutagenesis sites (see Table 1) are colored by tactic: (1) N-terminal variations, purple/blue/cyan, C-terminal variations, yellow/orange/red; (2) entropy-reducing surface mutations, light-green; (3) internal deletions, grey-green; and (4) pseudoactivation mutations, dark-green. Missing residues are represented as dots.

Our initial construct design tactic was alteration of the N- or C-termini of full-length MK2, thereby either removing flexible terminal segments that might hinder crystallization or simply providing different termini that might enable different crystal forms (via altered packing, isoelectric point, hydrophobicity, etc.). This approach has been used previously by several groups, and limited systematic N- and C-terminal truncations have been explored [18]; the construct used most often has been MK2(41–364), a form of the enzyme noted to be constitutively-active [8]. Given the poor behavior of the early constructs that included (part of) the proline-rich N-terminal domain, we deleted this domain in all subsequent constructs. We explored several termini: Gln41, His47, and Arg50; and Gln327, Leu342, Asp345, Thr357, Arg364, Asp366, and His400 (Figure 1). These residues delineate the kinase domain at the N-terminus, and the kinase domain, the autoinhibitory α-helix, and the NES/NLS at the C-terminus.

Our second tactic addressed the protein physical characteristic of surface entropy. Following Longenecker and colleagues [19], a series of point mutants was designed to alter flexible surface lysine and glutamate residues to minimize crystal protein entropy and entropic loss on crystallization. This approach was successfully used with RhoGD1, for which new crystal forms were identified that exhibited enhanced diffraction. Several MK2 mutants in which alanine was substituted for lysine or glutamate are shown in Table 1 (see also Figure 1). All mutants were constructed at one time, without iterative improvements.

Our third tactic addressed the internal flexibility of MK2. In several of the prior MK2 structures, the kinase activation loop (residues 220–238) engages in significant crystal contacts. We hypothesized that deletion of this long, flexible loop (Figure 1) might drive formation of alternate, better-diffracting crystal lattices. We thus examined two activation loop deletions: MK2(41–364, ΔL220-G238) and MK2(47–366, ΔL220-G238).

Our final tactic sought to reduce the chemical and conformational heterogeneity of MK2. All reported MK2 structures are of the unphosphorylated enzyme. Previous studies had shown that mutation of phosphothreonine residues to glutamate led to constitutive activation of the kinase [16,17]. We reasoned that by altering the activation state of the protein, we would not only access a more homogeneous enzyme but also distinct conformational states, and hence increase the likelihood of useful crystal forms. The pseudoactivated mutants were based on two truncated constructs, MK2(41–364) and MK2(47–366). Glutamate mutations T222E and T334E (both singly, and as the double mutant) replaced the threonines reported to activate MK2 when phosphorylated. We prioritized mutation of T222 and T334 over S272 because existing data suggested that the threonine residues were the more significant MK2 phosphorylation sites [16]. T334 is shown in Figure 1; T222 is located within the disordered portion of the activation loop.

Expression, Purification and Crystallization

The MK2 constructs (Table 1) were cloned using standard techniques and expressed in E. coli as glutathione S-transferase (GST) fusion proteins. The expression plasmids encoded GST followed by thrombin and tobacco etch virus (TEV) protease cleavage sites and the desired MK2 sequence. It proved important to develop a method for rapid generation and screening of multiple constructs in parallel. After plasmid construction (PCR, mutagenesis, ligation, etc.) and transformation, typically in parallel sets of 4–8 constructs, test expression was carried out on a small scale, to examine both yield and especially protein solubility. Representative results are shown in Figure 2 and Table 2. Low temperature induction (18°C) provided the optimal balance between expression yield and solubility for most constructs; higher temperatures increased the proportion of insoluble protein. One pseudoactivated construct, MK2(47–366, T222E), exhibited robust expression with both low and medium temperature induction; typical results are also shown in Figure 2. This systematic expression/solubility triage was used for all constructs. Constructs that expressed at high (soluble) levels were prioritized for small-scale purification.

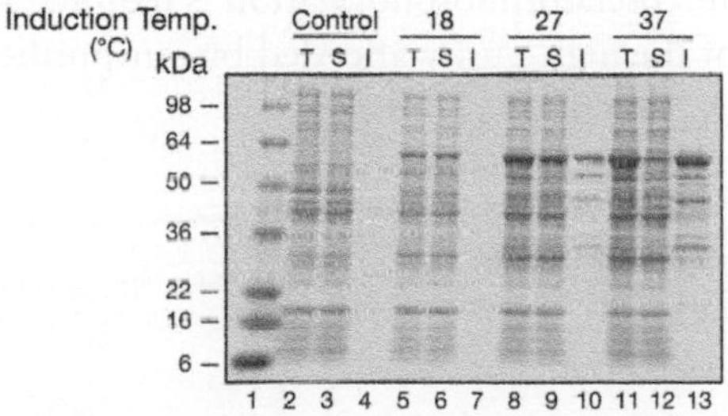

Figure 2. Analysis of MK2(47–366, T222E) expression as a function of temperature. Total, soluble, and insoluble fractions (see Methods) were separated by 4–20% SDS-PAGE; stained with Coomassie Blue. Lanes: (1) MW markers; (2–4) control (uninduced); (5–7) 18°C; (8–10) 27°C; (11–13) 37°C;. T, total; S, soluble; I, insoluble. All samples were taken 5 h after induction with IPTG. See also Table 3.

Table 2. Solubility assessment for representative MK2 constructs.

MK2 Construct		Solubility		
Backbone	**Mutation(s)**	**18°C**	**27°C**	**37°C**
41–364	--	Low	Low	Low
41–364	K64A	Low	Low	Low
41–364	Δ(L220-G236)	High	High	High
41–364	T222E	Low	Low	Low
41–364	K330A	Low	Low	Low
41–364	T334E	Low	Low	Low
41–364	K343A, K344A, K364A	Low	Low	Low
47–366	--	High	High	Medium
47–366	K64A	High	High	Medium
47–366	K84A	High	Med	Low
47–366	Δ(L220-G236)	High	High	High
47–366	T222E	High	High	Low
47–366	T222E, T334E	Low	Low	Low
47–366	T334E	Low	Low	Low

Expression tests were conducted at three temperatures for each construct. Low (<10%), medium (10–50%) and high (>50%) grades were assigned to each construct based on the ratio of soluble/insoluble MK2 as assessed by SDS-PAGE. An example gel is shown in Figure 2.

Routine procedures (glutathione affinity chromatography, TEV protease cleavage, cation exchange and finally size exclusion chromatography) were used to purify the MK2 constructs. Initially, limited attempts were made to purify proteins using parallel 24-well methods (filtration plates, etc.). But, the rapidity with which conventional purification could be performed made the use of small-scale plate methods unnecessary. Yields from the glutathione affinity chromatography capture step were 4–30 mg/L of culture; the parental constructs MK2(41–364) and MK2(47–366) had crude yields of 5 mg/L. Final yields for all constructs were 0.4–12 mg/L. Constructs that gave the highest yields were progressed first to large-scale purification and crystallization trials; only one construct, MK2(47–366, K56A), produced less than 0.5 mg/L and therefore was not progressed. Example yields for several pseudoactivated constructs are shown in Table 3. MK2(47–366, T222E) was found to be devoid of enzymatic activity, in agreement with previous reports that more than one (pseudo)phosphorylation is required to activate MK2 [16]. A typical example of the high purity afforded by our purification scheme is shown in Figure 3 for MK2(47–366, T222E).

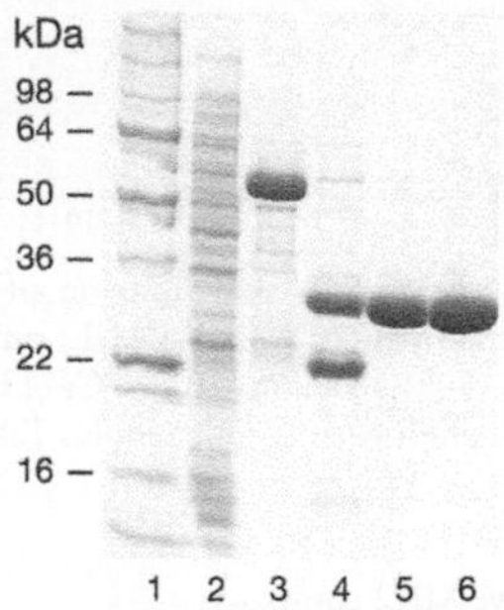

Figure 3. Representative MK2 purification: MK2(47–366, T222E). Protein fractions were separated by 4–20% SDS-PAGE; stained with Coomassie Blue. Lanes: (1) MW markers; (2) glutathione affinity column flow-thru; (3) glutathione affinity column eluate (40 mM glutathione, pH 8.0); (4) TEV protease cleavage; GST is present below MK2; (5) MonoS 10/10 eluate (~200 mM NaCl); (6) Superdex 75 10/60 peak fraction.

Table 3. Summary of pseudoactivated MK2 construct expression, enzymatic activity, and crystallization.

Construct	Yield (mg/L)		Enzymatic Activity (cts/nM)		Crystallography	
	Crude	Final	Without p38 activation	With p38 activation	Crystals Obtained?	Diffraction?
MK2(41–364, T222E)	3.0	1.0	0.30	0.34	Yes	No
MK2(47–366, T222E)	6.0	4.0	<0.02	<0.02	Yes	Yes
MK2(47–366, T334E)	4.5	2.0	1.5	1.2	Yes	No

Forty-three of forty-four constructs produced crystals using commercial crystallization screens (Hampton Crystal Screen I/II™ and Wizard Screen I/II™). The most common crystallization condition was 2 M (NH4)2SO4, 0.2 M Li2SO4, 0.1 M CAPS, pH 10.5. Crystals were optimized with an emphasis on finding novel conditions. We created a customized in-house crystallization grid to optimize co-crystallization conditions for a variety of MK2/inhibitor complexes (Table 4). Seven crystal forms were ultimately identified (Figure 4, Table 5); three of these have been independently identified by other investigators: Form IV [8-10,21]; Form V [9,22]; and Form VII [9]. Beyond MK2, we have since used this and similar grids to crystallize other kinase/inhibitor complexes.

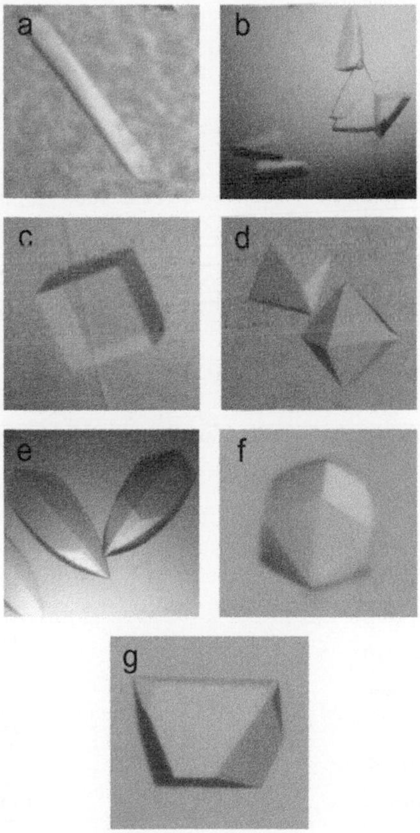

Figure 4. Seven MK2 crystal forms were identified (Table 5). (a) Form I: MK2(41–364, K343A, K344A, K364A); 2 M (NH4)2SO4, 35 mM Cymal®-3 (3-cyclohexyl-1-propyl-β-D-maltoside). (b) Form II: MK2(47–366); 2 M $(NH_4)_2SO_4$, 100 mM HEPES, pH 7.0, 100 mM Li_2SO_4. (c) Form III: MK2(47–366); 2 M $(NH_4)_2SO_4$, 0.1 M Na citrate, pH 5.0, 4% 1,4-butanediol. (d) Form IV: MK2(47–366, T222E), 2 M sodium malonate, pH 5.5, 0.01 mM Anapoe 80. (e) Form V: MK2(41–364); 1.5 M Na malonate, pH 8.0. (f) Form VI: MK2(41–364); 1.8 M Na malonate, pH 8.0. (g) Form VII: MK2(41–364); 1.75 M $(NH_4)_2SO_4$, 0.1 M Na citrate, pH 8.0. Crystals grew to a maximal size of ~0.1–0.2 × 0.1–0.2 × 0.2 mm (Forms V-VII); other crystals were ~0.1 × 0.1 × 0.1 mm (Forms III & IV) or smaller. See Table 5 for additional details.

Table 4. Customized MK2 robotic crystallization screen conditions.

Variable	Parameters
Buffer	Citrate HEPES
pH	7.0 – 8.5
Precipitant	Ammonium Sulfate Ammonium Malonate Na Malonate K Malonate Na/K Phosphate
Additives	DMSO 2-Propanol PEG 400 PEG 3350 Na Malonate Na Citrate

Table 5. Seven MK2 crystal forms.

Form	Construct(s)	Crystallization Conditions *	Space Group	Unit Cell (*a*, >*b*, *c*, Å)	Res. (Å) *N* †	Notes
I	MK2(41–364) & MK2(47–366) surface mutants ‡	AS, LS, & NP pH 5–8	N.D.	N.D.	>11 *N.D.*	Rods; high mosaicity
II	MK2(41–364) & MK2(47–366) surface mutants ‡	AS, LS, & NM pH 5–8	N.D.	N.D.	>11 *N.D.*	Plates; high mosaicity
III	MK2(47–366)	2 M AS 0.1 M Na citrate pH 5.0 4% 1,4-butanediol	$P2_13$	215	3.4 *4*	Cubes
IV	MK2(47–366, T222E)	2 M NM pH 5.5 10 μM Anapoe 80	$F4_132$	254	2.9 *1*	Bipyramids; inhibitor soaking req'd
V	MK2(41–364)	1.5–1.8 M NM pH 8.0	$P6_3$	158 158 138	2.6–3.3 *4*	Hex. bullets; co-crystals
VI	MK2(41–364)	1.5–1.8 M NM pH 8.0	$P6_3$	144 144 152	2.6–3.3 *4*	Hex. bullets; co-crystals; "collapsed" Form V
VII	MK2(41–364)	1.75 M AS 0.1 M Na citrate pH 8.0	$P2_12_12_1$	140 180 215	2.6–3.3 *12*	Sharp blocks; co-crystals

* AS – $(NH_4)_2SO_4$; LS – Li_2SO_4; NM – sodium malonate; NP – sodium phosphate.
† Resolution of diffraction. *N* – MK2 molecules/crystallographic asymmetric unit.

Three pseudoactivated constructs were analyzed for diffraction: MK2(41–364, T222E), MK2(47–366, T222E), and MK2(47–366, T334E) (Table 3). Although all yielded moderate amounts of protein and crystals suitable for diffraction testing, only one, MK2(47–366, T222E), diffracted well. The detergent Anapoe 80 proved to be extremely effective with these crystals, improving diffraction to 2.9-Å resolution (Form IV, Table 6). Crucially, this crystal form was used to solve our first MK2 structure in complex with a prototypical inhibitor chosen from our high-throughput screening lead chemotype. As the project progressed, however, it was supplanted by other crystal forms (especially Forms V-VII) that were more amenable to co-crystallization.

Table 6. Crystallographic statistics for an MK2/lead compound inhibitor complex.

Parameter	Value
Construct	MK2(47–366, T222E)
Crystal form	IV
Space group	$F4_132$
Resolution (Å)	20.0–2.9
Unit cell length (a; Å)	253.508
Unique reflections	29,178
R_{sym}	0.069
7 I/σ_I 8	10.9
Data completeness (%)	99.7
Mean multiplicity	8.2
R_{cryst} (%)	22.9
R_{free} (%)	26.8

Pseudoactivated MK2 Adopts the Conformation of Inactive MK2

The crystal structure of pseudoactivated MK2 in complex with a micromolar lead compound was determined in space group F4132 (Table 6). This crystal form was reported subsequent to our work by other investigators [8-10]. Superimposition with apo-MK2 (Protein Data Bank entry 1KWP[7]) illustrates a similar "closed" conformation Figure 5. One significant difference is observed in the arrangement of the glycine-rich loop, which assumes a non-canonical orientation by flipping away from the active site to form a short helix. This rearrangement increases the solvent exposure of the ATP binding pocket, making this crystal form a good candidate for inhibitor soaking. We successfully soaked an MK2-specific inhibitor into the ATP binding pocket, leading to our reference crystal structure (Figure 5). Due to the disorder of the activation loop, we were unable to resolve the T222E pseudoactivating mutation. Although the mutation was not directly involved in driving a novel crystal form, through altered crystal packing or (apparently) a significantly altered activation loop conformation, we believe that the enhanced solubility and stability MK2(47–366, T222E) facilitated crystallization.

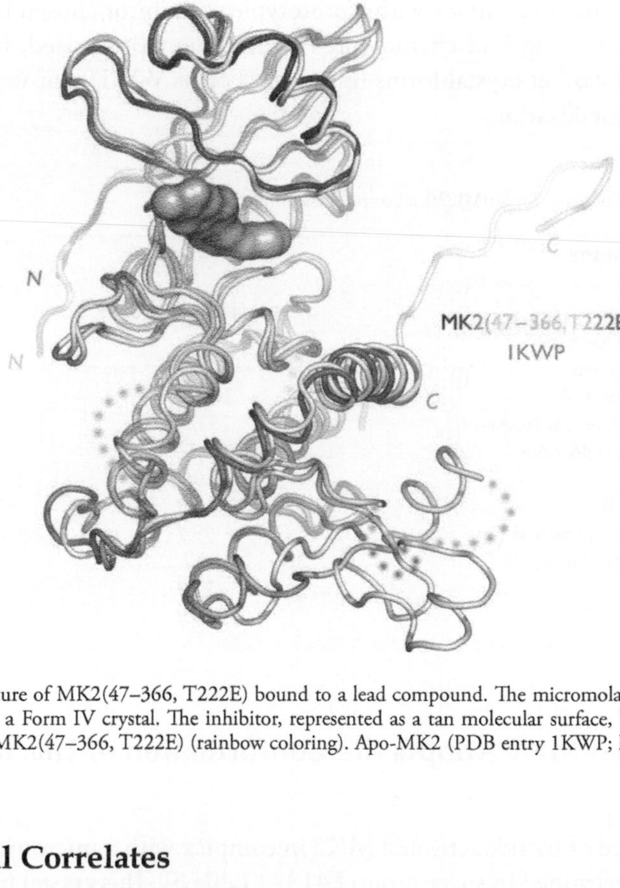

Figure 5. Structure of MK2(47–366, T222E) bound to a lead compound. The micromolar-potency inhibitor was soaked into a Form IV crystal. The inhibitor, represented as a tan molecular surface, binds deeply in the ATP pocket of MK2(47–366, T222E) (rainbow coloring). Apo-MK2 (PDB entry 1KWP; light-grey) is shown for comparison.

Structural Correlates

For reasons that are still unclear, MK2 has an overwhelming propensity to make specific, trimer-forming intermolecular contacts as it crystallizes. Some of our mutations abrogated these contacts; instead of shifting crystallization to new conditions and crystal forms, however, those MK2 constructs simply did not crystallize. Thus, a level of mutagenesis that would be sufficient for most proteins of this size was surprisingly less effective with MK2.

What are these trimers? As noted by Hillig et al. [18], two distinct packing interactions are present in both Form IV (space group F4132; PDB entry 2JBO[9]) and Form VII (P212121; PDB entry 2JBP[9]) MK2 crystals (Table 5). The Type

1 trimer is mediated by a draping of the N-terminus, beginning around residue 47, over the N-lobe of another MK2 subunit. Constructs beginning at residues 41 or 47 retain this contact and crystallize; those beginning at residue 50 lose the contact and do not form crystals. The Type 2 trimer is mediated by the C-terminal portion of the activation loop packing against helices F, G, and H. Constructs in which the activation loop was deleted, being unable to form these contacts, do not crystallize. Notably, Glu233-Arg313 and Glu238-Arg280 salt bridges mediate Type 2 contacts. Targeting of these glutamate residues in a second round of entropy-reduction mutagenesis might have altered the Type 2 contacts enough to spur formation of other crystal forms.

MK2 trimer formation is due entirely either to crystallographic symmetry (Form IV) or to non-crystallographic symmetry (Form VII). Form IV has one molecule/asymmetric unit, and the two types of trimers are formed by adjacent, non-intersecting crystallographic 3-fold symmetry axes. Conversely, since space group P212121 has no 3-fold axes, the 12 molecules/asymmetric unit in Form VII are arranged such that both trimer types are formed by two non-crystallographic 3-fold axes that nearly intersect in the center of the 12-subunit, virus-like MK2 shell.

Amazingly, all characterized MK2 crystal forms shown in Table 5 are composed of Type 1 and/or Type 2 trimers. Forms V and VI (both P63; e.g., PDB entry 1NXK[8]) have four molecules/asymmetric unit; three subunits form the Type 1 non-crystallographic trimer; the fourth, "odd man out" subunit forms, through crystallographic symmetry, the Type 2 trimer. And, the tenuous packing in Form III (P213; four molecules/asymmetric unit, 78% solvent) is mediated by trimer formation at two adjacent, non-intersecting crystallographic 3-fold axes, as in the other cubic (Form IV) crystal form. Only the first reported MK2 structure (R3; PDB entry 1KWP; two molecules/asymmetric unit) breaks the pattern [7]. Uniquely compared to all other MK2 crystals, the construct used in that study included the complete MK2 C-terminus. Packing in this crystal form is mediated by the Type 1 trimer (crystallographic 3-fold) and a novel (parallel, crystallographic) trimeric contact centered at residue 370 that positions the extended C-terminus (ordered to residue ~385) to pack against another MK2 subunit.

Although the biological relevance of MK2 trimer formation is unknown, we note that trimer formation is structurally incompatible with formation of the MK2-p38 complex [23,24]. Thus, MK2 trimer formation may be a form of self-regulation relevant in vivo.

In summary, much of the mutagenesis work reported here was, in the end, stymied by the unusually strong proclivity of MK2 toward trimer formation. Nonetheless, our systematic approach did indicate that truncations rather than mutations were more effective for MK2 crystallization. The exact position of the MK2 N-terminus is more important than is the C-terminus, and the activation loop could not be deleted. Surface arginine and glutamate residues drive crystallization more strongly than do surface lysine residues prevent it. The pseudoactivation constructs were helpful, likely due to slight modulation of protein surface properties rather than by producing a different protein conformation. These unusual MK2 properties result in many closely related (though at first glance apparently different) crystal forms (Table 5). Additional surface mutants (including untargeted glutamates, and also arginines) would seem to be required to drive MK2 into truly different crystal forms that can be produced under low ionic strength (e.g., polyethylene glycol) conditions rather than in high salt.

Conclusion

The systematic methods for the design, production and evaluation of MK2 protein constructs presented here allowed us to reproducibly obtain suitable crystals such that structure-based drug design could proceed. Our methods are robust and allowed rapid evaluation of about fifty constructs. This rapidity enabled the scale-up production of selected MK2 constructs in multi-milligram quantities proteins for structural studies. Using rational site-directed mutagenesis and in-house customized crystallization screens (Table 4), we were able to identify several novel crystallization conditions. Combined with variation of other parameters, such as surface side chain entropy and (pseudo)activation state, these high-throughput techniques produced five new MK2 crystal forms, most with improved diffraction characteristics (Table 5). One such crystal form, grown with a MK2 phosphorylation-site pseudoactivation mutant, was used to solve our first MK2/lead inhibitor complex (Figure 5).

Several key lessons were learned from this exercise. First, setting reasonable criteria for construct performance and prioritization is essential to identify constructs suitable for further evaluation and possible scale-up. In the case of MK2, most constructs gave satisfactory performance in expression and purification. Choices were necessary, however; we prioritized higher-yielding, more soluble constructs (Table 2) for scale-up and more extensive crystallization screening. Implicitly, we assumed that constructs that expressed or purified poorly were less likely to crystallize well. Our assumption appears to be supported by the expression yield, protein melting point, and crystallization data of Malawski et al. [18]. Second, a systematic crystallization screen (Table 4) was required to identify crystallization conditions in a robust manner. Indeed, the interplay between multiple constructs and multiple, customized crystallization solutions likely contributed greatly to our success. We note that the composition of customized crystallization screens is protein-specific – for MK2, we used a preponderance of high ionic strength conditions, since repeated screening with, for example, PEG solutions never provided crystals. Other proteins will behave differently. Third, reduction of surface side chain entropy can require several iterative rounds before productive mutation sites are identified. In retrospect, more than one round of surface residue mutations in this study, optimally informed by initial structural information, might have yielded additional crystal forms (e.g., mutation of Glu233 and Glu238 to disrupt the Type 2 trimer contact). Fourth, high throughput construct exploration must be initiated early in a drug discovery program in order to synchronize with hit-to-lead synthetic chemistry efforts, preferably in concert with initial target selection studies, i.e. before a high-throughput inhibitor screen is begun. This head start is especially critical for problematic crystallization targets like MK2. Applying these lessons learned from our experience with MK2 has helped us to accelerate many other structural programs, enabling us to impact lead discovery programs more rapidly and efficiently.

Methods

Cloning

Most human MK2 http://www.expasy.org/uniprot/P49137 constructs were engineered as fusion proteins with Schistosoma japonicum glutathione S-transferase (GST; http://www.expasy.org/uniprot/P08515), using the pGEX4T-1 vector (GE Healthcare). The sequence used was: GST-(SDLVPR↑GSENLYFQ↑G)-MK2. The linker sequence encodes thrombin and tobacco etch virus (TEV) protease cleavage sites ("↑"). Two protease sites were included for maximal flexibility in removal of the fusion tag after purification; we almost exclusively used the TEV protease site, resulting in an unnatural glycine residue N-terminal to MK2. A few constructs were also made as His6-FLAG-TEV-MK2 fusion proteins, using the pET21a+ vector (Invitrogen). The sequence used was: (MGHHHHHHG SGDYKDDDDKDYDIPTTENLYFQ↑G)-MK2. We refer to these vectors as "pGEX4T-1-GST-Thr-TEV" and "pET21a+-His6-FLAG-TEV," respectively.

Mutagenic primers were designed according to the QuikChange XL Site-Directed Mutagenesis Kit (Stratagene) instructions and were purchased from Invitrogen. Briefly, primer pairs were made for each mutation. Both primers in each pair contained the desired mutation and annealed to the same sequences on opposite strands of the plasmid template. The desired mutations (nucleotide replacements or deletions), in the middle of each primer, were flanked by 15–19 bases of wildtype sequence on both sides. Primer stocks and dilutions were arrayed in 96-well V-bottom microplates (MJ Research) and stored at -20°C.

Site-directed mutagenesis was performed using the QuikChange kit (Stratagene) using the plasmids pGEX4T-1-GST-Thr-TEV-MK2(47–366) and pGEX4T-1-GST-Thr-TEV-MK2(41–364) as templates. PCR was performed in 96-well V-bottom microplates using a DNA Engine Dyad (MJ Research) thermocycler complete with an ALS-1296, 96-well alpha unit (MJ Research). Reaction mixtures were cycled 18 times according to this schedule: 95°C, 50 s; 60°C, 50 s; 68°C, 14 min. Cycling was preceded by incubation at 95°C (1 min) and followed by incubation at 68°C (7 min).

Transformation of XL-10 Gold Ultracompetent E. coli cells (Stratagene) was performed in 24-well RB Blocks (Qiagen). Thawed cells (45 μL) were mixed on ice with 2 μL of 2-mercaptoethanol in each well. A 5 μL aliquot of each PCR reaction mix was added to the appropriate well and the mixture was incubated on ice for 10 min. The blocks were then heat-shocked by immersing the lower half of the blocks into a 42°C water bath (30 s), then placing the block back on ice (2 min). SOC medium (0.5 mL; preheated to 42°C) was then added to each well and the block was immediately placed in an incubator at 37°C for 1 hr with shaking. A 0.1 mL aliquot of each transformation was spread onto 100 mm LB agar plates containing 100 μg/mL ampicillin and incubated overnight at 37°C.

Plasmid DNA was prepared using the QIAprep 8 Turbo Miniprep Kit (Qiagen) in combination with the Qiavac 6S vacuum manifold (Qiagen) according to manufacturer's instructions. The DNA was quantitated spectrophotometrically and diluted to 100 μg/mL with water for sequence analysis. The coding sequence of all constructs was verified.

Expression

Plasmids encoding the MK2 constructs were transformed into BL21 (DE3) strain for expression studies. Competent cells were transformed with 0.5 μL of each plasmid in 24-well blocks as above (without 2-mercaptoethanol). The transformation mix (0.1 mL) was spread onto 100 mm LB agar plates containing 100 μg/mL ampicillin and incubated overnight at 37°C. Starter cultures grown at 37°C were used to inoculate 2.5 mL of LB medium containing 100 μg/mL ampicillin in 24 deep-well blocks. Once an OD600 of 0.4–0.6 was reached, the blocks were shifted to 18°C for 1 h and then induced with IPTG (0.4 mM) for 4 h. Cells were harvested by centrifugation, frozen at -80°C. Large-scale cultures were induced in the same way with the exception that the flasks were shifted to 4°C for 40 min prior to induction.

Analysis of the soluble and insoluble fractions was performed by SDS-PAGE on 4–20% gels (Invitrogen). Cells were thawed and resuspended for lysis in Bug Buster HT (Novagen; 0.25 mL per unit OD600). Bugbuster HT containing 60 kU rLysozyme (Novagen; 0.1 mL) was added to the cell suspension (0.1 mL). The mixture was incubated on ice for 15 min before addition of 0.2 mL of Bugbuster HT containing 20 mM DTT and Complete, EDTA-free Protease Inhibitor cocktail (Roche; 1 tablet/30 mL Bugbuster HT). An aliquot of the crude lysate was separated by centrifugation; the supernatant was analyzed as the soluble fraction. The pellet was resuspended in an equal volume of BugBuster HT and analyzed as the insoluble fraction. Proteins were visualized with SimplyBlue stain (Invitrogen)

Western blotting was used to confirm protein identity. Gels were transferred to PVDF membranes using a Protean II Mini Trans-Blot apparatus (BioRad). After blocking membranes with 5% nonfat dry milk in TBS-T buffer (50 mM Tris·HCl, pH 8.0, 150 mM NaCl, 0.1% Tween-20), the blot was probed with horseradish peroxidase-conjugated anti-GST antibody diluted 1:1000 in TBS-T/5% nonfat milk. Membranes were washed extensively with TBS-T, and then bound antibodies were visualized using the SuperSignal West Pico chemiluminescent substrate kit (Pierce).

Purification

MK2 variants were purified using the following procedure at 4°C: Cell pellets were thawed and resuspended in lysis buffer containing 50 mM Tris·HCl, 250 mM NaCl, 10% glycerol, 1 mM DTT, 1 mM EDTA, pH 7.5. The cell suspension was sonicated on ice for 20 s iterations and then centrifuged (22,000 × g) for 45 min. A 10 mL glutathione affinity column (GE Healthcare) was prepared by washing with ten column volumes of Buffer A (50 mM Tris·HCl, 250 mM NaCl, 10% glycerol, 2.5 mM DTT, 1 mM EDTA, pH 8.0) containing Complete, EDTA-free Protease Inhibitor cocktail. Soluble cell lysate was applied to the column and then extensively washed with Buffer A. MK2 was eluted from the column with Buffer A + 40 mM glutathione. The GST tag was cleaved using TEV protease, typically 4–16 h at 4–15°C. The sample was then diluted ten-fold with Buffer B (50 mM MES, 10% glycerol, pH 6.0) and loaded onto a MonoS 10/10 column (GE Healthcare) equilibrated with Buffer B + 20 mM NaCl. The protein was eluted at ≈200 mM NaCl using a stepwise Buffer B/NaCl gradient. MK2-containing fractions were pooled and concentrated to 10 mg/mL, and then loaded onto a Superdex 75 16/60 column (GE Healthcare) equilibrated with 50 mM Tris·HCl, 250 mM NaCl, 10% glycerol, 1 mM DTT, pH 7.5; protein was eluted at 1 mL/min. MK2-containing fractions were pooled and sample purity was assessed by SDS-PAGE; protein identity was confirmed using mass spectrometry.

Activity Assays

Enzymatic assays utilized a homogeneous time-resolved fluorescence method (CisBio-US, Inc.) to quantitate product formation. Reactions contained in 40 μL: varying amounts of enzyme, 4 μM peptide substrate (Biotin-Ahx-AKVSRS-GLYRSPSMPENLNRPR), 10 μM ATP, 20 mM MOPS, 10 mM MgCl2, 5 mM EGTA, 5 mM 2-phosphoglycerol, 1 mM Na3VO4, 0.01% Triton X-100, 5% DMSO, 1 mM DTT, pH 7.2. After 60 min at room temperature, reactions were quenched by adding 10 μL of 0.5 M EDTA. Phosphorylated peptide was measured by addition of 75 μL of 24 ng/mL Eu3+-cryptate-labeled anti-phospho-14--3-3 binding motif (CisBio-US, Inc.), 1.47 μg/mL SureLight™ allophycocyanin-streptavidin (CisBio-US, Inc.), 50 mM HEPES, 0.1% BSA, 0.01% Tween-20, 0.4 M KF. The developed reaction was incubated in the dark at room temperature for 10 min, then read in a time-resolved fluorescence detector (Perkin Elmer Discovery or BMG Rubystar) at 620 nm and 665 nm simultaneously, using a 337 nm nitrogen laser for excitation. The 665/620 emission ratio is proportional to the amount of phosphorylated peptide product. Since the HTRF method does not provide absolute quantities of product formed, specific activities were calculated as HTRF counts/nM MK2 protein.

Crystallization

MK2 constructs yielding more than 1 mg/L of culture were progressed to crystallization trials. Protein was concentrated to 10 mg/mL using an Ultrafree-15 Biomax 10 kDa molecular weight cut-off centrifugal filter device (Millipore). Various inhibitors were added individually to concentrated protein stocks to a final concentration of 1 mM. Complexed MK2 protein (0.5 μL) was mixed with 0.5 μL of various crystallization solutions from Crystal Screen 1™ and Crystal Screen 2™ (Hampton Research) and Wizard Screen 1™ and Wizard Screen 2™ (DeCode Biostructures). The resulting drops were dispensed into 96-well sitting drop trays (Greiner) using a Hydra II+1 liquid handler (Thermo Scientific Matrix). Trays were stored at 18°C and visualized manually. Crystallization was tested extensively at 4°C, uniformly without success.

Accumulated MK2 crystallization hits suggested parameters to be explored more closely in a customized robotic screen, the most effective being the precipitating reagent and pH range. A complete 96-well screen, designed for use with the Hydra II+1, consisted of four 4 × 6 grid screens: (1) (NH4)2SO4/sodium citrate, pH 7–8.5; (2) sodium malonate, pH 7–8.5; (3) sodium phosphate; pH 7–8.5; and (4) a randomized screen obtained by randomly mixing the above three precipitants with other additives (Table 4). Seven different crystal forms were identified from this comprehensive screen, as shown in Table 5 and Figure 4.

Crystals for inhibitor soaking were grown in sitting drops by the vapor-diffusion method using MK2(47–366, T222E). MK2 (1.5 μL, 10 mg/mL) was added to 1.5 μL of reservoir solution (2 M sodium malonate, pH 5.5, and 0.01 mM Anapoe 80 [Hampton Research]) and then the drop was sealed in vapor contact with 500 μL of reservoir solution. Crystals grew to about 0.2 mm in size in 3 days. For soaking, one MK2 crystal was added to 60 μL of 1 mM inhibitor dissolved in mother liquor and incubated at 18°C overnight.

Diffraction Testing and Structure Determination

MK2/inhibitor complex crystals were harvested into a cryoprotectant solution (20% glycerol plus mother liquor) using a fiber loop and flash-cooled in liquid nitrogen. Cystals were stored in liquid nitrogen until diffraction testing. X-ray diffraction testing was conducted in-house using a FR591 rotating anode generator (Bruker AXS) with a MAR345 image plate detector (MARResearch) and Osmic optics (Rigaku USA). A total of 535 crystals were tested, and over 80 crystals were selected for synchrotron data collection if diffraction reached at least 3.5-Å resolution. Advanced Photon Source (Argonne, IL) and National Synchrotron Light Source (Upton, NY) synchrotron beamlines were used primarily for data collection, although a few crystals were selected for in-house data collection.

Diffraction data was processed with the HKL2000 program suite [25]. After determining the crystal orientation, the data were integrated with DENZO, scaled and merged with SCALEPACK, and placed on an absolute scale and reduced to structure factor amplitudes with TRUNCATE [26]. Five percent of the unique reflections were assigned, in a random fashion, to the "free" set, for calculation of the free R-factor (Rfree) [27]; the remaining 95% of reflections constituted the "working" set, for calculation of the R-factor (Rcryst). The x-ray diffraction data for a representative inhibitor-soaked MK2 crystal (Form IV) are summarized in Table 6.

The CCP4 program suite was used to solve and refine the structure [28]. The cross-rotation function was calculated using MOLREP [29], using the apo MK2 structure reported previously (Protein Data Bank entry 1KWP; [7]) as the search model. Initial coordinates were generated based on the one solution apparent at 2.9 Å resolution. Refinement began with rigid-body refinement in REFMAC [30], resulting in an Rcryst of 37.0% (Rfree 40.0%) for all reflections with $F > 2.0\sigma F$, 20–2.9 Å. Manual rebuilding of the model was conducted using the molecular graphics program O [31] and examination of sigmaA-weighted 2FO-FC and FO-FC electron-density maps [32]. Restrained refinement using REFMAC converged at an Rcryst of 22.9% (Rfree 26.8%), 20–2.9 Å. The quality of the model was assessed with PROCHECK [33] and WHATCHECK [34].

Authors' Contributions

MAA led the MK2 structural biology sub-team; additionally, she participated in construct design, provided protein purification and crystallization oversight, collected diffraction data, and performed the structure determination. SS made the majority of the constructs and performed some of the protein expression. DB, DM, TX, JH, and MP purified protein, set up crystallizations, and collected diffraction data. MT and MD contributed protein characterization data. CH and SK contributed enzymatic data. JD and BN performed the bulk of the protein expression. LQ contributed the early template constructs. DWB provided structural biology oversight; additionally, he participated in construct design. JM provided construct construction and protein expression oversight and participated in construct design. RVT led the MK2 drug discovery project and participated in construct design. RS led the MK2 high-throughput construct construction and protein expression sub-team and participated in construct design. MAA, DWB and RS jointly prepared the manuscript, in consultation with all of the co-authors. All authors have read and approved this manuscript.

Acknowledgements

We thank Vincent Stoll and Kenton Longenecker (Abbott Laboratories) for helpful discussions regarding construct design, comments on the manuscript and assistance in data collection at APS. We also thank Eric Goedken (Abbott Bioresearch Center) for comments on the manuscript.

References

1. Stokoe D, Campbell DG, Nakielny S, Hidaka H, Leevers SJ, Marshall C, Cohen P: MAPKAP kinase-2; a novel protein kinase activated by mitogen-activated protein kinase. EMBO J 1992, 11(11):3985–3994.
2. Stokoe D, Caudwell B, Cohen PT, Cohen P: The substrate specificity and structure of mitogen-activated protein (MAP) kinase-activated protein kinase-2. Biochem J 1993, 296(3):843–849.
3. Engel K, Plath K, Gaestel M: The MAP kinase-activated protein kinase 2 contains a proline-rich SH3-binding domain. FEBS Lett 1993, 336(1):143–147.

4. Kotlyarov A, Neininger A, Schubert C, Eckert R, Birchmeier C, Volk H, Gaestel M: MAPKAP kinase 2 is essential for LPS-induced TNF-a biosynthesis. Nat Cell Biol 1999, 1:94–97.

5. Hegen M, Gaestel M, Nickerson-Nutter CL, Lin L-L, Telliez J-B: MAPKAP Kinase 2-Deficient Mice Are Resistant to Collagen-Induced Arthritis. J Immunol 2006, 177(3):1913–1917.

6. Kotlyarov A, Yannoni Y, Fritz S, Laass K, Telliez JB, Pitman D, Lin LL, Gaestel M: Distinct cellular functions of MK2. Molecular and cellular biology 2002, 22(13):4827–4835.

7. Meng W, Swenson LL, MJ F, Hayakawa K, ter Haar E, Behrens AE, Fulghum JR, Lippke JA: Structure of Mitogen-activated Protein Kinase-activated Protein (MAPKAP) Kinase 2 suggests a bifunctional switch that couples kinase activation with nuclear export. J Biol Chem 2002, 277:37401–37405.

8. Underwood KW, Parris KD, Federico E, Mosyak L, Czerwinski RM, Shane T, Taylor M, Svenson K, Liu Y, Hsiao C-L, et al.: Catalytically Active MAP KAP Kinase 2 Structures in Complex with Staurosporine and ADP Reveal Differences with the Autoinhibited Enzyme. Structure 2003, 11(6):627–636.

9. Hillig RC, Eberspaecher U, Monteclaro F, Huber M, Nguyen D, Mengel A, Muller-Tiemann B, Egner U: Structural Basis for a High Affinity Inhibitor Bound to Protein Kinase MK2. J Mol Biol 2007, 369(3):735–745.

10. Anderson DR, Meyers MJ, Vernier WF, Mahoney MW, Kurumbail RG, Caspers N, Poda GI, Schindler JF, Reitz DB, Mourey RJ: Pyrrolopyridine Inhibitors of Mitogen-Activated Protein Kinase-Activated Protein Kinase 2 (MK-2). J Med Chem 2007, 50(11):2647–2654.

11. Plath K, Engel K, Schwedersky G, Gaestel M: Characterization of the Proline-Rich Region of Mouse Mapkap Kinase2: Influence on Catalytic Properties and Binding to the c-abl-SH3 Domain in Vitro. Biochem Biophys Res Comm 1994, 203(2):1188–1194.

12. Engel K, Kotlyarov A, Gaestel M: Leptomycin B-sensitive nuclear export of MAPKAP kinase 2 is regulated by phosphorylation. EMBO J 1998, 17:3363–3371.

13. Ben-Levy R, Hooper S, Wilson R, Paterson HF, Marshall CJ: Nuclear export of the stress-activated protein kinase p38 mediated by its substrate MAPKAP kinase-2. Curr Biol 1998, 8(19):1049–1057.

14. Veron M, Radzio-Andzelm E, Tsigelny I, Ten Eyck LF, Taylor SS: A conserved helix motif complements the protein kinase core. Proc Natl Acad Sci USA 1993, 90(22):10618–10622.

15. Zu YL, Wu FY, Gilchrist A, Ai YX, Labadia ME, Huang CK: The Primary Structure of a Human Map Kinase Activated Protein Kinase 2. Biochem Biophys Res Comm 1994, 200(2):1118–1124.
16. Ben-Levy R, Leighton IA, Doza YN, Attwood P, Morrice N, Marshall CJ, Cohen P: Identification of novel phosphorylation sites required for activation of MAPKAP kinase-2. EMBO J 1995, 14:5920–5930.
17. Engel K, Schultz H, Martin F, Kotlyarov A, Plath K, Hahn M, Heinemann U, Gaestel M: Constitutive Activation of Mitogen-activated Protein Kinase-activated Protein Kinase 2 by Mutation of Phosphorylation Sites and an A-helix Motif. J Biol Chem 1995, 270(45):27213–27221.
18. Malawski GA, Hillig RC, Monteclaro F, Eberspaecher U, Schmitz AAP, Crusius K, Huber M, Egner U, Donner P, Muller-Tiemann B: Identifying protein construct variants with increased crystallization propensity – A case study. Prot Sci 2006, 15(12):2718–2728.
19. Longenecker KL, Garrard SM, Sheffield PJ, Derewenda ZS: Protein crystallization by rational mutagenesis of surface residues: Lys to Ala mutations. Acta Crystallogr D Biol Crystallogr 2001, 57:679–688.
20. Derewenda ZS, Vekilov PG: Entropy and surface engineering in protein crystallization. Acta Crystallogr D Biol Crystallogr 2006, 62(Pt 1):116–124.
21. Kurumbail RG, Pawlitz JL, Stegeman RA, Stallings WC, Shieh HS, Mourey RJ, Bolten SL, Broadus RM: Crystalline structure of human MAPKAP kinase-2. WO 2003/076333 A2 2003, 135.
22. Wu JP, Wang J, Abeywardane A, Andersen D, Emmanuel M, Gautschi E, Goldberg DR, Kashem MA, Lukas S, Mao W, et al.: The discovery of carboline analogs as potent MAPKAP-K2 inhibitors. Bioorg Med Chem Lett 2007, 17(16):4664–4669.
23. ter Haar E, Prabhakar P, Liu X, Lepre C: Crystal structure of the p38 alpha-MAPKAP kinase 2 heterodimer. J Biol Chem 2007, 282(13):9733–9739.
24. White A, Pargellis CA, Studts JM, Werneburg BG, Farmer BT 2nd: Molecular basis of MAPK-activated protein kinase 2:p38 assembly. Proc Natl Acad Sci USA 2007, 104(15):6353–6358.
25. Otwinowski Z, Minor W: Processing of X-ray diffraction data collected in oscillation mode. In Macromolecular Crystallography, Part A. Volume 276. Edited by: Carter CW Jr, Sweet RM. Academic Press; 1997:307–326.
26. French S, Wilson K: On the treatment of negative intensity observations. Acta Crystallogr A 1978, 34:517–525.

27. Brunger AT: The free R value: a novel statistical quantity for assessing the accuracy of crystal structures. Nature 1992, 355:472–475.
28. COLLABORATIVE COMPUTATIONAL PROJECT N: The CCP4 Suite: Programs for Protein Crystallography. Acta Crystallogr D Biol Crystallogr 1994, 50:760–763.
29. Vagin A, Teplyakov A: An automated program for molecular replacement. J Appl Crystallogr 1997, 30:1022–1025.
30. Murshodov GN, Vagin AA, Dodson EJ: Refinement of macromolecular structures by the maximum-likelihood method. Acta Crystallogr D Biol Crystallogr 1997, 53:240–255.
31. Jones TA, Zou JY, Cowan SW, Kjeldgaard M: Improved methods for building protein models in electron density maps and the location of errors in these models. Acta Crystallogr A 1991, 47:110–119.
32. Read RJ: Improved Fourier coefficients for maps using phases from partial structures with errors. Acta Crystallogr A 1986, 42:140–149.
33. Laskowski RA, MacArthur MW, Moss DS, Thornton JM: PROCHECK: A program to check the stereochemical quality of protein structures. J Appl Crystallogr 1983, 26:283–291.
34. Hooft RWW, Vriend G, Sander C, Abola EE: Errors in protein structures. Nature 1996, 381:272.

CITATION

Argiriadi MA, Sousa S, Banach D, Marcotte D, Xiang T, Tomlinson MJ, Demers M, Harris C, Kwak S, Hardman J, Petras M, Quinn L, DiMauro J, Ni B, Mankovich J, Borhani DW, Talanian RV, and Sadhukhan R. Rational Mutagenesis to Support Structure-Based Drug Design: MAPKAP Kinase 2 as a Case Study. BMC Structural Biology 2009, 9:16 doi:10.1186/1472-6807-9-16.

Nanotechnology Approaches to Crossing the Blood-Brain Barrier and Drug Delivery to the CNS

Gabriel A. Silva

ABSTRACT

Nanotechnologies are materials and devices that have a functional organization in at least one dimension on the nanometer (one billionth of a meter) scale, ranging from a few to about 100 nanometers. Nanoengineered materials and devices aimed at biologic applications and medicine in general, and neuroscience in particular, are designed fundamentally to interface and interact with cells and their tissues at the molecular level. One particularly important area of nanotechnology application to the central nervous system (CNS) is the development of technologies and approaches for delivering drugs and other small molecules such as genes, oligonucleotides, and contrast agents across the blood brain barrier (BBB). The BBB protects and isolates CNS structures (i.e. the brain and spinal cord) from the rest of the body, and creates a unique biochemical and immunological environment. Clinically, there are a number of scenarios where drugs or other small molecules need to gain access to the CNS following systemic administration, which necessitates being able to cross the BBB. Nanotechnologies can potentially be designed to carry out multiple specific functions at once or in a predefined sequence, an important requirement for the clinically successful delivery and use of drugs and other molecules to the CNS, and as such have a unique advantage over other complimentary technologies and methods. This brief review introduces emerging work in this area and summarizes a number of example applications to CNS cancers, gene therapy, and analgesia.

Introduction

Nanotechnologies are materials and devices that have a functional organization in at least one dimension on the nanometer (one billionth of a meter) scale, ranging from a few to about 100 nanometers. Nanoengineered materials and devices aimed at biologic applications and medicine in general, and neuroscience in particular, are designed fundamentally to interface and interact with cells and their tissues at the molecular level. The potential of nanotechnological applications to biology and medicine arise from the fact that they exhibit bulk mesoscale and macroscale chemical and/or physical properties that are unique to the engineered material or device and not necessarily possessed by the molecules alone. This supports the development of nanotechnologies that can potentially carry out multiple specific functions at once or in a predefined sequence, which is an important property for the clinically successful delivery of drugs and other molecules to the central nervous system (CNS).

An ability to cross the blood-brain barrier (BBB) to deliver drugs or other molecules (for example, oligonucleotides, genes, or contrast agents) while potentially targeting a specific group of cells (for instance, a tumor) requires a number of things to happen together. Ideally, a nanodelivery-drug complex would be administered systemically (for example, intravenously) but would find the CNS while producing minimal systemic effects, be able to cross the BBB and correctly target cells in the CNS, and then carry out its primary active function, such as releasing a drug. These technically demanding obstacles and challenges will require multidisciplinary solutions between different fields, including engineering, chemistry, cell biology, physiology, pharmacology, and medicine. Successfully doing so will greatly benefit the patient. Although this ideal scenario has not yet been realized, a considerable body of work has been done to develop nanotechnological delivery strategies for crossing the BBB.

Applications to Drugs and Other Molecules

A significant amount of work using nanotechnological approaches to crossing the BBB has focused on the delivery of antineoplastic drugs to CNS tumors. For example, radiolabeled polyethylene glycol (PEG)-coated hexadecylcyanoarcylate nanospheres have been tested for their ability to target and accumulate in a rat model of gliosarcoma [1]. Another group has encapsulated the antineoplasitc drug paclitaxel in polylactic co-glycolic acid nanoparticles, with impressive results. In vitro experiments with 29 different cancer cell lines (including both neural and non-neural cell lines) demonstrated targeted cytotoxicity 13 times greater than with drug alone [2]. Using a variety of physical and chemical characterization methods, including different forms of spectroscopy and atomic force microscopy, the investigators showed that the drug was taken up by the nanoparticles with very high encapsulation efficiencies and that the release kinetics could be carefully controlled. Research focusing on the delivery of many of the commonly used antineoplastic drugs is important because most of these drugs have poor solubility under physiologic conditions and require less than optimal vehicles, which can produce significant side effects.

In another example, various compounds – including neuropeptides such as enkephalins, the N-methyl-D-aspartate receptor antagonist MRZ 2/576, and the chemotherapeutic drug doxorubicin – have been attached to the surface of poly(butylcyanoacrylate) nanoparticles coated with polysorbate 80 [3-7]. The polysorbate on the surface of the nanoparticles adsorb apolipoproteins B and E and are taken up by brain capillary endothelial cells via receptor-mediated endocytosis. Nanoparticle-mediated delivery of doxorubicin is being explored in a rodent model of glioblastoma [3,8]. Importantly, recent work in a rat glioblastoma model revealed significant remission with minimal toxicity, setting the stage for potential clinical trials [8].

The delivery of other drugs is also being investigated. Dalargin is a hexapeptide analog of leucine-enkephalin containing D-alanine, which produces CNS analgesia when it is delivered intracerebroventricularly, but it has no analgesic effects if it is administered systemically, specifically because it cannot cross the BBB on its own [9,10]. [3H]Dalargin was conjugated to the same poly(butylcyanoacrylate) nanoparticles described above, injected systemically into mice, and demonstrated by radiolabeling to cross the BBB and accumulate in brain [10]. Other, similar studies have also demonstrated delivery of dalargin using polysorbate 80-coated nanoparticles [11]. Other polysorbate 80 nanoparticles have been chemically conjugated to the hydrophilic drug diminazenediaceturate (diminazene) and proposed as a novel treatment approach for second stage African trypanosomiasis [12]. In other work, PEG-treated polyalkylcyanoacrylate nanoparticles were shown to cross the BBB and accumulate at high densities in the brain in experimental autoimmune encephalomyelitis [13], a model of multiple sclerosis [14,15].

For other applications, molecules other than drugs must cross the BBB for therapeutic or diagnostic reasons, including oligonucleotides, genes, and contrast agents. Solid lipid nanoparticles consisting of microemulsions of solidified oil nanodroplets loaded with iron oxide and injected systemically into rats have been shown to cross the BBB and accumulate in the brain with long-lasting kinetics [16]. Iron oxides are classic superparamagnetic magnetic resonance imaging (MRI) contrast agents. Because iron oxides are insoluble in water, they must be delivered as modified colloids for clinical applications, which is usually achieved by coating them with hydrophilic molecules, such as dextrans [17]. Therefore, the delivery vehicle used is critical in determining the functional properties of the contrast agent. By taking advantage of the ability of these solid lipid nanoparticles to cross the BBB, nanoparticles complexed with iron oxides may provide new ways to image the CNS using MRI.

Other work has focused on the delivery of oligonucleotides in an in vivo mouse model and an in vitro endothelial cell model, with the aim being to develop novel treatments for neurodegenerative disorders [18]. The investigators synthesized a nanogel consisting of cross-linked PEG and polyethylenimine that spontaneously encapsulated negatively charged oligonucleotides. In their in vivo model, they demonstrated that intravenous injections resulted in a 15-fold accumulation of oligonucleotides in the brain after 1 hour, with a concurrent twofold decrease in accumulation in liver and spleen when compared with freely administered oligonucleotides (not encapsulated in nanogel particles).

A related area is the delivery of genes to the CNS for gene therapy. A tyrosine hydroxylase (TH) expression plasmid was delivered to the striatum of adult rats using PEG immunoliposome nanoparticles in order to normalize TH expression levels in the 6-hydroxydopamine rat model of Parkinson's disease [19]. Using specific antibodies to transferrin receptors conjugated to the nanoparticles, TH plasmids were shown to be expressed throughout the striatum.

Conclusion

Nanotechnology-based approaches to targeted delivery of drugs and other compounds across the BBB may potentially be engineered to carry out specific functions as needed. The drug itself – in other words the biologically active component being delivered, whatever that may be – constitutes one element of a nanoengineered complex. The rest of the complex is designed to carry out other key functions, including shielding the active drug from producing systemic side effects, being prematurely cleared or metabolized, crossing the BBB, and targeting specific cells after it has gained access to the CNS. Implicitly, all of this must be achieved by any drug intended to have CNS effects, regardless of whether it is part of a nanoengineered complex.

An important advantage of a nanotechnological approach, as compared with the administration of free drug or the drug associated with a nonfunctional vehicle, is that these critical requirements do not need to be carried out by the active compound, but by supporting parts of the engineered complex. This allows the design of the active drug to be tailored for maximal efficacy. Currently, most nanoengineered systems for crossing the BBB take advantage of drugs that are already in clinical use and therefore have greater potential for reaching the clinic relatively quickly.

In addition to the delivery of drugs and other compounds across the BBB for therapeutic purposes, the ability to cross the BBB selectively and efficiently in animal models using nanoengineered technologies will have a significant impact on research that focuses on the normal physiology of the CNS and its pathology, by allowing targeted in vivo studies of specific cells and processes using methods that take advantage of the intact live organism. Ideally, methods for crossing the BBB will complement other nanotechnological tools being developed to study the CNS, including quantum dot labeling and imaging [20].

List of Abbreviations Used

BBB: blood-brain barrier; CNS: central nervous system; PEG: polyethylene glycol.

Competing Interests

The author declares that he has no competing interests.

Acknowledgements

Parts of this paper were adapted from a more detailed review on nanotechnology approaches for crossing the BBB written by the author [21]. This work was supported by funds from NIH grant NINDS NS054736.

This article has been published as part of BMC Neuroscience Volume 9 Supplement 3, 2008: Proceedings of the 2007 and 2008 Drug Discovery for

Neurodegeneration Conference. The full contents of the supplement are available online at http://www.biomedcentral.com/1471-2202/9?issue=S3.

References

1. Brigger I, Morizet J, Aubert G, Chacun H, Terrier-Lacombe MJ, Couvreur P, Vassal G: Poly(ethylene glycol)-coated hexadecylcyanoacrylate nanospheres display a combined effect for brain tumor targeting. J Pharmacol Exp Ther 2002, 303:928–936.
2. Feng SS, Mu L, Win KY, Huang G: Nanoparticles of biodegradable polymers for clinical administration of paclitaxel. Curr Med Chem 2004, 11:413–424.
3. Gulyaev AE, Gelperina SE, Skidan IN, Antropov AS, Kivman GY, Kreuter J: Significant transport of doxorubicin into the brain with polysorbate 80-coated nanoparticles. Pharm Res 1999, 16:1564–1569.
4. Alyautdin RN, Petrov VE, Langer K, Berthold A, Kharkevich DA, Kreuter J: Delivery of loperamide across the blood-brain barrier with polysorbate 80 coated polybutylcyanoacrylate nanoparticles. Pharm Res 1997, 14:325–328.
5. Kreuter J, Alyautdin RN, Kharkevich DA, Ivanov AA: Passage of peptides through the blood-brain barrier with colloidal polymer particles (nanoparticles). Brain Res 1995, 674:171–174.
6. Alyautdin RN, Tezikov EB, Ramge P, Kharkevich DA, Begley DJ, Kreuter J: Significant entry of tubocurarine into the brain of rats by adsorption to polysorbate 80-coated polybutylcyanoacrylate nanoparticles: an in situ brain perfusion study. J Microencapsul 1998, 15:67–74.
7. Friese A, Seiller E, Quack G, Lorenz B, Kreuter J: Increase of the duration of the anticonvulsive activity of a novel NMDA receptor antagonist using poly(butylcyanoacrylate) nanoparticles as a parenteral controlled release system. Eur J Pharm Biopharm 2000, 49:103–109.
8. Steiniger SC, Kreuter J, Khalansky AS, Skidan IN, Bobruskin AI, Smirnova ZS, Severin SE, Uhl R, Kock M, Geiger KD, Gelperina SE: Chemotherapy of glioblastoma in rats using doxorubicin-loaded nanoparticles. Int J Cancer 2004, 109:759–767.
9. Rousselle C, Clair P, Smirnova M, Kolesnikov Y, Pasternak GW, Gac-Breton S, Rees AR, Scherrmann JM, Temsamani J: Improved brain uptake and pharmacological activity of dalargin using a peptide-vector-mediated strategy. J Pharmacol Exp Ther 2003, 306:371–376.
10. Alyaudtin RN, Reichel A, Lobenberg R, Ramge P, Kreuter J, Begley DJ: Interaction of poly(butylcyanoacrylate) nanoparticles with the blood-brain barrier in vivo and in vitro. J Drug Target 2001, 9:209–221.

11. Schroeder U, Sommerfeld P, Ulrich S, Sabel BA: Nanoparticle technology for delivery of drugs across the blood-brain barrier. J Pharm Sci 1998, 87:1305–1307.
12. Olbrich C, Gessner A, Kayser O, Muller RH: Lipid-drug-conjugate (LDC) nanoparticles as novel carrier system for the hydrophilic antitrypanosomal drug diminazenediaceturate. J Drug Target 2002, 10:387–396.
13. Calvo P, Gouritin B, Villarroya H, Eclancher F, Giannavola C, Klein C, Andreux JP, Couvreur P: Quantification and localization of PEGylated polycyanoacrylate nanoparticles in brain and spinal cord during experimental allergic encephalomyelitis in the rat. Eur J Neurosci 2002, 15:1317–1326.
14. Ercolini AM, Miller SD: Mechanisms of immunopathology in murine models of central nervous system demyelinating disease. J Immunol 2006, 176:3293–3298.
15. Kanwar JR: Anti-inflammatory immunotherapy for multiple sclerosis/experimental autoimmune encephalomyelitis (EAE) disease. Curr Med Chem 2005, 12:2947–2962.
16. Peira E, Marzola P, Podio V, Aime S, Sbarbati A, Gasco MR: In vitro and in vivo study of solid lipid nanoparticles loaded with superparamagnetic iron oxide. J Drug Target 2003, 11:19–24.
17. Dupas B, Berreur M, Rohanizadeh R, Bonnemain B, Meflah K, Pradal G: Electron microscopy study of intrahepatic ultrasmall superparamagnetic iron oxide kinetics in the rat. Relation with magnetic resonance imaging. Biol Cell 1999, 91:195–208.
18. Vinogradov SV, Batrakova EV, Kabanov AV: Nanogels for oligonucleotide delivery to the brain. Bioconjug Chem 2004, 15:50–60.
19. Zhang Y, Calon F, Zhu C, Boado RJ, Pardridge WM: Intravenous nonviral gene therapy causes normalization of striatal tyrosine hydroxylase and reversal of motor impairment in experimental parkinsonism. Hum Gene Ther 2003, 14:1–12.
20. Pathak S, Cao E, Davidson MC, Jin S, Silva GA: Quantum dot applications to neuroscience: new tools for probing neurons and glia. J Neurosci 2006, 26:1893–1895.
21. Silva GA: Nanotechnology approaches for drug and small molecule delivery across the blood brain barrier. Surg Neurol 2007, 67:113–116.

CITATION

Silva GA. Nanotechnology Approaches to Crossing the Blood-Brain Barrier and Drug Delivery to the CNS. BMC Neuroscience 2008, 9(Suppl 3):S4. doi:10.1186/1471-2202-9-S3-S4.

Copyrights

18. This work was performed at Sandia National Laboratories, supported by the U. S. Department of Energy under Contract # DE-AC04-94AL85000.

do not necessarily state or reflect those of the United States Government or any agency thereof.

Index

A

B

C

I

J

K

O

P

Q

R

S

T

U

V

W

Y

Z

Printed and bound by CPI Group (UK) Ltd, Croydon, CR0 4YY
23/10/2024
01777682-0004